Indirect calorimetry

Indirect calorimetry

Techniques, computations and applications

edited by:

W.J.J. Gerrits and E. Labussière

Wageningen Academic Publishers

EAN: 9789086862610
e-EAN: 9789086868094
ISBN: 978-90-8686-261-0
e-ISBN: 978-90-8686-809-4
DOI: 10.3920/ 978-90-8686-809-4

Cover design: Bram Dirven

First published, 2015

Table of contents

Preface

Indirect calorimetry has been a well-established technique to measure heat production of men and animals for many years. Measurement of the consumption of dioxygen and production of carbon dioxide are the basis for the measurement of heat production. Whereas some twenty years ago, quite a number of well-known calorimetry facilities were discontinued, in part for financial reasons, during the last decade, new facilities with innovative techniques and interpretation of data are constructed in many places. Combining straightforward measurement of gaseous exchange with accurate climate control, recording of physical activity and feed intake, use of stable isotopes and sophisticated modelling techniques allow scientists to make progress in various research areas.

When given the opportunity to enter this area of research, we were lucky to be embedded in an environment with good equipment, excellent researchers and highly skilled technicians at Wageningen University, the Netherlands and at INRA, St. Gilles, France. We are grateful to Martin Verstegen, Jean Noblet and Jaap van Milgen for sharing their expertise and for providing this environment. Continuing this line of research, it is important to be able to build on previous expertise. It is tempting to do so without questioning it or knowing its background.

This book provides a scientific basis for indirect calorimetry, dealing with its background, smart ways to design calorimeters, gas measurements and computational techniques to deal with complex data. Novel techniques allow the connection between short term changes in heat production, protein turnover and substrate oxidation, e.g. using stable isotopes. Various applications of indirect calorimetry are addressed, including heat production measurements in producing animals, hatching eggs, companion animals and in animals housed under heat stress conditions. In addition, various ways of measuring methane emissions are discussed. This book is intended for scientists working or interested in calorimetry or metabolism research, or people designing calorimetry systems, opening their eyes for applications they did not yet think of.

When approaching experts for a contribution to this book, they all responded with enthusiasm, at that time probably not realising what they were getting themselves into. We highly appreciate their contributions and the discussions that followed, also contributing to our research network. We highly value the independent reviews that were performed on all chapters by the scientists that are listed and acknowledged on page 285.

Walter Gerrits & Etienne Labussière,
Editors

Foreword

Calorimetry techniques that allow an evaluation of heat production and energy balance have then been implemented with great success over the last century and have produced bases for (net) energy evaluation of feeds, factorial energy requirements, prediction models and energy responses of animals to environment factors. The basic technique mostly used has been the so-called indirect calorimetry, usually in open-circuit, with measurement of O_2 consumption and CO_2 production. Some of the facilities were closed in the eighties and nineties of the previous century. However, in recent years, there has been a series of new developments in measurements of sometimes new traits related to energy and nutrients metabolism of farm animals in existing facilities. New facilities are under development in countries like China, Brazil, USA, Australia. Similar trends in new developments have been observed for human calorimetry. These new developments are usually described very briefly in publications and have been poorly documented.

The editors of this book have asked specialists to provide an update and survey on classical but also on new methods employed to measure heat production and associated metabolism, thus providing, in 14 chapters, a modern update on various aspects of energy metabolism.

This book starts with a chapter discussing the background and validity of deriving heat production from gaseous exchanges (i.e. indirect calorimetry). The authors survey techniques used to increase the robustness and accuracy of metabolic measurements, using as practical approaches as possible. Designing and building calorimetry systems is a challenging task. Adapting the size of a respiration chamber to the metabolic mass inside is one of the ways to increase accuracy of short term changes in heat production. In Chapter 2, authors illustrate how to design and build such calorimetry systems. The importance of the automated determination of incurrent in addition to excurrent gas concentrations during metabolic measurement, and the use of those concentrations for minimising analyser drift and maximising measurement accuracy, are described in detail in Chapter 3. Chapters 4, 6 and 8 describe how basal metabolism (or fasting heat production), feeding and physical activity are calculated and contribute to total heat production. These chapters also provide an overview of how physical activity can be measured within the setting of indirect calorimetry. They focus also on the dynamics of heat production by using a mathematical evaluation of data obtained from short-term gas exchange measurements and circadian variation of fat and carbohydrate oxidation in relation to feed intake. Modelling of the dynamics of gas exchanges has led to new developments of calorimetry concepts and techniques. A recent development is the use of stable isotopes to measure substrates oxidation including protein turnover in parallel with gas exchanges (Chapters 5 and 7). It is described how whole body nutrient oxidation can be measured when tracer amounts of the nutrient of interest are continuously infused and when $^{13}CO_2$ expiration and ^{13}C enrichment of the precursor for oxidation are measured in blood plasma. Pros and cons of various methods, including the doubly labelled water technique for energy expenditure, and ^{15}N labelled glycine for protein turnover are discussed. While generally considered costly, measurements in respiration chambers are still considered as the golden standard for the measurement of gas exchange, and are often used as a reference for the development of alternative methods that can be applied in animals under a wide range of conditions. Chapters 9 and 10 discuss various alternative techniques to measure enteric methane production. This is very important with regard to finding methods to reduce the contribution of animals to climate change. As climatic and thermal stress can affect energy metabolism, it is necessary to have a well-designed, accurate and reliable air conditioning system inside a respiration chamber to rule out the unwanted effects of an uncontrollable climate. Chapter 11 describes how this can be done and authors elaborate about possible bias from biological and physical factors and how to deal with them. The authors of Chapter 12 discuss methods which have been developed in recent years to measure the development of the embryo and the utilisation of nutrients in hatching avian eggs by using indirect calorimetry. Similarly, Chapter 13 deals with developments in energy metabolism measurements in companion animals with emphasis on sources of variability in results. One may regret that no special consideration is given to calorimetry measurements in fish. Finally, a quite practical calculation exercise of indirect calorimetry heat production is given in Chapter 14 with a detailed description of the many steps involved in the calculation of heat production from raw data. These steps are rarely published in full detail, nor are they well documented for the training of students, researchers or technical staff.

Overall, this book is not a textbook on the theory of energy metabolism but an update of new aspects and approaches of the measurement of energy metabolism. It gives a very good insight in new metabolism traits to consider and new possibilities to study simultaneously energy, protein and nutrient metabolism using indirect calorimetry facilities. It describes how the principles and methods used in various laboratories have led to applications of the findings in practice and how new developments can add to the relevance of these applications. Compared to other books on energy metabolism, it is very innovative in its approach with very valuable information for scientists, teachers, students and people working with calorimetry systems in various species.

Jean Noblet, Director of Research at INRA

Martin Verstegen, Professor emeritus in Animal Nutrition, Wageningen University

Chapter 1: Deriving heat production from gaseous exchange: validity of the approach

W.J.J. Gerrits[1], J.J.G.C. van den Borne[1] and E. Labussière[2,3]*
[1]Animal Nutrition Group, Wageningen University, P.O. Box 338, 6700 AH Wageningen, the Netherlands; [2]INRA-UMR Pegase, 35590, Saint-Gilles, France; [3]Agrocampus Ouest – UMR PEGASE, 35000 Rennes, France; walter.gerrits@wur.nl

Abstract

The use of indirect calorimetry as a means to quantify heat production (Q) and net substrate oxidation has increased rapidly since the pioneering work of Lavoisier, and today, indirect calorimetry is often used as a reference for other measures of Q. Simple equations were developed and widely adopted to calculate the production of heat from the measurement of the production of CO_2 and CH_4, the consumption of O_2 and urinary nitrogen loss. The coefficients in these equations were derived from stoichiometry of the complete oxidation of carbohydrates, fat and protein. In this chapter, taking the Brouwer equation as an example, these calculations are explained and their validity to compute Q is discussed. Particular attention is paid to anaerobic fermentation and *de novo* lipogenesis from carbohydrates. It is concluded that Q can be predicted satisfactorily from the O_2 consumed and CO_2 produced using factors derived from oxidation of carbohydrates, fat and protein. Caution is required for corrections for CH_4, particularly in ruminants, although quantitatively, the impact of these corrections on the estimates of Q seems limited. The approach is also valid when *de novo* lipogenesis from carbohydrates is substantial. As it is assumed that O_2 and CO_2 pools within the body do not change within a measurement period, a critical attitude towards this approach is required whenever pushing a biological system with e.g. a diet or infusate while using it to estimate short term variation in Q. Examples of these include strenuous exercise or an environmental challenge, e.g. heat stress, during which energy containing intermediates can be temporarily stored in the body and/or the bicarbonate pool fluctuates. Lastly, the measurement of retained energy based on indirect calorimetry in growing subjects is compared with other techniques. Comparison with the comparative slaughter technique in growing subjects reveals that measurement based on indirect calorimetry tends to lead to higher estimates of retained energy, but treatment differences within studies are quantified accurately. It can be argued that the bias reflects a true difference in Q, related to differences in e.g. housing conditions. Comparison of indirect calorimetry with the carbon-nitrogen balance technique and with direct calorimetry, reveal a very close match between these techniques. In conclusion, although a critical attitude is appropriate, Q can be accurately derived from gaseous exchange measurements using the simple equations developed in the previous century, with coefficients based on the complete oxidation of carbohydrates, fat and protein.

Keywords: indirect calorimetry, Brouwer equation, comparative slaughter, carbon-nitrogen balance

DOI 10.3920/978-90-8686-261-0_1, © Wageningen Academic Publishers 2015

1.1 Introduction

As recently summarised by Frankenfield (2010), the use of indirect calorimery as a means to quantify substrate oxidation and heat production (Q) has increased rapidly since the discovery that air was a mixture of gases by John Mayow in 1674. Pioneering work of Lavoisier and De Laplace (1783) demonstrated that Q, measured by a triple chambered ice calorimeter coincided with the consumption of O_2 and the simultaneous production of CO_2. The widespread acceptance of the first law of thermodynamics (conservation of energy in a closed system) in the nineteenth century and the rapid technological development over the last two and a half centuries has led to indirect calorimetry being a commonly accepted technology to measure the production of heat following the oxidation of substrates in living subjects. Although the strong correlation between O_2 consumption, CO_2 production and Q was clear since the work of Lavoisier and De Laplace, the development of coefficients to predict Q from gaseous exchange required considerable research effort. Based on work conducted in the late 19th and early 20th century by scientists like Max Rubner, Nathan Zuntz, Adolf Magnus-Levy, Wilbur Atwater, Francis Benedict, John Haldane (see historic overview by Frankenfield, 2010), simple approaches were developed in the forties and fifties of the last century, with coefficients being derived from the complete oxidation of carbohydrates, fat and protein. The idea was based on the notion developed in 1838 by G.H. Hess that the heat produced in a chemical reaction is independent of the pathway between the initial and final states (Hess's law, from Blaxter, 1989). This implies that it does not make a difference whether a substrate is directly oxidized completely, or whether intermediate products such as lactic acid, fatty acids or ketone bodies, are produced which are subsequently transformed and oxidized at a later stage. In this chapter, the calculation of Q, from volumes of CO_2 produced (VCO_2, l) and O_2 consumed (VO_2, l) is explained and discussed, and independent validations of the technique are presented.

1.2 Calculating heat production from gaseous exchange

1.2.1 The principle

There are many variations on the theme, but for simplicity, we follow the formula published by Brouwer in 1958 and later adapted by the 'Sub-committee of Constants and Factors', published in 1965. This formula is used quite often in indirect calorimetry studies with livestock, and closely resembles approaches used by others (Blaxter, 1989), including humans (Livesey, 2002). The approach is based on measurements of VO_2, VCO_2 and Q, released upon combustion of 1 g of fat, carbohydrate and protein (Table 1).

From these data, it follows that total VO_2, VCO_2 and Q can be expressed as a function of the quantities of protein, fat and carbohydrates oxidized. After re-arranging of equations, Q (kJ), carbohydrate (g) and fat (g) oxidation can all be expressed as a function of VO_2 (l), VCO_2 (l) and oxidized protein (P, in g; urine-N × 6.25) (Equation 1, 2 and 3).

$$Q\ (kJ) = 16.175\ VO_2 + 5.021\ VCO_2 - 0.958\ P \quad (1)$$

$$\text{Carbohydrate oxidation (g)} = -2.965\ VO_2 + 4.170\ VCO_2 - 0.390\ P \quad (2)$$

$$\text{Fat oxidation (g)} = 1.718\ VO_2 - 1.718\ VCO_2 - 0.315\ P \quad (3)$$

Equation 1 is the basis for the Brouwer equation (uncorrected for CH_4, see Equation 4 below). Equations 2 and 3 provide a basis for the estimation of oxidation of carbohydrates and fats. Throughout the years it has become clear that these equations provide an approach to net, rather than true, rates of oxidation (e.g. Frayn, 1983). The potential and limitations of indirect calorimetry to provide indications of rates of substrate oxidation is dealt with in Chapter 5 (Van den Borne *et al.* (2015). Note that the combustion heat of proteins exceeds the heat released upon the biological oxidation of proteins listed in Table 1 (23.8 kJ/g vs 18.4 kJ/g). This is to account for the formation of urea as an energy containing end-product, excreted in urine.

1.2.2 Influence of composition of proteins, carbohydrates and fats on the coefficients

Minor variations on Equation 1 were adopted earlier by others. Differences in the coefficients adopted by various groups mainly originate from different monomer composition of proteins, fats and carbohydrates (Appendix 1A). Most constants were based on early studies by Zuntz (1897), Rubner (1901), Magnus-Levy (1907), Benedict and Talbot (1914), Benedict and Tompkins (1916) and Lusk (1917, 1918) as reviewed by Brockway (1987). For carbohydrates, the amount of oxygen required (in l per g) is 0.829 for starch and glycogen, 0.786 for sucrose and lactose, and 0.746 for glucose (Appendix 1B). For sugar alcohols, such as sorbitol and glycerol, these values may deviate. For protein, the amount of oxygen required (in l per g) is 1.008 for beef muscle and casein, and 1.052 for zein. The amino acid composition of proteins determines the stoichiometry of protein oxidation (Appendix 1C). For example, O_2 consumption varies from 2.2 l per g N for arginine to 16.01 l per g N for phenylalanine. In practice, 1 g of urinary N is associated with 0.854 to 1.581 l of O_2 consumption when comparing the major protein sources, and estimates of protein oxidation from N excretion can be in error from +14 to -39% of the true value (Livesey and Elia, 1988; Weir, 1949). For fat, the

Table 1. Gaseous exchange and heat production of oxidized substances (from Brouwer, 1965).[1]

Substance	Consumed with oxidation of 1 g O_2 (l)	Released with oxidation of 1 g		RQ
		CO_2 (l)	Q (kJ)	
Starch	0.829	0.829	17.57	1.00
Fat	2.013	1.431	39.75	0.71
Protein	0.957	0.774	18.41	0.81

[1] Q = heat production; RQ = respiratory quotient.

amount of oxygen required (in l per g) is 2.015 for maize oil and 2.037 for lard. Frayn (1983) used palmitoyl-stearoyl-oleoyl-glycerol ($C_{55}H_{104}O_6$) as a typical fat, which requires 2.029 l of oxygen to oxidize 1 g. In general, variation in the respiratory quotient (RQ) between long chain fatty acids is rather small, but RQ increases with increasing degree of unsaturation (Appendix 1D). For C18 fatty acids, for example, RQ increases from 0.70 to 0.74 with increasing degree of unsaturation, and even higher RQ's are reported for arachidonic acid, eicosapentaenoic acid and docosahecaenoic acid. However, when comparing 40 common dietary fat sources, RQ ranged between 0.71 and 0.73, indicating that the assumption of a constant RQ is acceptable for fat (Livesey and Elia, 1988).

1.3 Validity of the approach

The coefficients published by Brouwer (1965) are based on the measurement of the heat of combustion, VO_2 and VCO_2 upon the complete oxidation of starch, fat and protein. Obviously, when indirect calorimetry is part of the measurement of a complete energy balance, energy rich excretion products have to be accounted for. This is common practice for urinary excretions and methane emissions, but not always for components like H_2 and ketone bodies that can also occur in breath. In recent studies in dairy cattle, increased emissions of both H_2 (Van Zijderveld *et al.*, 2011) and N_2O (Petersen *et al.*, 2014) were measured in response to feeding nitrate as a methane mitigation strategy. Although the contribution to biology or greenhouse gas emissions can be considerable, the energetic losses are thought to be very low. In the study of Van Zijderveld *et al.* (2011), the energy emitted as H_2 amounted less than 0.2% of the measured Q.

Following the logic of the approach used by Brouwer and others, the net, rather than true rates of substrate oxidation are required, as the energy liberated when glucose is first converted into an intermediate product, and subsequently oxidised is the same as for its direct oxidation. It then is tempting to focus on the simplified stoichiometry of the overall reactions, and compare the heat of combustion of a substrate with the Q predicted by the Brouwer equation (e.g. Livesey, 2002). However, several metabolic pathways exist that are important to consider. These may be pathways in which the stoichiometry deviates from that of a complete oxidation, and during which these fixed relationships between oxidative metabolism and Q may not apply. Some relevant examples are discussed below.

1.3.1 Anaerobic fermentation

Fermentation of dietary carbohydrates may, depending on the species, be an important source of energy, typically being largely anaerobic, and yielding microbial biomass, short chain fatty acids and fermentation heat. Whereas in hindgut fermenters, microbial biomass production largely appears in faeces and can be regarded as an energetic loss, in ruminants, 80-90% of the microbial biomass produced in the rumen may be digested in the intestinal tract, yielding substrates for sequestration or later oxidation. For ruminant animals, the subcommittee led by

Brouwer added a negative correction factor proportional to the methane production (VCH_4, l) to Equation 1, yielding Equation 4, often referred to as the Brouwer equation:

$$Q = 16.175\ VO_2 + 5.021\ VCO_2 - 0.958\ P - 2.167\ VCH_4 \qquad (4)$$

This correction factor actually alters the relationship between the measured gas exchange and Q as presented in Equation 1, and is based on the aerobic oxidation of CH_4, i.e. the quantities of O_2 consumed and CO_2 produced during the complete oxidation of CH_4, known to yield 39.5 MJ/l. As the aerobic oxidation of CH_4 does not occur in the ruminant, it can be questioned to what extent this assumption is justified. Maybe the approach to also assume oxidative metabolism of an energy containing end-product is taking the Brouwer approach one step too far. Applying Equation 4, it appears that for every l of VCH_4 reduced, Q is increased by 2.167 kJ. As VCO_2 will simultaneously increase by 1 l ($CO_2 + 4H_2 \rightarrow CH_4 + 2H_2O$), it follows that 18% ([5.021+2.167]/39.5) of the energy spared from CH_4 reduction is inevitably lost as Q, leaving 82% to be used for anabolism. As recently discussed by Van Zijderveld (2011), this approach can be questioned, as the alternative use of hydrogen, the energetic precursor of methane, when not used to produce methane, may well be very different.

In analogy to Equations 1 to 3, Brouwer's equation, including the methane correction (Equation 4), was also re-arranged to enable the calculation of the net rates of oxidation of carbohydrates and fats. These equations, originating from Andersen (1922), presented by Brouwer (1958), have the same coefficients for VO_2 and VCO_2 as presented in Equation 2 and 3. This causes these equations to predict net rates of substrate oxidation, not adding up to the total Q. As argued by Van den Borne *et al.* (2015) in Chapter 5, the equations for predicting the net rates of substrate oxidation, including the coefficients for $VCH_{4,}$ should therefore not be used.

In hindgut fermenters, anaerobic fermentation would lead to increased faecal loss of microbial biomass. In analogy to the correction for the oxidation of proteins (Equations 3 and 4) this would imply that there is an energy containing end product to account for. In the estimation of Q from gaseous exchange in hindgut fermenters, this has to be accounted for. In addition, hydrogen sinks may be inadequate to capture all H_2 produced, and when carbohydrate fermentation is substantial, N excretion may be diverted from urine to faeces. The latter may lead to an underestimate of protein oxidation, as actually the urea is recycled into the colon (pigs, e.g. Gerrits *et al.*, 2012) or into the rumen (e.g. Berends *et al.*, 2014). This influences the estimation of Q by using the Brouwer or similar equations. Livesey (2002) has listed a number of potential correction factors for Q with increasing carbohydrate fermentation in humans.

1.3.2 *De novo* lipogenesis from glucose

The *de novo* synthesis of fatty acids from glucose is an example often used to discuss the validity of the approach by Brouwer. In rapidly growing monogastric animals as well as glucose infused patients, an RQ exceeding 1 has been reported. As discussed by Elia and Livesey

(1988), *de novo* fatty acid synthesis from glucose can be considered to consist of two parts: (1) the conversion of glucose into fatty acids, a process in which 4 of the 6 carbon atoms of glucose can be sequestered in a fatty acid whereas the 2 others are excreted as CO_2, without a need for O_2. The RQ of this step would be ∞; (2) the use of glucose to provide part of the NADPH required for fatty acid synthesis, a process with an RQ of 1. The variation in overall stoichiometry of *de novo* lipogenesis among literature sources is quite large, but was analysed by Elia and Livesey (1988) to be related to variations in the estimation of the amount of glucose needed to support fatty acid synthesis and providing the required glycerol. An example of the overall stoichiometry of the formation of 1 mol of triacylglycerol (TAG; containing oleate, stearate and palmitate esterified to glycerol) from glucose would be according to Equation 5 (taken from Frayn, 1983):

$$83 \text{ glucose} + 30\ O_2 \rightarrow 6\ \text{TAG} + 168\ CO_2 \quad (5)$$

The RQ for this overall process would be 168/30 = 5.6. According to Blaxter (1989) and Elia and Livesey (1988), the Q calculated from VO_2 and VCO_2 is valid during both net oxidative and net lipogenic states. Therefore, it appears that the heat produced during lipogenesis can be predicted by factors derived from substrate oxidation (Equation 5), and this approach is also valid in rapidly growing subjects with a high starch intake. The proof of Elia and Livesey (1988) for this approach was provided by algebraic equations derived from theory, whereas Blaxter based it on the comparison of the ΔH of the conversion of glucose to palmitic acid with the Q calculated from VO_2 and VCO_2. Direct *in vivo* proof, however, is lacking, as the heat produced from *de novo* fatty acid synthesis cannot be separated from other heat produced. Indirect evidence may be obtained from a comparison of the indirect calorimetry approach and the carbon-nitrogen (CN) balance technique (Section 1.4).

1.3.3 Other metabolic processes

In the relationship between gaseous exchange and Q, the net, rather than the true substrate oxidation rates are relevant. When glucose, for example, is converted into fat and then oxidized, the same heat arises as when the glucose is directly oxidized. Depending on the kinetics of such conversion and the duration of the measurements, this phenomenon may be accurately quantified or it may disturb the outcome of a calorimetry measurement. There are many other metabolic pathways that deserve attention, some of them covered by Frayn (1983). For example, gluconeogenesis from amino acids is a process with a very low RQ. Anaerobic glycolysis yields heat and lactate, without any effect on VCO_2 or VO_2. Resynthesis of glucose from lactate in the liver would require ATP, likely generated from oxidative phosphorylation, thus involving CO_2 production and O_2 consumption. This kind of substrate cycling is therefore not expected to yield large errors in the estimation of Q from VCO_2 and VO_2. A possible exception is the short term measurement of Q when lactate or other intermediate products are accumulating in the body. In addition, it may disturb the estimation of the net rates of carbohydrate and fat oxidation from gas exchange (Chapter 5; Van den Borne *et al.*, 2015).

1.4 Comparison with other methods

In the literature, indirect calorimetry is quite often used as a reference for other methodologies for the measurement of Q. However, several authors have challenged its validity by comparing an energy and protein balance with its heat component measured by indirect calorimetry against other techniques, e.g. direct calorimetry, the carbon-nitrogen balance technique (CN), or against the comparative slaughter (CS) method. Comparisons between direct and indirect calorimetry are rare. In a study with sheep performed in 1967 at the Rowett Research Institute, indirect calorimetry was compared with direct calorimetry using a gradient layer, direct calorimeter. Results of 65 measurements using direct calorimetry were compared with 60 observations performed using indirect calorimetry on the same sheep, at varying levels of feeding, fed a mixture of concentrates. Results indicated a close match between the two techniques (well within 5%) (W.P. Flatt, personal communication). In adult humans, Q estimated from indirect calorimetry was slightly higher than heat loss measured by direct calorimetry (Dauncey, 1980; Seale *et al.*, 1990; Table 2).

When performing a CN balance, the N balance is measured by the quantified difference of N inputs and outputs via faeces, urine, and aerial NH_3, the latter following volatilisation of NH_4^+ originating from urea. Similar methodology is often used in combination with indirect calorimetry. Measurement of the C balance requires quantitative measurement of all C inputs (food) and losses (urine, faeces, VCH_4, VCO_2), using similar equipment as indirect calorimetry, with the exception of O_2 analysers. Energy and protein balances can be estimated, and Q derived by difference between metabolisable energy intake and retained energy (RE). In this way, the assumptions underlying the Brouwer equation or equivalents are avoided. Within-day patterns of Q, however, cannot be examined. The latter is also impossible using the CS technique, typically performed in growing subjects, preferably over a rather large weight range. The CS technique measures protein, fat and RE by difference between their respective quantities at the end and start of the measurement. For estimating the starting protein, fat and energy mass, a representative initial group of animals is sacrificed. Metabolisability of the ingested energy has to be measured and considered representative for the entire growth trajectory. Production of combustible gases is usually ignored. It is generally acknowledged that the CS technique is most reliable, but any errors made would typically lead to underestimation of the ER, whereas errors in the methods based on gaseous exchange would point in the opposite direction (Table 2).

From the comparison of indirect calorimetry with CS it appears that there may be a systematic overestimate of energy retained by indirect calorimetry, but experimental contrasts are generally well predicted. In a study by Noblet *et al.* (1987) in growing pigs, indirect calorimetry overestimated energy retained measured by CS by 10%, but changes in energy retained due to differences in lysine and protein intake were quantified accurately. Likewise, in a study by Quiniou *et al.* (1995) in growing pigs, indirect calorimetry overestimated energy retained by 4% when compared with CS, but differences, induced by dietary protein content were quantified accurately. A study by Le Dividich *et al.* (2007) in milk fed piglets, incremental changes in

Table 2. Comparison of the estimation of retained energy (RE) by indirect calorimetry when compared with comparative slaughter methods (CS) or carbon-nitrogen balances (CN) and direct calorimetry (DC).[1]

Reference	Species	No balances	No subjects	Experimental contrasts	Reference method	RE calorimetry/RE reference method[2]	Note
Just *et al.*, 1982	pigs 20-90 kg BW	480	80	dietary protein (18-20%), crude fibre (3.1-9.7%)	CS	-0.3%, ranging from -13.3 to 12.2%	large range in differences, not clearly diet dependent; correlations may have been strong, but are not presented
Noblet *et al.*, 1987	pigs 20-55 kg BW	24	24	dietary lysine (0.67-0.81%) and protein (15-18%)	CS	10.2%	treatment differences well predicted, difference in Q may be real, related to differences in housing conditions
Quiniou *et al.*, 1995	pigs 20-50 kg BW	12	12	dietary protein (18-24%)	CS	4.4%	good match treatment difference
Le Dividich *et al.*, 2007	pigs 1.5-2.5 kg BW	18	18	3 levels of milk feeding	CS	5.7% (incremental)	similar incremental response to metabolisable energy intake from milk
Christensen *et al.*, 1988	pigs 20-120 kg BW	326	56	6 BW classes, feeding levels, different diets	CN	-4% (incremental)	larger differences may occur when high body fat gain or at high fibre diets (not substantiated)
Thorbek, 1975	pigs 20-90 kg BW	367	48	BW classes, diet composition	CN	-2.4% (incremental)	
Zhao *et al.*, 1997	rats 90-250 g	72	48	low and high fibre diet, three ambient temperatures	CN	<2% (incremental)	differences small and negligible, similar responses to ambient temperature and fibre level
Vermorel *et al.* 1980	calves 60-250 kg BW	57	57	4 BW classes, diet composition (milk replacer/solid feed)	CN	<1%	residual variability is lower with results from indirect calorimetry
Seale *et al.* 1990	humans 85 kg BW	4	4	-	DC	-0.6%	four healthy adult men
Dauncey *et al.* 1980	humans 60 kg BW	24	8	3 levels of energy intake	DC	-1.2%	four healthy males and 4 healthy females

[1] BW = body weight; Q = heat production.

[2] (100-100×RE measured by indirect calorimetry/ RE measured by the reference method); whenever incremental is indicated, the difference refers to the increase in RE calorimetry/increase in RE the reference method, imposed by the experimental contrasts referred to.

energy retained measured by indirect calorimetry overestimated energy retained by 6% when compared with CS, but differences related to different feeding levels and diet compositions were quantified accurately. The exception is the (large) dataset by Just *et al.* (1982), from which this information cannot be extracted. On average, energy retained measured by indirect calorimetry was virtually the same as RE measured by CS. However, differences induced by dietary protein and crude fibre contents, measured by indirect calorimetry were poorly related to those measured by CS (variation from -13 to +12%). As noted by Noblet *et al.* (1987), CS has to be conducted over a large range in body weights whereas indirect calorimetry is usually performed over a week or less (quite often respiration measurements are done over a 24 or 48 h period). It is therefore likely that part of the difference presented in literature, attributed to methodological errors, may actually reflect a real difference in Q related to slightly different or varying housing conditions (Noblet *et al.*, 1987).

Indirect calorimetry and CN methods show quite comparable results, and a much closer absolute resemblance of measured RE than do indirect calorimetry and CS. The measurement of VCO_2 is common for both techniques, and VO_2 is not needed for CN. Experimental contrasts are predicted accurately. In most cases, the slightly higher estimates of Q by indirect calorimetry lead to lower estimates of RE. Thorbek (1975, pigs), Christensen *et al.* (1988, pigs) and Zhao *et al.* (1997, rats) concluded from their systematic comparison of indirect calorimetry and CN methods, from a discussion of sources of errors, that there is no reason to prefer one over the other. It should be noted, however, that the obvious advantage of indirect calorimetry over CN is that indirect calorimetry methods allow the estimation of within day variation in Q, and provide a good lead to estimate the net rate of substrate oxidation. In addition, there are indications that indirect calorimetry produces results with lower individual variability (Vermorel *et al.*, 1980; calves).

Overall, data reported in the literature to compare techniques measuring or calculating heat produced by living organisms indicate a close agreement between results obtained from the different techniques. Nevertheless, it should be noted that as most of the papers on this topic aimed at validating one particular technique, disagreeing results are less likely to be reported.

In conclusion, Q can be predicted satisfactorily from the O_2 consumed and CO_2 produced using factors derived from the oxidation of carbohydrates, fat and protein. Caution is required for corrections for CH_4. A critical attitude towards this approach is required whenever pushing a biological system with a diet, infusate or when using it to estimate short term variation in Q. Examples of these include strenuous exercise or an environmental challenge, e.g. heat stress, during which energy containing intermediates can be temporarily stored in the body and/or the bicarbonate pool fluctuates. Comparison with other methods, i.e. the comparative slaughter technique, direct calorimetry and the carbon nitrogen balance technique generally reveal a close match between these techniques. Therefore, although a critical attitude is appropriate, Q can be accurately derived from gaseous exchange measurements using the simple equations developed in the previous century, with coefficients based on the stoichiometry of the complete oxidation of carbohydrates, fat and protein.

References

Berends, H., Van den Borne, J.J.G.C., Røjen, B.A., Van Baal, J. and Gerrits, W.J.J., 2014. Urea recycling contributes to nitrogen retention in calves fed milk replacer and low-protein solid feed. Journal of Nutrition 144 (7): 1043-1049.

Blaxter, K.L., 1989. General principles. In: Blaxter K.L. (ed.) Energy metabolism in animals and man. Cambridge University Press, Cambridge, UK, pp. 6-22.

Brockway, J.M., 1987. Derivation of formulae used to calculate energy expenditure in man. Human Nutrition: Clinical Nutrition 41C: 463-471.

Brouwer, E., 1958. On simple formulae for calculating the heat expenditure and the quantities of carbohydrate and fat metabolized in ruminants, from data on gaseous exchange and urine N. In: Thorbek, G. and Aersoe, H. (eds.) 1st Symposium on Energy metabolism, Copenhagen, DK. Principles, methods and general aspects. EAAP publication no. 8. EAAP, Rome, Italy, pp. 182-194.

Brouwer, E., 1965. Report of the Sub-committee on Constants and Factors. In: Blaxter, K.L. (ed.). Energy metabolism, proceedings of the 3rd Symposium, Troon, Scotland, EAAP Publication no 11. Academic Press, London, UK, pp. 441-443.

Christensen, K., Chwalibog, A., Henckel, S. and Thorbek, G., 1988. Heat production in growing pigs calculated according to the RQ and CN methods. Comparative Biochemistry and Physiology 91A: 463-468.

Dauncey, M.J., 1980. Metabolic effects of altering 24 h energy intake in man, using direct and indirect calorimetry. British Journal of Nutrition 43: 257-269.

Elia, M. and Livesey, G., 1988. Theory and validity of indirect calorimetry during net lipid synthesis. American Journal of Clinical Nutrition 47: 591-607.

Frayn, K.N., 1983. Calculation of substrate oxidation rates *in vivo* from gaseous exchange. Journal of Applied Physiology: Respiration Environmental Exercise Physiology 55(2): 628-634.

Frankenfield, D.C., 2010. On heat, respiration and calorimetry. Nutrition 26: 939-950.

Gerrits, W.J.J., Bosch, M.W. and Van den Borne, J.J.G.C., 2012. Quantifying resistant starch using novel, *in vivo* methodology and the energetic utilization of fermented starch in pigs. Journal of Nutrition 142: 238-244.

Just, A., Fernandez, J.A. and Jorgensen, H., 1982. Nitrogen balance studies and nitrogen retention. In: Laplace, J.P., Corring, T. and Rerato, A. (eds.) Digestive physiology in the pig. Institut National de la Recherche Agronomique, Paris, France, pp. 111-119.

Lavoisier A.L. and De Laplace, P.S., 1783 [1994]. Memoir on heat, read to the Royal Academy of Sciences, 28 June 1783. Obesity Research 2(2): 189-202.

Le Dividich, J., Marion, J. and Thomas, F., 2007. Energy and nitrogen utilization of sow colostrum and milk by the piglet. Canadian Journal of Animal Science 87: 571-577.

Livesey, G., 2002. Thermogenesis associated with fermentable carbohydrate in humans, validity of indirect calorimetry, and implications of dietary thermogenesis for energy requirements, food energy and body weight. International Journal of Obesity 26: 1553-1569.

Livesey, G. and Elia, M., 1988. Estimation of energy expenditure, net carbohydrate utilization, and net fat oxidation and synthesis by indirect calorimetry: evaluation of errors with special reference to the detailed composition of fuels. American Journal of Clinical Nutrition 47: 608-628

Noblet, J., Henry, Y. and Dubois, S., 1987. Effect of protein and lysine levels in the diet on body gain composition and energy utilization in growing pigs. Journal of Animal Science 65: 717-726.

Petersen, S.O., Hellwing, A.L.F., Brask, M., Hojberg, O., Poulsen, M., Zhu, Z. and Lund, P., 2014. Nitrous oxide emission partly alleviates the methane mitigation effect of nitrate intake by dairy cattle. In: Livestock, Climate Change and Food Security Conference, May 19-20, 2014 Madrid, Spain, p. 46 [abstract].

Quiniou, N., Dubois, S. and Noblet, J., 1995. Effect of dietary crude protein level on protein and energy balances in growing pigs: comparison of two measurement methods. Livestock Production Science 41: 51-61.

Seale, J.L., Rumpler, W.V., Conway, J.M. and Miles, C.W., 1990. Comparison of doubly labelled water, intake-balance and direct- and indirect-calorimetry methods for measuring energy expenditure in adult men. American Journal of Clinical Nutrition 52: 66-71.

Thorbek, G., 1975. Studies on energy metabolism in growing pigs. II Protein and fat gain in growing pigs fed different feed compounds. Efficiency of utilization of metabolizable energy for growth. Publication no 424, Report from the National Institute of Animal Science. Landhusholdningsselskabet, Copenhagen, Denmark, pp. 80-92.

Van Zijderveld, S.M., 2011. Dietary strategies to reduce methane emissions from ruminants. PhD thesis, Wageningen University, Wageningen, the Netherlands, pp. 95-115.

Van Zijderveld, S.M., Gerrits, W.J.J., Dijkstra, J., Newbold, J.R., Hulshof, R.B.A. and Perdok, H.B., 2011. Persistency of methane mitigation by dietary nitrate supplementation in dairy cows. Journal of Dairy Science 94: 4028-4038.

Van den Borne, J.J.G.C., Heetkamp, M.J.W., Alferink, S.J.J. and Gerrits, W.J.J., 2015. Moving from a complete energy balance towards substrate oxidation: use of stable isotopes. In: Gerrits, W.J.J. and Labussière, E. (eds.) Indirect calorimetry. Techniques, computations and applications. Wageningen Academic Publishers, Wageningen, the Netherlands, pp. 87-114.

Vermorel, M., Bouvier, J.C. and Geay, Y. 1980. Utilisation de l'énergie des aliments par le veau ruminant. Influence de l'apport de lait, du niveau d'alimentation et de l'âge du veau. Annales de Zootechnie 29: 65-86.

Weir, J.B., 1949. New methods of calculating metabolic rate with special reference to protein metabolism. Journal of Physiology 109: 1-9.

Zhao, X.Q., Jorgensen, H., Just, A., Du, J.Z. and Eggum, B.O., 1997. Energy expenditure and quantitative oxidation of nutrients in rats (*Rattus norvegicus*) kept in different thermal environments and given two levels of dietary fiber. Comparative Biochemistry and Physiology 116A (4): 351-359.

Appendix 1A.

Constants for protein, fat and carbohydrate when oxidized in the body.

Oxidized substrate	VO_2 (l/g)	VCO_2 (l/g)	Heat (kJ/g)	RQ	Reference
Carbohydrate[2]	0.746	0.746	15.65	1.00	Lusk, 1917
Fat	2.019	1.427	39.58	0.70	
Protein	0.966	0.774	18.07	0.80	
Carbohydrate[1]	0.829	0.829	17.50	1.00	Lusk, 1928
Fat	2.019	1.427	39.58	0.70	
Protein	0.967	0.775	18.06	0.80	
Carbohydrate[1]	0.829	0.829	17.57	1.00	Carpenter, 1948
Fat	2.013	1.431	39.75	0.71	
Protein	0.957	0.774	18.41	0.81	
Carbohydrate[1]	0.829	0.829	17.57	1.00	Brouwer, 1965
Fat	2.013	1.431	39.75	0.71	
Protein	0.957	0.774	18.41	0.81	
Carbohydrate[2]	0.746	0.746	-	1.00	Frayn, 1983
Fat	2.029	1.430	-	0.70	
Protein	0.966	0.782	-	0.81	
Carbohydrate[2]	0.746	0.746	15.64[3]	1.00	Jéquier and Felber, 1987
Fat	2.031	1.427	39.83[3]	0.70	
Protein	0.966	0.744	18.60[3]	0.81	
Carbohydrate[2]	0.746	0.746	17.5	1.00	Livesey and Elia, 1988
Fat	2.019	1.438	39.5	0.71	
Protein	1.002	0.836	19.68	0.83	
Carbohydrate[1]	0.829	0.829	21.10	1.00	Westerterp, 2008
Fat	2.019	1.427	19.60	0.71	
Protein	0.967	0.775	18.70	0.80	

[1] As starch.

[2] As glucose.

[3] Calculated from publication.

Appendix 1B.

Oxygen consumption (VO_2), carbon dioxide production (VCO_2), heat of combustion ($-\Delta H_c$), and respiratory quotient (RQ) for carbohydrates and polyols (Blaxter, 1989; Livesey and Elia, 1988; Lusk, 1928).

Oxidised substrate	VO_2 (l/g)	VCO_2 (l/g)	$-\Delta H_c$ (kJ/g)	RQ
Starch	0.829	0.829	17.48	1.00
Glycogen	0.829	0.829	17.49-17.51	1.00
Sucrose	0.786	0.786	16.48-16.57	1.00
Lactose	0.786	0.786	16.50	1.00
Glucose	0.746	0.746	15.56-15.65	1.00
Fructose	0.747	0.747	15.61	1.00
Glycerol	0.851	0.732	17.98-18.03	0.86
Xylitol	0.849	0.772	16.96	0.91
Sorbitol	0.800	0.736	16.71	0.92
Maltitol	0.814	0.781	16.98	0.96
Lactitol	0.814	0.781	16.98	0.96

Appendix 1C.

Oxygen consumption (VO_2), carbon dioxide production (VCO_2), heat of combustion ($-\Delta H_c$), and respiratory quotient (RQ) for amino acids and conventional food protein (modified and calculated from Livesey and Elia, 1988).

Oxidised substrate	VO_2 (l/g)	VCO_2 (l/g)	$-\Delta H_c$ (kJ/g)[1]	RQ
Alanine	0.75	0.63	18.28	0.83
Aspartic acid	0.50	0.59	11.25	1.17
Asparagine	0.51	0.51	11.37	1.00
Glutamic acid	0.69	0.69	14.99	1.00
Glutamine	0.69	0.61	15.11	0.89
Glycine	0.45	0.45	11.47	1.00
Proline	1.07	0.88	24.84	0.82
Serine	0.53	0.53	13.00	1.00
Cysteine	0.83	0.47	20.05	0.56
Tyrosine	1.18	1.05	25.27	0.89
Isoleucine	1.28	0.94	28.81	0.73
Leucine	1.28	0.94	28.82	0.73
Lysine	2.15	1.52	23.79	0.71
Methionine	1.13	0.68	25.34	0.60
Phenylalanine	1.36	1.15	29.44	0.85
Threonine	0.75	0.66	17.16	0.88
Tryptophan	1.32	1.09	26.86	0.83
Valine	1.15	0.86	26.26	0.75
Arginine	0.71	0.52	15.83	0.73
Histidine	0.72	0.65	17.64	0.90
Conventional food proteins			17.43-21.15	0.81-0.87

[1] Corrected for heat of combustion of urinary urea.

Appendix 1D.

Oxygen consumption (VO_2), carbon dioxide production (VCO_2), heat of combustion ($-\Delta H_c$), and respiratory quotient (RQ) for fatty acids, triglycerides, fats and oils (modified and calculated from Livesey and Elia, 1988).

Oxidized substrate	Carbon chain length:double bonds	VO_2 (l/g)	VCO_2 (l/g)	$-\Delta H_c$ (kJ/g)	RQ
Acetic acid	2:0	0.75	0.75	14.56	1.00
Propionic acid	3:0	1.06	0.91	20.62	0.86
Butyric acid	4:0	1.27	1.02	24.78	0.80
Triglycerides[1]	6:0	1.60	1.21	31.48	0.76
	8:0	1.74	1.29	34.19	0.74
	10:0	1.84	1.34	36.08	0.73
	12:0	1.91	1.38	37.47	0.72
	14:0	1.97	1.40	38.54	0.71
	16:0	2.02	1.41	39.38	0.70
	16:1	1.99	1.43	39.05	0.72
	18:0	2.05	1.44	40.07	0.70
	18:1	2.03	1.44	39.77	0.71
	18:2	2.00	1.46	39.48	0.73
	18:3	1.98	1.47	39.18	0.74
	20:0	2.08	1.46	40.63	0.70
	20:4	1.99	1.50	39.56	0.75
	20:5	1.97	1.50	39.29	0.76
	22:0	2.11	1.45	41.11	0.69
	22:1	2.09	1.46	40.87	0.70
	22:6	1.97	1.50	39.34	0.76
Conventional food fats				37.69-40.09	0.71-0.74

Chapter 2: Design of climate respiration chambers, adjustable to the metabolic mass of subjects

M.J.W. Heetkamp[1], S.J.J. Alferink[2], T. Zandstra[2], P. Hendriks[3], H. van den Brand[1] and W.J.J. Gerrits[2]*

[1]Wageningen University, Department of Animal Sciences, Adaptation Physiology Group, P.O. Box 338, 6700 AH Wageningen, the Netherlands; [2]Wageningen University, Department of Animal Sciences, Animal Nutrition Group, P.O. Box 338, 6700 AH Wageningen, the Netherlands; [3]Wageningen University, Technical services; marcel.heetkamp@wur.nl

Abstract

Open-circuit respiration chambers can be used to measure gas exchange and to calculate heat production (Q) of humans and animals. When studying short-term changes in Q, the size of the respiration chamber in relation to the subject of study is a point of concern. The washout time of a chamber, defined as the proportion of the chamber size to the rate of ventilation, needs to be minimised for accurate measurement of short term changes in Q. To date, most respiration chambers have a fixed size, limiting their use for different species, sizes and number of subjects, thus hampering studying the short term dynamics of Q. This chapter presents various approaches to the design, construction and testing of respiration chambers, adjustable to the metabolic mass inside. As investment costs for constructing respiration chambers are high, flexibility in the use of chambers can contribute substantially to an efficient use of resources. Furthermore, an outline is given to sensor criteria and calibration and finally, the validation of a whole indirect-calorimetric system is described. Air leak tolerance is defined and attention is paid to caretaking of animals, excreta collection and animal and personnel welfare and safety. Respiration facilities, recently constructed at Wageningen University are presented as an example. Briefly, four 45 m^2 climate chambers can be used, e.g. for heat or cold stress experiments, to incubate eggs or as a hygiene barrier. Within each chamber, one or two smaller airtight, size adaptable respiration rooms, can be built in where ambient temperature, humidity and ventilation rate can be controlled independently. In each respiration room a wide range of ventilation flow rates can be accomplished and both hypobaric and hyperbaric air pressure control can be established, allowing energy metabolism experiments with specific pathogen free animals (hyperbaric) or trials with infectious agents (hypobaric).

Keywords: indirect calorimetry, calorimeter, short-term accuracy, multi-dimensional, validation

2.1 Introduction

Open-circuit respiration chambers are frequently used to measure heat production (Q) of human and animal subjects. Although the technique is also applied to study short-term

DOI 10.3920/978-90-8686-261-0_2,

changes in Q, accuracy of measurement of short term changes is hampered by the size of the respiration chamber in relation to the size of the subject of study (Van Milgen *et al.*, 1997; Lighton, 2008). An instantaneous change in O_2 consumption (VO_2) or CO_2 production (VCO_2) inside the respiration chamber results in a small and prolonged response at the gas analyser level. Relatively large volumes of the respiration chambers related to the metabolic mass of the subject of study and short intervals between measurements leads to noisy data when Q is calculated (Van Klinken *et al.*, 2013). The extent of the dampening effect is directly proportional to the washout time of the chamber, which is defined as the proportion of the chamber size to the rate of ventilation. Typically, the washout time can vary, e.g. from 60 min for two cows in a 80,000 l chamber, to 45 min for 20 hatching eggs in a 260 l chamber, to 20 min for a group of three 15 kg piglets in a 1,800 l chamber (calculations based on Wageningen chambers). Despite sophisticated mathematical techniques to deal with such problems (Labussière *et al.*, 2013; Van Klinken *et al.*, 2012, 2013), maximising the metabolic mass of subject(s) inside a fixed sized chamber is important for accurate measurement of short term changes in Q. This is, for example, essential when estimating activity related Q or in studies where indirect calorimetry is combined with measurements of stable isotope ($^{13}CO_2$ / $^{12}CO_2$) ratios to study within day variation in substrate oxidation (Chapter 5; Van den Borne *et al.*, 2015). For example, ventilated hoods have a washout time near zero, but free movement of the subjects is then restricted and group housing of animals is impossible. Therefore, adaptability of the size of the chamber to the metabolic activity inside (i.e. number and size of the subjects) is the preferred way to improve the accuracy of measurement of short term changes in Q. In addition, as investment costs for constructing respiration chambers are high, flexibility in the use of chambers in various configurations can contribute substantially to an efficient use of resources.

The combination of climate control and respiration chambers allows the study of metabolic responses to changes in climate conditions, e.g. variation in temperature and/or relative humidity (Aerts *et al.*, 2000; Huynh *et al.*, 2005; Renaudeau *et al.*, 2013), or to create an environment free of specific pathogens (Verstegen *et al.*, 1993) or even with specific pathogens (Van Dam *et al.*, 1997; Van Eerden *et al.*, 2006).

The investment costs of the construction of climate respiration chambers are largely determined by the costs related to climatisation in all its facets as central cooling and heating systems, heat exchangers, air distribution systems, sensors, control valves and controllers, multiplexing meters, climate control software, etc. Constructing multiple respiration rooms within one climatised chamber can increase the number of experimental units that can be measured simultaneously, thereby greatly contributing to an efficient use of experimental resources. When constructing separate rooms in a chamber of course consequences on the accuracy of the climate control should be anticipated. In addition, animal welfare issues have to be considered, for example, the possibility of avoiding social isolation during the measurements of energy metabolism (Bourguet *et al.*, 2011; Cimadevilla *et al.*, 1997; Van Adrichem and Vogt, 1993) by maintaining visual and/or auditory contact during measurements.

In this chapter, important issues related to the design of size adjustable, climatised respiration systems, particularly for large animals are described. The facilities constructed at Wageningen University, the Netherlands, will be used as an example throughout this chapter. Technical data, used materials, and testing protocols are presented to allow the reader to use and apply information selectively.

2.2 Techniques to reduce the size of a respiration chamber

A rather simple and cheap technique to decrease volume in a fixed sized chamber is to insert airtight objects. Plastic or metal boxes (Figure 1A) or balloons (like PVC coated polyester tissue gas bags, Figure 1B), can be used as long as air (diffusion) tightness of these objects is secured.

Another way to reduce size can be the use of a single movable wall, as presented in Figure 2. By repositioning the movable wall, floor size and thus volume of the chamber can be reduced in a continuous range, meaning that short term measurements of different species of different size can be performed accurately.

When changing the internal volume of a chamber, one should consider potential changes in air circulation near the measured subject. High airspeed should be avoided and therefore the internal air circulation rates need to be adjusted. The measurement of the total chamber volume is more complicated when objects are added or walls are repositioned, meaning that it need to be done accurately as this is an important part of the calculation of total gas exchange.

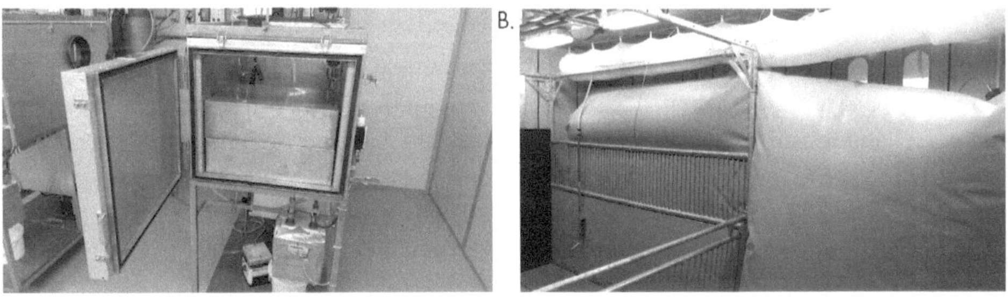

Figure 1. (A) An example of the use of two metal boxes inside a fixed sized chamber of 267 l. A maximum of four stainless steel boxes of 28 l can be used to decrease the volume of the chamber to 155 l. At an average ventilation rate of 10 l/min, this reduces the washout time from 27 to 16 min. (B) An example of the use of a big balloon as an airtight gas bag, normally used for storage of bio gas, constructed from PVC coated polyester tissue. It can be made in any dimension to fit in an empty space within a chamber. It is used to decrease the volume of a 34.2 m^3 chamber by 8.0 m^3. At a ventilation rate of 500 l/min, this reduces the washout time from 68 to 52 min (Photos Wageningen University, Wageningen, the Netherlands).

A B

Figure 2. Two respiration chambers (A) both 1.5×3.5 m animal floor space with one flexible wall (B, width 1.5 m) to fit one calf, cow, pig or groups of smaller animals. Inflatable tubing around this wall with 2 bar (= kg/cm^2) air pressure secures airtight closure at various positions (Photo Wageningen University, Wageningen, the Netherlands).

In case of flexible gas bags, variation in their volume in time should be kept to a minimum. Therefore inside bag air pressure should be controlled at a constant differential level against room pressure (in the gas bag in Figure 1b for example +150 Pa is used). To measure the volume of a bag is another challenge. A calculation can be made by measuring its dimensions and correcting for all round shapes. Alternatively, the bag can be pressurized, measure its internal pressure, add a known volume of air and measure its internal pressure again. The volume then can be calculated by difference. Finally, a room volume can be estimated by washing out a high level of a specific gas, like CO_2, which is described later in this chapter.

2.2.1 Design of size adaptable systems

Chambers are surrounded by analysis and control equipment and areas to handle feed, products (like milk) and excreta. Some general considerations for a complete indirect calorimetry setup are: (1) minimize the length of air sample tubing between chambers and gas analysers; (2) when working with large animals and different housing systems, consider making a large entry directly from outside to facilitate transport and to keep internal walking areas clean; (3) keep climate control equipment away from animals and excreta and consider easy access for testing and maintenance; (4) when conducting a complete energy balance in large animal species or group-housed animals, it may be necessary to collect excreta quantitatively with cleaning water. In that case, it can be considered to construct facilities for collecting, weighing and mixing these large quantities underneath the respiration chambers.

2.2.2 General chamber design and compartmentation

The maximum size of a chamber depends on the type of housing and subject(s) to be measured in its largest setting. For creating multiple airtight rooms within one climatised chamber, the use of movable walls are the preferred option. For multi-purpose use (respirometry, varying climatic conditions, hygiene statuses), it is desirable that these movable walls can be completely removed from a chamber. In addition, the force that an over or under pressurised room exerts on the walls and ceiling should be taken into account. Materials and constructions used also should take forces from animal movements into account. Top layers of materials used should be strong and easy to clean. The height of the chamber is depending on climate control equipment and the amount and size of equipment that need to be mounted under the ceiling out of animal reach and without hampering personnel working inside.

The Wageningen calorimetry unit comprises of four climate chambers, each of which can contain up to two airtight rooms for gas exchange measurements. At its largest size, a chamber measures (l × w × h) 6.5×7.0×2.5 m. The volume then is 114.5 m^3 and it can house four cows or a group of 20 pigs of 100 kg body weight. A top view of one of these chambers is shown in Figure 3. At all dotted lines on the left chamber, wall panels can be positioned to create one or two airtight rooms. As an example, in the right chamber 45 cm wide wall panels are drawn to create two similar maximum sized internal respiration rooms within one chamber. Rooms A and B measure 4.5×2.7 m (excluding the air conditioning unit) and the remaining area (room C) is only used as corridor to reach room B via in-building door 4.

Each chamber contains five independent climate control units (6) bounded by the possible flexible wall position marked by the dotted lines. Flexible air socks can be connected to each unit to distribute the air homogenously within rooms A, B and C (like the one visible at the top of Figure 1B). Four, separate stainless steel slurry pits (7) are present for the separate collection of excreta. All outer fixed walls (1) and the ceiling are built from three layer 'sandwich' panels (layer 1 and 3: coated metal; layer 2: Styrofoam) and covered with plywood plates and a polyester-layer on the inside. The concrete floor is constructed with a Styrofoam core and a polyester top coating. Sand added to the topcoat reduces slipping hazard. Furthermore, each chamber has an extra airtight sliding door (3) for transportation of large housing constructions or animals directly from outside. The outside walls and ceiling of each chamber are fixed by a metal construction and extra attention was paid to constructing and fastening the large ceiling. Air pressure differences and the use of flexible walls with pressurized inflatable tubing exert large force on especially the large ceiling and therefore, it was fastened to metal H-shaped beams, 180×405 mm (w × h) positioned just above the ceiling.

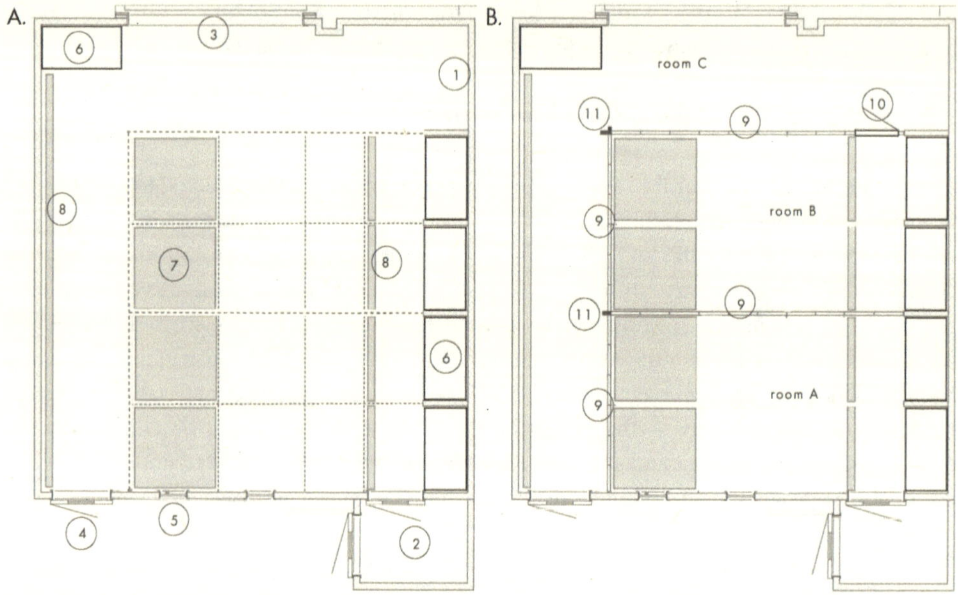

Figure 3. Two top views of one chamber with movable wall panels. Each chambers contains: (1) outer fixed walls; (2) an airlock with two doors (each w × h = 85×200 cm including a 20×20 cm window); (3) a sliding door in the outer wall (w × h = 270×230 cm; (4) an extra door in the fixed wall towards a central corridor in the building; (5) two windows (w × h = 40×20 cm); (6) a total of five air-conditioning units; (7) four, separate stainless steel slurry pits, each with a volume of approximately 0.75 m^3; l × w = 120×120 cm with metal lattice; (8) one long and four separate small drainage gutters; (9; wall panels positioned in part of the 4 by 4 area raster indicated by the dotted lines in the left picture. A total of 16 different sized rooms or eight different sized double rooms can be made; (10) a double width wall panel with door; (11) one or two corner posts fastened to floor and metal ceiling grid. Airtightness of this element is accomplished by an inside pressurised air-cylinder with will press this post, top and bottom coated with a flexible rubber layer, against floor and ceiling (Wageningen University, Wageningen, the Netherlands).

2.2.3 Adjustable size of respiration rooms

Within a chamber (Figure 3) one or two airtight rooms can be created using removable wall panels of 426 mm wide and 2,486 mm high (Figure 4) giving a 450 mm wide and 2,500 mm high surface with inflated surrounding tubing. With the use of a series of these wall panels one airtight animal room of 0.90, 1.80, 3.15 or 4.50 m long by 1.35, 2.70, 4.05 or 5.40 m wide can be created. In this way, 16 different configurations can be realized by placing the flexible walls in different positions (dotted lines in Figure 3). In the case of one airtight room, floor area can vary between 0.90×1.35 = 1.2 m^2 and 4.50×5.40 = 24.3 m^2. Consequently, airtight room volume, including the 0.67×1.35 m sized air-conditioning units, can vary between 5.3 and 69.8 m^3.

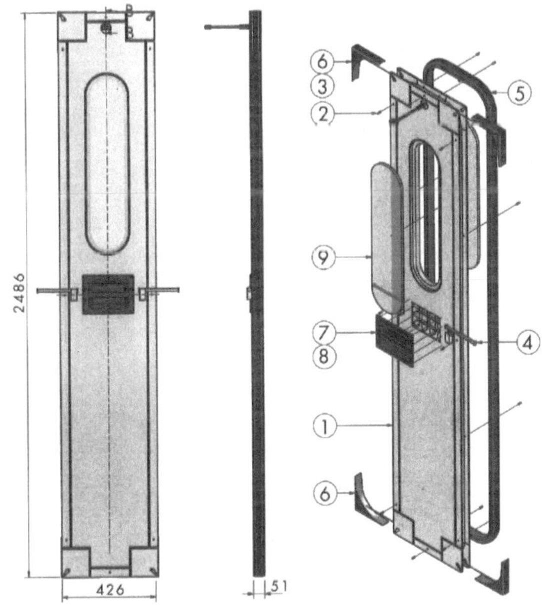

Figure 4. Wall panels for flexible positioning. Walls (1) are made of 5 cm grid hollow polypropylene plates and windows (9) of a double layer polycarbonate. Airtight closure is made possible with inflatable tubing (5) and rubber corner elements (6). A hand grip (7) can help to move the approximately 23 kg weight construction. Each wall part can be moved into a special metal ceiling grid and positioned against a fixed wall or against each other. Metal pins (4) assist to keep wall parts aligned when inflating the air tube to approximately 1.5-2.0 bar (Wageningen University, Wageningen, the Netherlands).

By placing extra flexible wall panels, two airtight animal rooms can be built with the same variation in length (0.90, 1.80, 3.15 or 4.50 m) and a width of 1.35 or 2.70 cm. Building two rooms would give a variation in floor area between 0.90×1.35 = 1.2 m^2 and 4.50×2.70 = 12.2 m^2 with a volume of 5.3 to 34.9 m^3, respectively for each room.

Within a two hour time frame, two people can alter the complete configuration inside a chamber. For ease of handling and a financial point of view a body of lightweight hollow poly-propylene plate is suggested. Although poly-propylene can be welded with simple tools when broken, care should be taken in their use as they are fragile when compared with fixed, permanent walls.

Doors and windows from the chamber and wall panels should be designed for efficient caretaking and cleaning. In the positioning of the walls to create airtight rooms, windowed panels can be used to allow visual contact between rooms if desired.

Figure 5 shows a 3D top view of a 6.5×7.0 m chamber with wall panels placed at positions for two maximum sized airtight rooms. In practice, half of the panels have two windows which in total provide approximately 22% window surface to allow visual contact between animals in different rooms.

The construction of animal housing systems is not elaborated on in this chapter. Some examples used in the Wageningen chambers can be seen in Figures: e.g. Figure 1B shows a 5 m^2 pen for pigs and Figure 7B shows a crate to house a cow fastened with a belt around her neck. Housing and feeding systems can be used according to agricultural practice or of special design. They should preferably only be fastened onto the floor and therefore its construction should be extra stabilised. A system for milking dairy cows also can be mounted.

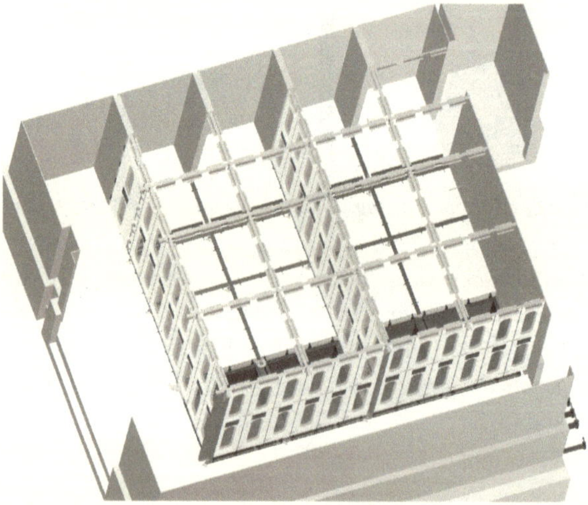

Figure 5. Three dimensional top view of a climate chamber, with flexible wall panels at positions as shown in Figure 3 right, to create two maximum equal sized airtight rooms. The sliding door in the left outer wall, the doors in the upper right airlock and lower right position are not shown. The raster of possible flexible wall position is shown in the floor as well as in the ceiling (Wageningen University, Wageningen, the Netherlands).

2.3 Climate control design

Important aspects to consider for the climate control design are: (1) the temperature and relative humidity (RH) set point range; (2) the accuracy of the temperature and RH sensors; (3) the maximum allowed variation in temperature and RH in time and place since there will be variation on the same spot in time and among different spots within a room at the same moment in time; (4) the number of temperature and RH sensors per chamber. Sensors need to be wired with enough length to make it possible to place them at an optimal position for any species, housing system and animal(s) position. As the sensors near an animal are rather susceptible to sudden changes in animal position, we recommend to use approximately five temperature sensors at various places near animals in each climatised room. Using the median value of all temperature sensors rules out disproportional effects of extreme values.

Because of adjustable room sizes and the possibility to build more airtight rooms in one chamber, the air-conditioning unit needs to be compartmented to a total of at least one more than the number of respiration rooms within one chamber: one for each of the airtight respiration rooms and one for the surrounding area of these rooms within a chamber. The circulation rate of air to and from the respiration room through the air-conditioning units need to be adjustable to assure airspeed to be at the desired level. Continuous maximal mixing of inside air is important for air conditioning as well as for gas exchange measurements. Circulating air needs to be filtered continuously before entering the air-conditioning units. From an animal and human welfare point of view noise level of air conditioning and filter units is a point of concern. We suggest that inside a room a noise level of 65 dB(A) is not exceeded.

Because of adjustable airtight room sizes in the Wageningen chambers, five separate independent air-conditioning units are constructed in each chamber. In a configuration with one airtight room in one chamber, one to four of these units can be incorporated. Each of two airtight rooms can include one or two air-conditioning units. Animal mass and its corresponding sensible and latent heat loss and the capacity of one air-conditioning unit determine how many of these units need to be in one room. The surrounding area has at least one unit to create the same environmental temperature inside and outside the airtight rooms. In this way condense formation is avoided and wall panels do not have to be insulated. The air circulating rate within each room is adjustable in a way that required climate conditioning at all room sizes can be covered, maintaining a maximum allowed air speed at animal level. The complexity of the climate control system with a varying number of air conditioning units in a different number of flexible sized rooms suggest that, especially while testing new chambers, software and algorithms need to be developed in a way that adaptability is rather easy. This complexity but also the variety of experiments and implementation of new ideas and equipment in time may frequently necessitate changes or expanding this software. Software need to be developed in a way that multiple settings, checks and outgoing alarms on preferably all measured values can be set by means of a clear user interface. In a real time situation data need to be presented in tables as well as in graphs for a regular check of behaviour and tuning of climate control in various rooms and room sizes, climate set points and animal mass.

Further technical details on the design of climate control of these chambers are given in Chapter 11 (Gaughan *et al.*, 2015).

2.3.1 Ventilation setup

The range of the ventilation rate can be estimated based on the expected animal mass and the maximum allowed CO_2 level. A maximum CO_2 concentration of 1.0% is suggested by Lighton (2008), which is considered a level not affecting metabolism of most animal species.

Depending on the expected range of ventilation and the desired accuracy (better than ±0.5%), gas flow meters need to be selected. The most suitable accurate gas flow meter for use with a wide range of different numbers of large animals, like pigs and cows is suggested to be a rotary gas flow meter. Due to the volumetric measuring principle of a rotary gas flow meter its metrology is not influenced by installation conditions. They are still very accurate when gas flows are low or irregular[1]. A rule of thumb for the desired amount of ventilation is 4-6 l/min per $kg^{0.75}$ animal mass, of course depending on production level.

Short air sampling lines contribute to fast responses and thereby to accurate measurement of short term changes in gas concentrations. For normal respiration measurements, either hypobaric or hyperbaric air pressure control can be used. The principle of both systems is illustrated in Figure 6. When constructing chambers for multiple applications, the two systems can be combined, allowing switching from hypobaric to hyperbaric systems. For example, in experiments with specific pathogen free animals, a hyperbaric setting is recommended, whereas for experiments with infectious agents, a hypobaric setting is preferred or even compulsory. Normal set point pressure difference between chamber or room and the environment is recommended to be between 50-100 Pa. To protect the construction from damaging, due to extreme air pressure reasons, mechanical maximum over and under pressure safety valves are essential. Depending on properties of the ventilation pump or fan, a chamber or room could implode or explode if air pressure is not secured. The sequence of devices in the incoming and outgoing airlines is critical for good gas exchange measurements. This sequence depends on the chosen air pressure set point (Figure 6). In a hypobaric setting a fan at the end of the ventilation air pipe sucks air from the flow control, through the gas flow meter out of the (pull

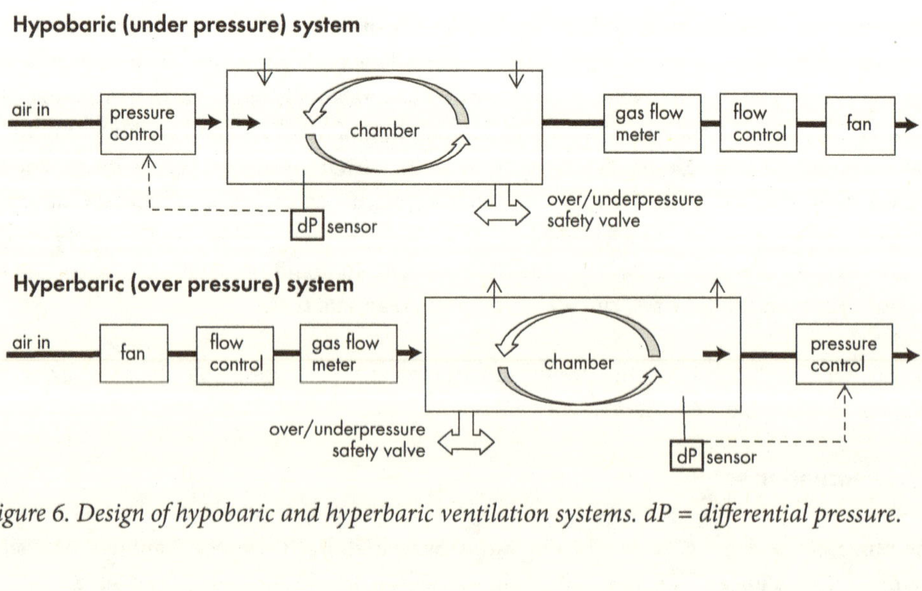

Figure 6. Design of hypobaric and hyperbaric ventilation systems. dP = differential pressure.

[1] Wikipedia (http://tinyurl.com/p76gzsd) and http://www.itron.com.

type) chamber or room. When the chamber or room meets airtightness criteria (see below), virtually all air enters the ingoing air channel from outside through the pressure control valve. In case of enough under-pressure (<-50 Pa) small leakages (thin arrows) will not have an effect, because air leaking into the chamber or room is assumed to have gas concentrations identical to those of analysed incoming outside air. It is important that this is checked, especially if leaking air is taking from an inside building area were people work or other animals live.

In the hyperbaric setting, a fan at the start of the ventilation air pipe blows fresh outside air through the flow control, through the gas flow meter into the (push type) room. When the room meets airtightness criteria, virtually all air flows outside through the pressure control valve in the outgoing air channel. In case of enough over-pressure (>+50 Pa) small leakages (thin arrows) will not have an effect, when assuming leaking air gas concentrations to be equal to measured concentrations in outgoing air.

Regardless the ventilation design chosen, a high level of circulation inside the airtight room is essential for a well-controlled homogeneous climate and for a good mixing of air gas components. As a rule of thumb, a minimal circulation rate of approximately once the total volume of the room per minute can be used. For example, if an airtight room is 35 m^3 then minimum air circulation rate needs to be close to 35×60 = 2,100 m^3/h. With adjustable sized airtight rooms air circulation rate need to be adjustable to ensure that at all possible room sizes circulation level can be at the required level. Attention has to be given to prevent too high air speeds, because it can cause draught at animal level. A maximum airspeed of ≤0.2 m/s at animal level is recommended (Verstegen *et al.*, 1987). In the example with the 35 m^3 room and a minimum circulating rate of 2,100 m^3/h, assuming an inside chamber height of 2.5 m and 10% surface needed for an inside air conditioning unit, a free floor surface of 35/2.5×(1-0.1) = 12.6 m^2 can be calculated. Average airspeed can then be calculated as 100/12.6/60/60 = 0.05 m/s. From this example it is clear that even a circulation level of twice the total volume per minute would not cause draft problems although it is clear that airspeed variation within a room can be expected. This variation can probably best be visualised by using smoke.

Differential air pressure (dP) control can be automated by measuring air pressure difference between the inside and direct outside of a room continuously and controlling a valve in the incoming (hypobaric) or outgoing (hyperbaric) airline. At a constant ventilation level this valve can also be operated manually but the dP then has to be checked regularly and adjusted if necessary. In both automated and hand controlled dP arrangement an alarm to warn if dP diverting too far from its set point needs to be set. This alarm also draws immediate attention to unwanted door opening, inaccurate door closure and other leaks. Furthermore, as there is a big inside airflow, it is good to realize that inside absolute pressure differences can be expected. In a hypobaric setting, the highest absolute pressure is usually recorded just downstream the climate control air circulation fan, preferably dP <-50 Pa. In a hyperbaric setting, the lowest absolute pressure is usually upstream of this fan, preferably dP >+50 Pa.

In the Wageningen chambers three ventilation systems are constructed per chamber of which two have an integrated gas flow meter. For the expected range of ventilation rates required, a rotor type meter Itron Delta 2050 MP series, type G100, with a range of 3-160 m^3/hour was chosen. Length of ventilation air and air sampling lines were minimised and by using a multiplexing gas analysis system where gas in all air sample lines is flowing continuously, time lag between air leaving a chamber or room and analysis is minimised to roughly 1 min. Both hypobaric and hyperbaric air pressure control were implemented by adding extra air pipes and hand valves by which a switch can be made within minutes. Mechanical maximum over and under pressure safety valves (+ and -150 Pa) for every possible airtight configuration were incorporated. At design time a maximum value of 2,000 m^3/h was calculated (based on the expected maximum heat load, water vapour load, temperature and relative humidity range and desired maximum speed of climate change) for each air-conditioning unit, whereas air temperature difference between in and outgoing air of an air-conditioning unit was set at a maximum of 3 °C. With this data the required capacity of fans and heat exchangers was calculated. Differential air pressure control is automated by valves in the incoming (pull type) or outgoing (push type) airline. Extra in-pipe fans were implemented in the in- and out-going airlines to realize sufficient air pressure control in the whole ventilation range.

Regular (e.g. once yearly) calibration of at least one 'non process lab standard' gas flow meter against an official standard is recommended, to ensure its accuracy within 0.5%. All process gas flow meters used within the respiration systems can then be calibrated in line with this 'lab standard' gas flow meter, preferably twice a year. In this way the absolute error of and the mutual error between process gas flow meters is minimised. Gas flow meter error curves are by definition non-linear and the error depends on the flow rate. Therefore, calibrations have to be conducted at least at three flow levels, including the flow rate to be used in the upcoming experiment. After a calibration test, a derivative error curve can be produced. To calibrate a diaphragm type gas volume meter, a stepwise increasing flow, followed by a stepwise decreasing flow is necessary because of hysteresis effects. Furthermore, for other types of gas flow meters it is advised to perform this test to at least have an idea about hysteresis. Securing the airtightness of the connection between two gas flow meters during a calibration test deserves special attention.

On-site in Wageningen, twice a year process gas flow meters (diaphragm type and rotor type) are tested in line with a 'lab standard' gas flow meter of the same type to get derivative error curves. On advice of the Dutch National Metrology Institute once a year the 'lab standard' gas flow meter is tested. After in series testing a process gas flow meter, a new error curve is calculated and compared to earlier ones. In case a gas flow meter has a clearly changed curve since last time, a more detailed inspection is considered. Larger changes predict technical errors or sometimes accumulation of dirt. For this reason it can be advised that all gas flow meters are to be preceded by dust filters even if filters are inside the air conditioning units.

2.3.2 Leak tolerance

In the construction of respiration facilities, criteria for maximum leak tolerance need to be met and obviously, these depend on the dP applied. The leak tolerance at a dP of -100 and +100 Pa used in the development of the facilities in Wageningen, based on long term experience is:

- for the airtight rooms, constructed within the fixed sized chamber using flexible walls (Figure 5): *0.05 × room volume (l) / hour*
- for the airtight fixed sized chamber surrounding the airtight rooms: *0.1 × chamber volume (l) / hour*

Air leakage (l/h) can be tested in a closed chamber or room with all inside systems (e.g. climate control, ventilation, lights) shut down and incoming and outgoing air lines sealed. A dP-meter and a flow meter with an accurate range in series with a hand-valve and fan are needed. dP can be controlled with a fan and hand-valve at +100±10 Pa and in a later test at -100±10 Pa. At a constant dP, leakage equals the measured gas flow during 60 s. In case -100 Pa and/or +100 Pa cannot be reached, leaks need to be detected and fixed prior to leak testing. Detecting large leakages is possible using smoke, with possible risks damaging electronic RH sensors, which can be seriously damaged by fatty smoke. Small leakages can best be localized using soap water.

Experience learnt that at the lower volumes, airtight closure at the rubber corner elements of the flexible walls (Figure 4) is rather critical and extra attention is necessary every time the tube around the wall panels is inflated. With soap water it is easy to do a quick scan and if necessary these elements should be re-positioned by hand. Leaks can be minimised by well-cleaning of the inflatable tubing and especially by regularly cleaning the rubber corner elements. This is of special importance when a flexible wall separates 2 measuring rooms because gas exchange through leakages would really negatively affect results.

2.3.3 Temperature, relative humidity, barometric pressure and dP sensors

It is recommended that sensors used comply with the following criteria: temperature: 4-wire type 1/3DIN-B Pt100 with accuracy ≤0.1 °C at 0 °C; RH with accuracy ≤2% absolute; differential pressure: range -500 to 500 Pa with accuracy ≤1% full scale.

The Wageningen choices in temperature and RH sensors are given in Chapter 11 (Gaughan *et al.*, 2015). In the Wageningen setup to measure differential air pressure in each of two rooms and a surrounding chamber three differential pressure transmitters Beck 984M, ranging from -500 to + 500 Pa with digital display are used.

Calibration of temperature and humidity sensors is described in Chapter 11 by Gaughan *et al.* (2015). Barometric pressure measurements can be compared with on-line data of nearby weather stations. An accurate mercury barometer can also help to check levels in time. Differential pressure sensors can be checked with a water column U-tube, where 100 Pa is equal to 10 mm of water column difference.

2.4 Gas analysis system

Analysing air content of a single respiration room can be done by a gas analysis system in the direct proximity. On the other hand, in a multi-chamber or multi-room setting, a multiplexing system can alternately prepare multiple air samples and calibration gasses on a regular basis. Multiplexing systems require air samples from inlet and exhausted air from all used rooms to be taken in a continuous flow through dust filters and tubing by airtight sample pumps to a central analysing system. Sample tubing needs to be isolated and heated well above the maximum expected dew point to keep water vapour from condensing inside the tubing. Length of gas lines between switching valves and analysers should be minimised and before analysis gas samples need to be conditioned for humidity, flow and preferably also for pressure.

Alternatively, analysis can be done as a differential measurement between incoming air and chamber air content. Gas samples can be either analysed in air as sampled from the chamber (i.c. containing moisture) or in dried air or air standardized for low humidity. The analysis in humid air is discussed in Chapter 3 by Lighton (2015). For analysis in dried air or air standardized for low humidity, air humidity has to be conditioned, using a desiccant (e.g. Drierite, W.A. Hammond Drierite Co. Ltd. Ohio, USA) or an air cooling system, respectively. Air also needs to be standardised in preferably absolute pressure and airflow to establish the number of gas molecules inside the measuring unit to be constant. Gas analysers normally have their own temperature control to prevent water to condensate inside and to standardise temperature at the measuring point.

If the measurement of short term changes in Q is of interest, adjusting the volume of the respiration room to the size of the metabolic mass inside is important, and the cycle time of the multiplexing unit, i.e. the time interval between two subsequent analyses of the same room, should be minimized (see e.g. Van Klinken *et al.*, 2012). The minimal cycle time should be tested thoroughly. It depends on the number of respiration rooms connected to the multiplexing unit, but also on the flow and volume of an analysis system itself. Even the last analyser in line should be flushed well enough to deliver a steady state signal at least for incoming air and calibration gas samples. Gas concentrations in air coming from rooms with living subjects are per definition not constant, complicating the estimation whether or not a stable reading can be achieved from the gas analyser. Air from calibration cylinders and inlet air can be used to test the minimal cycling time and CO_2 recovery tests, described below, should prove the validity of the chosen cycle time.

In the Wageningen facilities air samples are taken by gas pumps in a continuous flow through a dust filter, through heated fluorinated ethylene propylene tubing with outer/inner dimension of 6/4 mm tubing at a rate of about 1.5 l/min, to a gas analysis system, located in a room with minimised distance from all respiration rooms. Heating of air tubing is done by electrical tracing (Bartec HSB25 self-limiting parallel heating cable) and controlled at approximately 50 °C with a thermostat switch. The sequence of air streams to be analysed in the multiplexing unit is arranged by electronically operated valves, controlled by tailor made software. Before analysis, sampled gas is again dust filtered and conditioned for humidity at a level of 2.5-3.0 °C dew point (M&C TechGroup gas cooler series EC 4-channel dew point drier). Dry calibrating gasses, analysed once daily, and sometimes dry outside air are humidified at wet spots inside this drier to exactly the same humidity level as long as condensed water is in the drier. Air from all animal rooms does normally contain enough water to be condensed in the drier. Tests revealed that at the analysis gas flow level of approximately 1.0 l/min the amount of earlier condensed water, which is still available in the drier, will evaporate into dry gas samples to the 2.5-3.0 °C dew point level and is enough for at least 15 min humidification of dry gasses. In this way the dilution and interference effect of water vapour in all analysed gasses is equal and can therefore be neglected. Gas analysers are setup in series for ease of control of pressure and sample flow between a downstream air pressure controller before the first analyser and an upstream air pressure controller after the last analyser. Both precision pressure controllers (Brooks Model 5866 compact pressure controller; range 1000-1100 mbar) are hand tuned to create an analysis airflow of approximately 1.0 l/min. In front of the first (downstream) precision controller an extra mechanical downstream pressure controller is positioned to level all offered gas types (outside, chamber and cylinder gas) to a pressure of approximately 1.3 bar. Both precision controllers regulate the sample air pressure well above environmental air pressure to prevent air coming in the analysis system by small leaks. After the last pressure controller in line an extra air pump can be used, to create a lower than environmental pressure at the controllers outlet which will improve the performance of this controller especially at high barometric pressure. Pressure control inside the analysis line will be independent from environmental pressure. In this conditioned gas flow control setup analysis of CO_2 (total and ^{13}C labelled CO_2), CH_4 (all with non-dispersive infrared (NDIR) technique) and O_2 (paramagnetic) concentrations are performed with Advance Optima modular systems (ABB formerly known as Hartmann & Braun, Frankfurt, Germany). Concentrations of all analysed gases from inlet and exhaust air of each respiration room connected to the gas analysis system are stored in a computer database. In addition, zero and span concentrations from two calibration gas mixtures, analysed at least once daily are stored.

To check and/or adjust gas analyser's zero and span point deviation, calibration gas mixtures need to be used. Avoid using outside air for calibration purpose since content is not steady! After streaming calibration gasses and registering its different gas contents, correction values for each analyser can be calculated. Between moments in time when calibrations are performed, linear zero and span drifts can be assumed. When purchasing calibration gas mixtures be sure that a certificate with accurate gas analysis results is obtained. Linearity of NDIR analysers (CO_2, $^{13}CO_2$, CH_4) has to be checked and if necessary adjusted once in a while. Paramagnetic oxygen measurement on the other hand is a very linear measuring technique.

We use two calibration gas mixtures from 50 l cylinders to daily check the gas analysers. One containing approximately 1.0% CO_2 and 20.0% O_2, filled up to 100% with N_2. The other approximately 21.0% O_2 and 1000 ppm CH_4, also filled up to 100% with N_2. To prevent sudden changes in Q or respiratory quotient (RQ) levels within or between experiments as a result of replacing new calibration gas cylinders, the content of new cylinders are tested on all analysis systems, which have just been calibrated with the old set of gas cylinders that are still in use. In case measured gas concentration data of new cylinders differ from content given at the certificate of the new cylinder, a thorough check is done to find out potential causes of this difference. The actual gas concentrations are calculated from the measured gas concentrations and the concentrations measured in the calibration gas mixtures, thereby accounting for drift in the analysing system, using linear interpolation over time.

Probably the best way to handle air leakages while someone enters or leaves a chamber to take care of the animals is to take those moments in time out of the final datasets. While milking, be aware that the 'vacuum' air pipe takes out some air from a chamber because of leakages between teats and teat cups. The vacuum pipe should be closed very accurately (think of using a double closure) after milking so milking in other chambers or cleaning the milking system is not affecting gas exchanges measurements. Taking data away from the final dataset also takes into account the gas exchange of a human being working inside a chamber for a certain amount of time. In the Wageningen chambers all doors are equipped with switches. If switched it gives an electrical pulse to a counter and for each door every half a minute an opening or closure is recorded. While computing the final dataset this data is used to automatic rule out data.

2.5 Full system checks

Full system tests, like a CO_2 recovery test, an ethanol burning test and a nitrogen injection test are the ultimate tests for an indirect calorimetric system. It includes testing of chamber and piping airtightness, flow measurement, measurement of temperature, RH and barometric pressure to gas sample control and gas analysis. The frequency of full system tests need to be considered and it is advised that they should always be preceded by a gas flow meter test. It is a good habit to perform such tests before and after each experiment. Typically, a recovery $\leq$99% is indicative of problems and the test needs to be repeated. If persistent, then there is probably some kind of systematic error in the system. If a gas analysis system is used for more chambers/rooms (multiplexing) then comparison of recoveries between chambers can be used to pinpoint problems to either the gas analysing system or to airtightness or flow measurement of one (or more) chambers. To get an idea where the problem might be, it can sometimes help to change over equipment like gas flow meters, from a room with good recovery to a room with poor recovery and then repeat a recovery test. Recovery and repeatability between tests in one chamber need to be high to be sure about the accuracy of a whole system and therefore to trust its quality. When performing a CO_2 recovery test by releasing CO_2 inside a respiration chamber O_2 concentrations are affected in the opposite direction by dilution. As no O_2 is consumed nor produced, the actual O_2 exchange should be calculated zero or at least $\leq$1% of the measured VCO_2, and this should be verified.

We perform CO_2 recovery tests using a 10 l pure (>99.5%) high pressure CO_2 cylinder on an accurate weighing scale (0-30±0.001 kg). The amount of CO_2 needed can be estimated and checked with the amount available in the cylinder. A 10 l cylinder contains approximately seven kg liquid CO_2 (CO_2 becomes liquid >50 bar). As the molar mass of CO_2 is 44.01 g and the molar volume of CO_2 at 273.15 K and 101.3 kPa is 22.26 l/mol, 1.000 gram of liquid CO_2 gives 0.5058 l of STPD CO_2 gas. STPD stands for Standard Temperature (273.15 K) and Pressure (101.3 kPa), Dry gas. The amount of CO_2 (in g) to be released into a chamber's airtight room can be estimated from room volume and ventilation flow rate, and depends on the duration of the test. Typically such tests last overnight for 15-20 h. The amount of CO_2 required for such test consists of a part needed to reach the desired CO_2 concentration, depending on room volume, and a part related to maintaining this concentration over time, the latter depending on the duration of the test and flow rate:

$$Y = \frac{\text{Room volume (l)} \times [CO_2]/100 + \text{Ventilation rate (l/min)} \times [CO_2]/100 \times \text{time (min)}}{0.5058} \qquad (1)$$

in which Y = the quantity of CO_2 needed for the recovery test (g); $[CO_2]$ = the desired stable CO_2 concentration throughout the test (%); time = the duration of the test (min). For example in a room of 40,000 l, a desired CO_2 concentration of 0.7%, a test duration of 1000 min and a flow rate of 400 l/min, 6,053 g of CO_2 is required.

The weighing scale with cylinder on top is put inside a respiration room to be sure that small leakage from the cylinders pressure control does not affect the test. Then the room is closed, air pressure difference is checked and subsequently it is checked whether all gas analysis and other systems are running. CO_2 gas then is injected at a rate that CO_2 level in a room becomes somewhere between 0.5 and 0.9%, which is approximately the same amount of CO_2 as an animal would produce. For safety reasons, before the start of a test when CO_2 injection is tuned by hand inside a room, it is advised that two persons have to be around. After CO_2 injection level is set, the test can be started by writing down the time and CO_2 cylinder weight. The amount of CO_2 released is calculated by the weight change of the gas cylinder during the test, corrected for possible impurities of the CO_2 gas. Be aware that gas expansion takes a lot of energy, meaning that the cylinder and/or pressure controller will become cold and therefore water vapour can condense or even freeze on it. At the end of a test the cylinder need to be closed and dried before taking the weight. Measured gas concentrations, airflow and also temperature, RH and barometric pressure are used to calculate the measured quantity of injected CO_2. The CO_2 recovery then can be calculated:

$$CO_2 \text{ recovery (\%)} = \frac{CO_2 \text{ measured (l)} \times 100}{CO_2 \text{ from cylinder (l)}} \qquad (2)$$

At a desired overall whole system accuracy level of ≥99%, the CO_2 recovery should also give a value between 99 and 101%. If not, something in the system is wrong and it needs to be checked for causes of the poor recovery.

Another test more suitable for smaller chambers is to burn ethanol from a small lab ethanol burner for some time. The ethanol container should be sealed thoroughly to prevent alcohol from evaporating. Weight loss of the ethanol container can be recorded accurately in time and estimated VO_2 and VCO_2 can be calculated. If ethanol is burned completely (with a blue rather invisible flame) as in the chemical formula $C_2H_6O + 3\ O_2 \rightarrow 2\ CO_2 + 3\ H_2O$, one gram of ethanol with a molar mass of 46.0688 and a molar volume of 22.414 l/mol uses 1.4596 l STPD O_2 and produces 0.9731 l STPD CO_2. The VO_2 and VCO_2 measured by the analysis system can also be calculated and finally the recovery of CO_2 and the usage of O_2 can be calculated:

$$CO_2 \text{ recovered } (\%) = \frac{CO_2 \text{ measured (l)} \times 100}{\text{ethanol burned (g)} \times 0.9731 \text{ (l/g)}} \quad (3)$$

$$O_2 \text{ used } (\%) = \frac{O_2 \text{ measured (l)} \times 100}{\text{ethanol burned (g)} \times 1.4596 \text{ (l/g)}} \quad (4)$$

At a desired overall whole system accuracy level of ≥99%, the CO_2 recovery as well as the O_2 use should give a value between 99 and 101%. Another thing to check is whether the RQ, calculated as CO_2 measured / O_2 measured, is equal to 0.9731/1.4596 = 0.667 (is equal to 2 CO_2 / 3 O_2 from the chemical formula). If burning was incomplete, then probably RQ is wrong as well. If both CO_2 and O_2 factors are wrong, but still there is an acceptable RQ measured, this may indicate that gas analysis is okay but ventilation was measured with an error or affected by leakages in the system.

In some labs a nitrogen injection test is performed to mimic the decrease in O_2 concentration. In this test the amount of injected N_2 gas is measured gravimetrical or by accurate mass flow measurement and compared to the calculated amount of injected N_2 gas calculated from the dilution of oxygen (Mclean and Tobin, 1987). As mentioned above this oxygen dilution is also tested in a CO_2 recovery test were VO_2 should be calculated zero because actually no O_2 is consumed.

2.6 Estimating room volume

The volume of a respiration room is an important parameter in the calculations of Q. This volume can be calculated by the measurement of its inner dimensions. In a flexible wall setup however, volumes at all possible wall positions have to be measured and the current volume should of course be accurately used in calculations. Alternatively, it can be estimated from a CO_2 dilution test. This can be particularly useful when irregular shaped objects like balloons are inserted in the room to reduce its volume. In addition, comparison of the two methods can be a good check of the overall accuracy of the indirect calorimetry system. In the CO_2 dilution test, the CO_2 concentration in the room to be tested is increased well above 1%, but still within the measuring range of the CO_2 gas analyser. When this CO_2 concentration is reached, the injection is stopped by closing and removing the CO_2 cylinder. Then doors are closed and a test starts by reading at a shortest time interval the running gas flow meter start value and the current CO_2 concentration of the chamber or room. When the CO_2 concentration has decreased to approximately 0.1%, the test is stopped by reading the exact current CO_2 concentration value and at the same moment the gas flow meter value. After correcting ventilation flow to STPD, correcting the CO_2 and O_2 content of ingoing and outgoing air dependent on analyser drift the chamber/room STPD volume can be calculated as:

$$\text{STPD volume (l)} = \text{STPD ventilation (l)} \times \left(\frac{1}{(\text{nLog} \left(\frac{(\text{start point } CO_2\% - \text{mean incoming } CO_2\%)}{(\text{end point } CO_2\% - \text{mean incoming } CO_2\%)} \right))} \right) \quad (5)$$

To convert the STPD volume to chamber/room volume, it needs to be corrected for temperature, moisture content and barometric pressure under the conditions in which the measurements are obtained. The water pressure (pH_2O) inside a chamber/room can be calculated using the dry bulb temperature (Td) and relative humidity (RH) as (Verstegen *et al.*, 1987):

$$pH_2O \text{ (mmHg)} = \frac{RH \times (3.999 + 0.455417 \times Td + 0.001708 \times Td^2 + 0.000468 \times Td^3)}{100} \quad (6)$$

Finally, chamber/room volume at current conditions can be calculated using the barometric pressure (pBaro; in mmHg) as (derived from the combined gas laws in http://en.wikipedia.org):

$$\text{Volume (l)} = \frac{\text{STPD volume}}{((pBaro - pH_2O) \times 273.15) / ((273.15 + Td) \times 760)} \quad (7)$$

Because of the sensitivity of this test, where measuring noise on only two gas analysis points can have a rather big effect, the difference between measured and calculated chamber or room volume should not exceed 2%. When found to be repeatable, this test can be used as a quick check of a whole system with all its calibrated parts.

2.7 Collection of excreta

For conducting a complete energy balance, slurry pits or other reservoirs in the floor should be large enough for the collection of excreta for at least a week. Because of compartmentation abilities of the whole animal area, reservoirs also need to be assigned to different compartments and of course airtightness also here is a point of concern. Slatted floors for each species at each age need to be available to prevent leg injuries.

In the Wageningen setup per chamber four stainless steel slurry pits (Figure 7), each with a size of approximately 1.2×1.2×0.5 m (l × w × d) are created, which can contain in total approximately 2.8 m^3 manure. Piping from slurry pits are flexibly connected to systems in the central basement to collect, weigh, mix and sample the excreta.

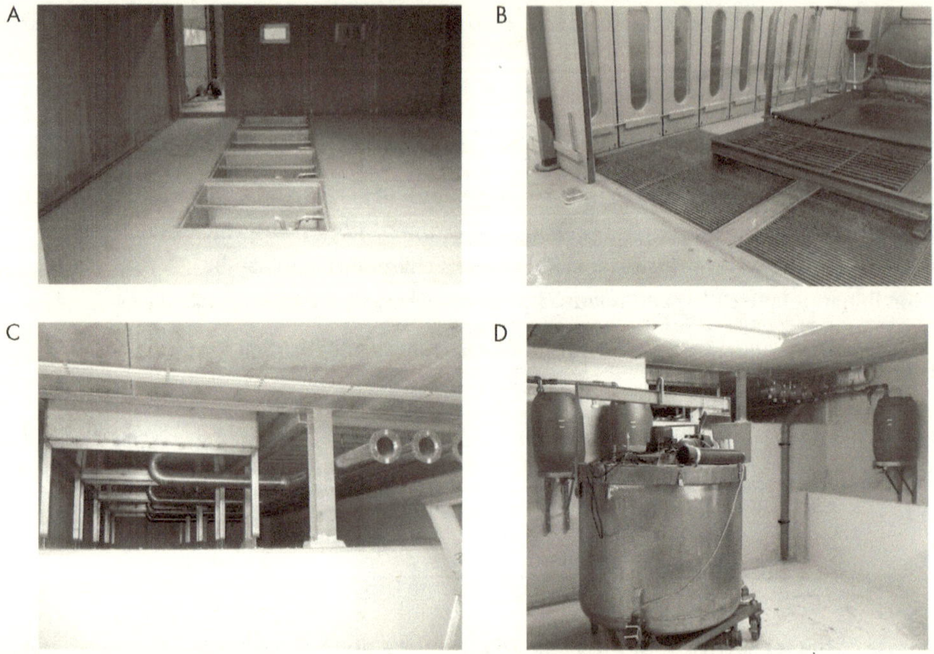

Figure 7. Top view (A, B) and bottom view (C, D) of four stainless steel slurry pits and a weighing and mixing vessel (D) to be connected to the waste pipe of each slurry pit with flexible tubing (Photos Wageningen University, Wageningen, NL).

2.8 Safety measures

Confining animals and personnel in airtight climate rooms brings the risk of suffocation and therefore the responsibility for adequate safety measures, meeting local requirements, which may strongly vary among countries. The continuous recording of various measured parameters (e.g. gas concentrations, ventilation level, climate and air pressure) opens the possibility to use these to trigger alarms for securing the safety for animals and personnel. Furthermore, logged data can be used to develop web applications to give direct access to alarm logging and recent measurements in case of an alarm call.

2.9 Summary

Respiration chambers with an adjustable volume have many advantages with regard to flexibility in the choice of size and/or number of animals used in experiments above fixed sized chambers. In this chapter, important issues related to the design of respiration chambers with special attention to size adjustable, climatised respiration systems, particularly for large animals are addressed. The facilities, recently constructed at Wageningen University, are used as an example throughout this chapter. The information presented on design issues, gas analysis systems, ventilation and climate control, leak tolerance and full system checks can be used selectively if renovation is required or new chambers are developed. In the end only a repeatable full system check can provide enough evidence that Q can be measured with a high level of accuracy.

References

Aerts, J.M., Berckmans, D., Saevels, P., Decuypere, E. and Buyse, J., 2000. Modelling the static and dynamic responses of total heat production of broiler chickens to step changes in air temperature and light intensity. British Poultry Science 41: 651-659.

Bourguet, C., Deiss, V., Boissy, A., Andanson, S. and Terlouw, E.M.C., 2011. Effects of feed deprivation on behavioral reactivity and physiological status in Holstein cattle. Journal of Animal Science 89: 3272-3285.

Cimadevilla, J.M., GonzalezPardo, H., Lopez, L., GarciaMoreno, L.M. and Arias, J.L., 1997. Social isolation and energy metabolism in rat hippocampus. Psicothema 9: 399-405

Gaughan, J.B., Heetkamp, M.J.W. and Hendriks, P., 2015. Indirect calorimetry: assessing animal response to heat and cold stress. In: Gerrits, W.J.J. and Labussière, E. (eds.) Indirect calorimetry. Techniques, computations and applications. Wageningen Academic Publishers, Wageningen, the Netherlands, pp. 213-230.

Huynh, T.T.T., Aarnink, A.J.A., Verstegen, M.W.A., Gerrits, W.J.J., Heetkamp, M.J.W., Kemp, B. and Canh, T.T., 2005. Effects of increasing temperatures on physiological changes in pigs at different relative humidities. Journal of Animal Science 83: 1385-1396.

Labussière, E., Dubois S., Van Milgen, J. and Noblet J., 2013. Partitioning of heat production in growing pigs as a tool to improve the determination of efficiency of energy utilization. Frontiers in Physiology 4: 146.

Lighton, J.R.B., 2008. Measuring metabolic rates, a manual for scientists. Oxford University Press, Oxford, UK.

Lighton, J.R.B., 2015. Metabolic measurement techniques: baselining, mathematical correction of water vapour dilution and response correction. In: Gerrits, W.J.J. and Labussière, E. (eds.) Indirect calorimetry. Techniques, computations and applications. Wageningen Academic Publishers, Wageningen, the Netherlands, pp. 57-72.

McLean, J.A. and Tobin, G., 1987. Animal and human calorimetry. Cambridge University Press, Cambridge, UK.

Renaudeau, D., Frances, G., Dubois, S., Gilbert, H. and Noblet, J., 2013. Effect of thermal heat stress on energy utilization in two lines of pigs divergently selected for residual feed intake. Journal of Animal Science 91: 1162-1175.

Van Adrichem, P.W.M. and Vogt, J.E., 1993. The effect of isolation and separation on the metabolism of sheep. Livestock Production Science 33: 151-159.

Van Dam, J.T.P., Schrama, J.W., Vreden, A., Verstegen, M.W.A., Wensing, T., Van der Heide, D. and Zwart, D., 1997. The effect of previous growth retardation on energy and nitrogen metabolism of goats infected with *Trypanosoma vivax*. British Journal of Nutrition 77: 427-441.

Van den Borne, J.J.G.C., Heetkamp, M.J.W., Alferink, S.J.J. and Gerrits, W.J.J., 2015. Moving from a complete energy balance towards substrate oxidation: use of stable isotopes. In: Gerrits, W.J.J. and Labussière, E. (eds.) Indirect calorimetry. Techniques, computations and applications. Wageningen Academic Publishers, Wageningen, the Netherlands, pp. 87-114.

Van Eerden, E., Van den Brand, H., Heetkamp, M.J.W., Decuypere, E. and Kemp, B., 2006. Energy partitioning and thyroid hormone levels during *Salmonella enteritidis* infections in pullets with high or low residual feed intake. Poultry Science 85: 1775-1783.

Van Klinken, J.B., Van Den Berg, S.A.A., Havekes, L.M. and Van Dijk, K.W., 2012. Estimation of activity related energy expenditure and resting metabolic rate in freely moving mice from indirect calorimetry data. PLoS One 7: e36162.

Van Klinken, J.B., Van den Berg, S.A.A. and Willems Van Dijk, K., 2013. Practical aspects of estimating energy components in rodents. Frontiers in Physiology 4: 94.

Van Milgen, J., Noblet, J., Dubois, S. and Bernier, F., 1997. Dynamic aspects of oxygen consumption and carbondioxide production in swine. British Journal of Nutrition 78: 397-410.

Verstegen, M.W.A., Van der Hel, W. and Pijls, F.J.M., 1993. Growth-rate and growth composition of artificially reared piglets from specific pathogen free sows. Animal Production 56: 217-223.

Verstegen, M.W.A., Van der Hel, W., Brandsma, H.A., Henken, A.M. and Brandsen, A.M., 1987. The Wageningen respiration unit for animal production research: a description of the equipment and its possibilities. In: Verstegen, M.W.A. and Henken, A.M. (eds.) Energy metabolism in farm animals. Martinus Nijhoff Publishers, Dordrecht, the Netherlands, pp. 21-48.

Chapter 3: Metabolic measurement techniques: baselining, mathematical correction of water vapour dilution and response correction

J.R.B. Lighton[1,2]

[1]*Sable Systems International, 6000 S Eastern Ave # 1, Las Vegas, NV 89119, USA;* [2]*School of Life Sciences, University of Nevada at Las Vegas, Box 4004, 4505 S. Maryland Parkway, Las Vegas, NV 89154-4004, USA; john@johnlighton.org, lighton@sablesys.com*

Abstract

In this chapter, I discuss techniques for increasing the robustness and accuracy of metabolic measurements, using as practical an approach as possible. After a short introduction to flow-through respirometry, the importance of the automated determination of incurrent in addition to excurrent gas concentrations during metabolic measurement, and the use of those concentrations for minimising analyser drift and maximizing measurement accuracy, are described in detail. Methods for reducing or eliminating disruptions in metabolic recordings caused by measuring incurrent gas concentrations are also described. In addition, the measurement of water vapour partial pressure and barometric pressure, and the mathematical correction of water vapour dilution, are proposed as an alternative to chemical or thermal desiccants to eliminate the dilution effect of water vapour on gas concentrations and flow rates. Techniques for extracting data on individual animals from multiplexed recordings, and correcting those data to minimise the response time distortions caused by the time constant of the respirometry chamber housing each animal, are also described.

Keywords: respirometry, metabolism

3.1 Introduction

First, a brief introduction to flow-through respirometry is in order. This technique, as implemented in this chapter (see Lighton, 2008 for a more general treatment), flows air past an animal at a known flow rate. The O_2 consumption (VO_2) and CO_2 production (VCO_2) rates of the animal are calculated from the concentration changes that the animal's aerobic catabolism causes to these gas species. The animal is usually confined to a chamber into which the air is pushed, or from which the air is pulled. Alternatively, the animal may wear a mask or hood from which air is pulled. In the case of push systems, a well-sealed chamber is essential because any leaks will cause variable and significant errors. Pull systems are much more forgiving in this regard and are widely used for large animals; only an exceedingly tiny pressure gradient from the outside to the inside of a chamber is required for essentially complete recovery of

DOI 10.3920/978-90-8686-261-0_3, © Wageningen Academic Publishers 2015

interior gas concentrations (Xu, 2007). Increasingly, pull systems are also being deployed for smaller animals, such as mice and rats used in biomedical research.

It is obvious that accurate flow rate measurement is of crucial importance. For this reason, mass flow meters, which report flowrates corrected to standard temperature and pressure, are almost universally used. For push systems, the flow rate entering the chamber (referred to as the incurrent flow rate) is measured; for pull systems, the flow rate leaving the chamber, musk or hood (referred to as the excurrent flow rate) is measured. A gas analyser chain, typically comprising O_2 and CO_2 analysers (in more modern systems, including H_2O analysers as well) analyses a small subsample of air that can be sourced, typically under computer control, from either the incurrent or excurrent flow streams. This ability to select between the incurrent or excurrent flow streams for analysis is essential, as discussed below.

The subsampled gas stream selected for analysis will contain a variable amount of water vapour, which dilutes the gas species being measured. In legacy metabolic measurement systems, the water vapour is removed, typically by chemical or thermal scrubbing. More advanced systems measure the amount of water vapour present, and mathematically correct the concentrations of other gas species to remove its dilution effect (Lighton, 2008). The gas analysers responsible for measuring O_2, CO_2 and, where relevant, CH_4 concentrations, should have the lowest possible drift and noise. This is especially relevant if fine temporal resolution of metabolic signals is required, as further explained below.

3.2 The importance of baselining

The measurement of baseline, or 'incurrent', gas concentrations interacts in an interesting way with response distortions caused by flowrates and respirometry chamber volumes. When measuring the metabolic rate of an animal in a chamber, the volume of the chamber and the flow rate of air leaving the chamber combine to limit the temporal resolution of metabolic data. However, it is possible to use the temporal distortion caused by the chamber to advantage, in order to measure incurrent gas concentrations frequently, thus improving the accuracy of metabolic measurement. The disruptions caused by repeated incurrent gas concentration measurements can be reduced or eliminated, and the primary metabolic vectors (or channels of data) can be reconstituted for correction of the temporal distortion caused by the time constant of the chamber (see definition below).

In all that follows, a single and pivotally important fact must be borne in mind. The equations that transform gas concentrations to metabolic measurements are vitally dependent upon concentration differences, rather than upon absolute concentrations. The concentration difference (or delta, Δ) of interest is that between the incurrent and excurrent fractional concentrations of a given respiratory gas species, e.g. ΔO_2 or ΔCO_2. Both incurrent and excurrent concentrations must be accurately known in order to calculate this delta parameter.

The reasonable, human perceptions of the researcher can be misleading and even pernicious. This causes excurrent gas concentrations to be regarded as more important than their incurrent equivalents, leading, in turn, to an indifference regarding the importance of measuring incurrent, or 'baseline' gas concentrations. Perhaps this is because the excurrent air stream embodies all of the expense and inconvenience of placing the animal in the chamber, and obtaining and setting up all of the complex apparatus required for metabolic measurement. The goal after all is to obtain an optimally accurate time-series data-stream of the gas exchange characteristics of one or more animals. It follows that any interruption to the continuous measurement of gas exchange parameters from a given animal will disrupt the metabolic time-series, so it seems best to interrupt the measurements as seldom as possible. This sounds reasonable, yet is entirely mistaken. At least, it is mistaken if the interruptions are thoughtfully and expertly implemented.

There are two key issues to understand here. The first and more important is that interruptions need not disrupt the time-series if, and only if, they occur at intervals substantially shorter than the time constant of the respirometry chamber. The second is that interruptions are a practical requirement in order to compensate for gas analyser drift and changes in the concentrations of incurrent gases.

Let us address the first issue in some detail. For the sake of argument, assume that we are measuring the metabolic parameters of a 750 kg animal housed within a 10,000 l respirometry chamber from which we are pulling air at a rate of 500 l per min. To a first approximation, the time constant of the chamber (neglecting the volume of the animal) is 10,000 l divided by 500 l/min = 20 min.

$$TC = \frac{V}{FR} \tag{1}$$

Where TC is the time constant of a chamber of volume V, from which air is pulled (or into which air is pushed) at the flow rate FR. Obviously, V and FR must be in volumetrically consistent units (for example, l and l/min), and TC inherits the reciprocal of the time units of FR (for example, min for l/min). In practice, the volume of the animal should be subtracted from the volume of the chamber. In most cases, assuming an animal density of 1.0 (1 l/kg) is sufficiently accurate.

Metabolic measurement chambers are first-order, linear time invariant systems (Lipták, 2003). Thus, even if the animal were to undergo a step change in metabolic rate, the gas concentrations leaving the chamber, assuming good mixing within the chamber, will not change by more than a factor of $(1 - 1/e)$ or ~63% over the duration of the chamber's time constant. Gas concentrations are, in other words, grossly lagged and distorted indicators of the actual metabolic rate of the animal within the chamber.

The situation appears dire, but in fact we can use the relative slowness of chamber response to our advantage, as will be shown below. Let us turn to the second of the two issues – the requirement for baselining – for a moment, before returning to ways in which the disruptions caused by baselining can be minimised.

We live in an imperfect world, and in celebration of this fact, gas analysers tend to drift. This is most noticeable in the case of O_2 analysers, which must resolve tiny depletions of the order of 0.02 to 1% from a massive standing concentration of roughly 21%. This is a serious problem because in metabolic measurement, we are generally not concerned with absolute gas concentrations, but rather the difference in gas concentrations between the incurrent and excurrent air-streams entering and leaving the respirometry chamber. Differential gas analysers may appear to mitigate this problem, and indeed, if used with care are capable of resolving concentration differences down to 0.3 ppm (Stephens *et al.*, 2007). However, such analysers still suffer from drift, and will still require the 'sanity check' of periodic checks of their true differentiality, which is to say, baselining. Differential gas analysers mitigate, but do not eliminate, the requirement for periodic baselining.

If we measure incurrent concentrations at the start of a recording and then measure them again several hours later, we have no way of knowing what creative excursions the measured incurrent concentrations may have taken during the course of the recording. It should be obvious that in order to measure the all-important delta concentrations of O_2 and CO_2, we must be meticulous in measuring incurrent as well as excurrent concentrations, and must do so over a time-interval that is short relative to the rate of drift. (That time interval will obviously depend on the gas concentration and analyser stability characteristics that are unique to a given installation.)

The perceived necessity for continuous metabolic measurement stands in contradiction to the necessity of frequent incurrent gas concentration measurements. Yet this contradiction is more apparent than real, as we shall see.

3.3 Minimizing disruptions from baselining

In practice, it is necessary at regular intervals to direct a sample of the incurrent air-stream through the gas analysis chain, thus unavoidably disrupting the measurement of excurrent gas concentrations. The benefits outweigh the disadvantages because each incurrent, or baseline, measurement allows for the correction of O_2 analyser drift, by spanning the analyser to ambient O_2 concentrations, and also allows the tracking of incurrent water vapour (where applicable) and CO_2 concentrations. Given the necessity of regular incurrent measurements, it is obvious that we must make them as brief as possible, and in every possible way minimize their disruptive effect on delta gas concentration measurements.

To make the incurrent measurements as brief as possible, the obvious approach is to use the fastest possible gas analysers. However, this is a sub-optimal solution because very fast gas analysers are also noisy (see Lighton, 2008 for a survey of gas analyser technologies). This makes them unsuitable for mathematical treatment of the gas exchange record to minimise the temporal distortions caused by the time constant of the chamber. Rather, the volume of the sub-sampling and gas analysis chain must be made as small as possible in order to minimise its time-constant, and the response time of the gas analysers should be reduced by mathematical compensation, for example by using the 'instantaneous transform' (Bartholomew *et al.*, 1981; see also Lighton and Halsey, 2011 and Lighton, 2012). This technique is further explained below. To minimise the volume of the gas analysis chain, it should be miniaturized to the extent possible, and preferably the various gas analysers should be tightly integrated into a single enclosure with short, thin plumbing interconnections. In addition, eliminating desiccators from the gas analysis chain achieves a significant reduction in volume and improvement in response time, at the cost of requiring a more sophisticated gas analysis chain that includes water vapour pressure and barometric pressure measurement.

The extent to which this approach can improve response times when switching from excurrent to incurrent gas concentrations is shown in Figure 1. See below for more information on the response correction technique employed there.

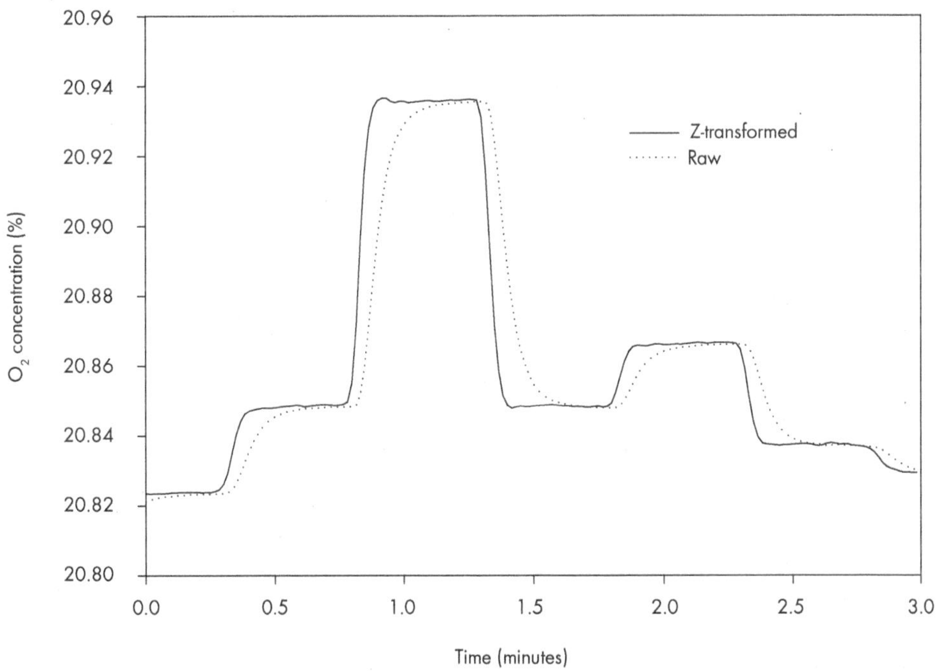

Figure 1. Improvement in analyser response times caused by application of the instantaneous transform (see text). The data vector is of O_2 concentration in a multiplexed Promethion metabolic measurement system, switched from various excurrent samples to incurrent air from ~0.75 min to ~1.25 min. The measurement duration for all gas samples is 30 s. The dotted line shows the original, raw data. The solid line shows the same data after instantaneous or 'Z' transformation. It can be seen that the data vector reaches an acceptably stable plateau value after only ~10 s.

Having minimised the duration of the incurrent sample, it is now possible, if we are dealing with metabolic measurement of a single experimental animal, to make the necessary corrections and then simply interpolate across the incurrent sample, effectively removing it from the corrected metabolic time-series. In my experience, with suitable equipment and software it is almost always possible to reduce the duration of an incurrent measurement to <1 min, and with some additional care and fine-tuning, to <15 s.

Each incurrent measurement gives us the opportunity to adjust the gain of the O_2 channel so that incurrent O_2, after correction for water vapour dilution and barometric pressure, yields a concentration of 20.94% STPD (standard temperature and pressure, dry). This is referred to as 'spanning' the analyser. For some inexplicable reason, perhaps historical inertia, bottled O_2 span gas is widely used to calibrate O_2 analysers. In fact, the atmosphere is almost invariably a much more accurate source of O_2 span gas (Tohjima *et al.*, 2005; Stephens *et al.*, 2007). Even though heavy traffic, etc., in urban areas may slightly perturb local O_2 concentrations, such minor variations are generally negligible compared to the accuracy envelope of typical span gases, which vary from 1% to as much as 5% of the nominal span concentration, and in addition may vary substantially from lot to lot. (In the interests of ensuring maximally stable incurrent O_2 and CO_2 concentrations, excurrent air pulled or pushed from masks or chambers should, of course, be evacuated away from the area of measurement, for example, to the outside of a building.) Thus, each incurrent gas measurement yields the opportunity to re-calibrate the O_2 analyser under software control or during the later analysis of the raw data. It also allows the spot measurement of incurrent CO_2 concentrations, which as we will see below, must be known for accurate VCO_2 data.

The frequency of incurrent measurements is dependent on the stability of the gas analysers, the stability of the environment in which the gas analysers are kept, and the degree of variability in incurrent gas concentrations, particularly water vapour and CO_2. As a rule of thumb, I recommend taking baselines every 15 min when measuring a single animal. The situation when measuring multiple animals is more complex and will be dealt with in detail below.

Returning at last to our first issue, and to our hypothetical example, we see that the long-time constant created by the volume of the chamber and the flow rate of the air pulled from it, now works to our advantage. The time constant of the chamber operates as a massive low-pass filter, but our incurrent gas concentration measurements take place after this filter. Thus, provided the incurrent measurements are as brief as possible compared to the time constant of the chamber, they can be removed by extrapolation without doing undue violence to the metabolic signal. This is shown in Figure 2, for a more extreme case with a much shorter chamber time constant. In the case of animals measured in chambers with long time-constants, it is then possible to correct the final data to reduce or eliminate the effect of the chamber time-constant, as further discussed below.

Thus, in the vast majority of cases when dealing with animals in chambers, a well-implemented incurrent measurement schema need not distort the metabolic data being measured, but to

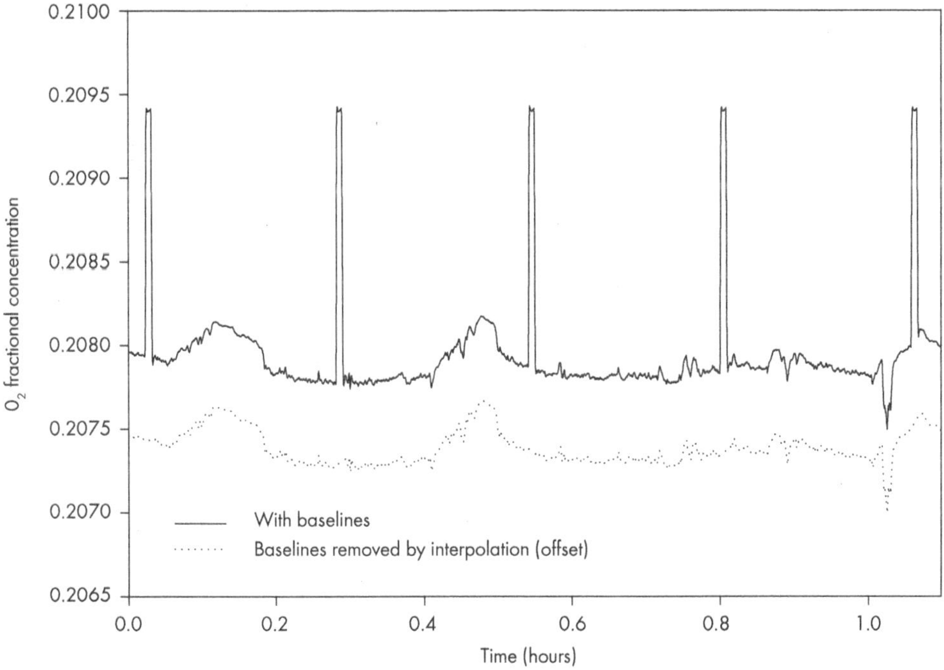

Figure 2. Removal of brief incurrent measurements (baselines) by interpolation. Data are from an O_2 vector in a continuous Promethion metabolic measurement system. The solid line shows O_2 data interrupted every 15 min for a 30 sec baseline measurement (fractional incurrent O_2 concentration 0.2094). In this system, the chamber time constant was 4.5 min, significantly less than for most large-animal systems. The dotted line shows the same data with the baselines removed by interpolation. The dotted trace is slightly displaced downwards for visibility.

the contrary, will substantially increase the accuracy of the measurements being made. This is especially true for O_2 analysers, which need to resolve tiny changes against a substantial offset.

3.4 Mathematical compensation for water vapour dilution

In order to measure gas concentrations for use with metabolic measurement calculations, it is necessary to remove the dilution effect of water vapour. It is also necessary to subtract the contribution of water vapour to the overall flow rate of the air stream pulled from the respirometry chamber. Traditionally, chemical or thermal desiccants had been used for the former, and calculated approximations based on relative humidity and barometric pressure for the latter (see later). However, chemical desiccants are short-lived, present disposal problems, and may interact with CO_2. Thermal desiccants are more mechanically complex, and do not in fact remove all water vapour from the air passing through them. They can only reduce the dew point of the air to approximately 1-2 °C, corresponding to a relative humidity of approximately

25% at an ambient temperature of 25 °C. In many locations, ambient relative humidity can drop substantially below this figure, rendering thermal scrubbers useless and causing them to generate misleading metabolic data (Melanson *et al.*, 2010). Although an accumulated reservoir of condensed water can buffer short-term dips below the condenser's dew point, such a reservoir will be rapidly exhausted by prolonged episode to dry air (as described by Melanson *et al.*, 2010). In addition, CO_2 is highly soluble in water, ~30× more so than O_2 (Prosser, 1991), and a significant volume of condensate will act as a buffering CO_2 capacitor. This may result in slow responses to shifts in measured CO_2 concentrations, especially when switching from incurrent to excurrent air streams, and potentially lead to deleterious phase lags between the CO_2 and O_2 concentrations signals if their rates of change are significant.

Basic physics comes to the rescue. By Dalton's law of partial pressures, we can exactly compensate for the dilution effect of water vapour provided that we know water vapour partial pressure and the total pressure of all gas species present, also known as barometric pressure (see Lighton, 2008 for a discussion). Instrumentation now exists that can measure water vapour partial pressure with sufficient accuracy to make this technique feasible, and in fact, improve upon the accuracy of traditional techniques (Melanson *et al.*, 2010). Water vapour pressure is measured using instrumentation that simultaneously measures relative humidity and the temperature of the relative humidity sensor. By calculating saturated water vapour pressure corresponding to the temperature of the sensor, and multiplying that figure by the water vapour activity (= relative humidity/100), water vapour partial pressure can be measured with good accuracy. Calibration of such a sensor can be automated by switching in a chemical desiccant column under computer control, which accomplishes two things: (1) it allows the water vapour pressure analyser to be zeroed against a known zero concentration of water vapour, and (2) by simultaneously measuring O_2 concentration, it allows the O_2 analyser to be spanned against dry air, and when the column is switched out, and the O_2 concentration falls as a result of water vapour dilution, it allows the precise ambient water vapour pressure to be calculated from the dilution effect of water vapour (Lighton, 2008).

$$WVP = BP \times \frac{(O_2' - O_2)}{O_2'} \tag{2}$$

Where O_2' is dry O_2, O_2 is O_2 measured in the presence of water vapour partial pressure WVP, and BP is the barometric pressure in units compatible with WVP, e.g. kPa. Of course, care must be taken to avoid condensation of water in the gas analyser chain (regulating its temperature at, e.g. 5-10 °C above the maximum anticipated animal or environmental temperature, where condensation is unlikely to occur, is a good idea), and analysers that are not damaged by the presence of water vapour should be used. In the case of O_2, paramagnetic or fuel cell analysers may be used. However, the diamagnetic properties of water vapour, which may interfere with paramagnetic O_2 detection, suggest that fuel cell analysers are a better choice if water vapour is present.

This allows the water vapour pressure analyser to be calibrated against the actual water vapour pressure it is experiencing. Such calibrations for zero and span can be carried out by direct interaction with the water vapour and O_2 analysers, or in software, provided that all raw data are acquired. In general, software correction (where the calibration is derived directly from measured concentration changes recorded when a high quality chemical desiccant is switched in and out of the analyte air stream under computer control) is to be preferred because it offers greater traceability.

To correct any gas concentration for water vapour dilution, use the following equation:

$$G' = G \times \frac{BP}{(BP - WVP)} \tag{3}$$

Where G' is the concentration of the mathematically dried gas species, and G is the concentration of the gas species in the presence of water vapour.

To correct a flow rate to a dry equivalent, use the following equation:

$$FR' = FR \times \frac{(BP - WVP)}{BP} \tag{4}$$

Where FR' is the dried flow rate, and FR is the flow rate in the presence of water vapour. It should be noted that if water vapour pressure and barometric pressure are not measured, correcting flowrates (whether pushed or pulled) to their dry equivalents is not possible. Applying a correction factor based on a consensus relative humidity, ambient temperature, and altitude-based barometric pressure is unacceptably inaccurate.

If water vapour is measured, the water loss rate of the animal in each respirometry chamber can likewise be measured. This may or may not be of interest. Two potential benefits can be derived from knowledge of this metric, which is a portmanteau sum of all the water produced by the animal, including respiratory and insensible water loss plus evaporation from urine and faeces. First, if the water intake of the animal is known, for example by gravimetric monitoring, and the water of hydration in any food intake is likewise known, it will be found that the overall water loss rate of the animal is significantly higher than the sum of those two figures. That increase is caused by metabolic water production, and this technique allows that figure to be directly measured. In addition, if the energy equivalent of food is derived from the mass of the food, that mass is significantly affected by varying hydration levels caused by ambient water vapour. It is possible in principle to remove this source of unwanted noise in energy equivalent calculations if the hydration characteristics of the food as a function of ambient water vapour pressure are suitably quantified.

In a pull system, it is necessary to know the incurrent concentration of CO_2 in order to calculate VCO_2 correctly. This is in order to compensate VCO_2 for the slight exaggeration caused by the removal of consumed oxygen from the excurrent air (Lighton, 2008). As mentioned above, each incurrent measurement yields a spot data point of incurrent CO_2. It is then possible to create, by interpolation, a pseudo-continuous channel of incurrent CO_2 concentrations for calculation purposes, as shown in Figure 3. This is opposite to the approach described above, in that incurrent CO_2 concentrations are retained, after moving to a separate data vector, and interpolated between, whereas the incurrent concentrations are removed from the main data vectors (O_2, CO_2) by interpolation. This approach yields a far more accurate and dynamically tracked data vector of incurrent CO_2 compared with the common practice of assuming an invariant incurrent CO_2 concentration, typically 0.03%, which is now substantially in error and becoming less accurate with each passing year.

3.5 Baselining, calibration and compensation: all together now

The following basic steps are recommended in processing data containing baseline concentration information:

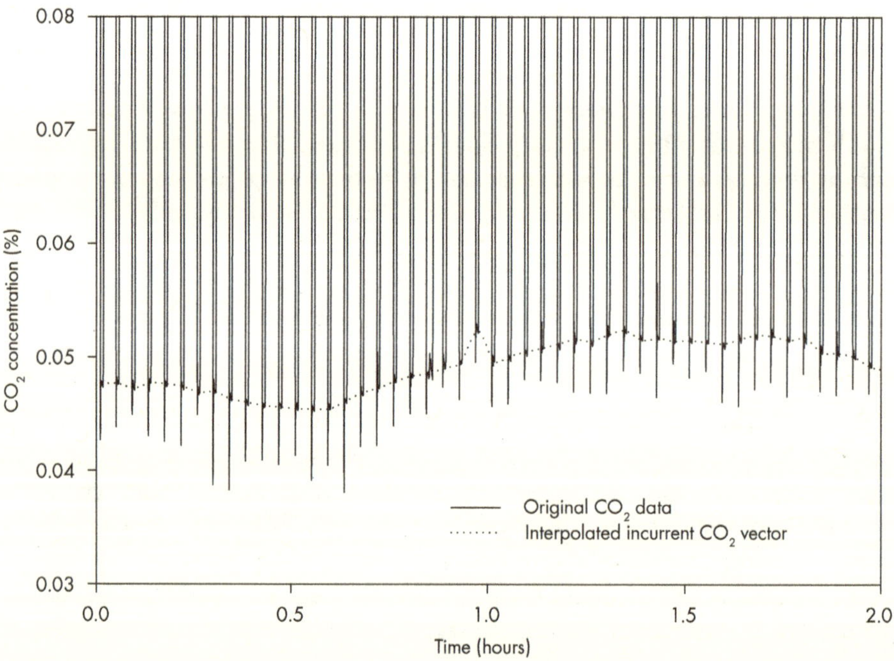

Figure 3. Creation of a pseudo-continuous data vector (dotted line) corresponding to incurrent gas concentrations (in this case CO_2) measured during baselining. Incurrent CO_2 levels were fairly variable, as the recording was made in an enclosed space with several people present. Only the lower concentration range of the original CO_2 vector is shown, for greater visibility of the incurrent interpolation. Non-first-order kinetics during the first few seconds of each switch to incurrent air cause brief downward spike artefacts, but these are ignored when producing the interpolated trace. The interpolation algorithm is set to use only the mean of the most level 25% within each baseline for calculation purposes.

1. Ensure that the location of the baselines in the data time series is known. This allows analysis algorithms to detect the position of the baselines and apply the corrections listed below, automatically. The location of the baselines can be denoted, for example, by markers or by a numerical value in a data vector specifically set aside for allowing automated detection of the gas analysis path.
2. Apply the instantaneous transform to the O_2, CO_2 and, if applicable, H_2O vectors to minimize the effect of response time of the gas analysers. Each analyser for a given gas species is likely to have a unique instantaneous transform constant (Z).
3. Slide the O_2 and CO_2 concentration vectors and, if applicable, the H_2O partial pressure vector, leftwards, i.e. backwards in time, so that their baseline initiations are coincident. Good temporal synchronization between the vectors is important, as the calculations that will follow require it.
4. Remove the dilution effect of water vapour from the O_2 and CO_2 vectors using Equation 3.
5. Correct the O_2 and CO_2 vectors to standard barometric pressure.
6. Span the O_2 baselines to 20.94%.
7. Ensure that the O_2 and CO_2 vectors are converted to fractional concentrations, i.e. percent over 100.
8. Subtract the O_2 baselines from the O_2 vector, and invert the subtraction by multiplying the resulting delta vector by -1, creating the delta O_2 vector.
9. Extract the baseline CO_2 concentrations from the CO_2 vector and interpolate between them to create a vector containing the incurrent CO_2 concentration. This will be required in step 12, below.
10. Subtract the CO_2 baselines from the CO_2 vector, creating the delta CO_2 vector.
11. Remove the water vapour from the flow rate using Equation 4.
12. Finally, apply the requisite respirometry equations. These are selected according to the flow modality of the system (push or pull).

For convenience, the key equations, all of which use dry gas fractional concentrations and flow rates, are listed below. FR = flow rate, F = fractional concentration, i = incurrent, e = excurrent. VO_2 and VCO_2 inherit the units of FR. For further details and derivations, see Lighton (2008).

For pull systems (see Equations 11.7 and 11.8 in Lighton, 2008):

$$VO_2 = \frac{FRe[(\Delta O_2) - FiO_2\,(\Delta CO_2)]}{(1 - FiO_2)} \tag{5}$$

$$VCO_2 = \frac{FRe[(\Delta CO_2) + FiCO_2\,(\Delta O_2)]}{(1 + FiCO_2)} \tag{6}$$

For push systems (see Equations 10.6 and 10.7 in Lighton, 2008):

$$VO_2 = \frac{FRi[(\Delta O_2) - FeO_2\,(\Delta CO_2)]}{(1 - FeO_2)} \tag{7}$$

$$VCO_2 = \frac{FRi[(\Delta CO_2) + FeCO_2\,(\Delta O_2)]}{(1 + FeCO_2)} \tag{8}$$

To calculate the heat production or energy equivalent corresponding to a given VO_2 and VCO_2, calculate the respiratory quotient or RQ (VCO_2/VO_2), and use the following oxyJoule equivalent in conjunction with VO_2 (Equation 9.13 in Lighton, 2008; many alternative equations exist, such as that of Weir, 1949).

$$J\ mlO_2^{-1} = 16 + 5.164(RQ) \tag{9}$$

3.6 Dealing with multiplexed systems

The situation is somewhat more complex when measuring multiple animals or chambers in succession. It is relatively easy to ensure that the duration of incurrent measurements is substantially less than the chamber time-constant in the case of a single animal. Where one gas analyser chain is shared between multiple chambers, the need to minimize the duration of incurrent measurements, and indeed the duration of measurements of excurrent air from any given chamber, becomes more critical.

When dealing with N chambers in a multiplexed metabolic measurement system, the critical number for the system is its cycle time, which is defined:

$$C = (N + \frac{N}{I}) \times D \tag{10}$$

Where C is the cycle time of the system, D is the dwell time required to obtain accurate concentration measurements when switching to a new gas source, and I is the incurrent interleaving ratio. This metric is determined by how frequently the incurrent gas concentrations are measured. If incurrent gas concentrations are measured after each excurrent measurement, the interleaving ratio is 1; if measured after every two excurrent measurements, the ratio is 2, and so on.

Clearly, the desirable situation is:

$$C << TC \tag{11}$$

In other words, where possible, the cycle time of the multiplexed system should be minimised in comparison to the time constant of the chambers.

Returning to our hypothetical example, we have TC = 20 min. Let us assume that we are measuring four chambers, each with similar inhabitants and time constants. Of course, the excurrent flow pulled from the chambers is not affected by the multiplexing operation, which concerns itself only with small sub-sampled flows from the main flows – in practice, 100 to 500 ml/min. If we assume a dwell time of 1 min, and an interleaving ratio of 4, then the cycle time of the system is 5 min, which is only ca. 25% of the time constant of each chamber. Thus, each chamber is sampled approximately 4 times during each time constant. During those samples, a 'snapshot' measurement of excurrent gas concentrations pulled from each chamber is made. Provided that the gas analysers do not add substantial noise to the concentration signal, this 'snapshot' measurement gives accurate results because the time constant of the chamber guarantees that real concentration changes in the excurrent air within that brief dwell time will be relatively small. It should be noted that excellent mixing within the chamber is required, without which the excurrent air is subject to random fluctuations in respiratory gas concentrations. Such fluctuations may add an unacceptably large random component if accurate resolution of short-term metabolic signals is of interest.

Analysis of multiplexed gas concentration readings is accordingly more complex. Corrections for analyser drift, and measurement of incurrent concentrations, take place during each incurrent measurement. The drift-corrected excurrent concentrations are converted to delta concentrations, resulting in a series of data points separated by the cycle time. This is shown in Figure 4.

3.7 Correcting for the chamber time-constant

We are now faced with minimising the impact of each chamber's time-constant. The instantaneous transform requires calculation of the first derivative of the gas exchange signals (most other correction protocols require this value as well but are not covered here). Although it is possible to use individual measurements spaced one cycle time apart, it is also possible to interpolate intelligently between individual spot readings and generate a pseudo-continuous metabolic record at a smaller time-interval (Figure 4, dotted line). This allows synchronization between metabolic records and data acquired at a higher frequency, for example activity, position or intake data.

If interpolation is used, the interpolation algorithm should be chosen with care. Arguably the best such algorithm was originally designed for computerized graphics generation, and is known as the Catmull-Rom spline (Catmull and Rom, 1974). It is unusual because it passes exactly through each sample point, and utilizes tangent vectors to approximate a 'best guess' at the underlying non-linear continuity of the data.

The suitably reconstituted gas exchange records from each chamber can now be smoothed preparatory to applying the instantaneous transform. Smoothing the data vector prior to applying the instantaneous transform significantly reduces the noisiness of the resulting data.

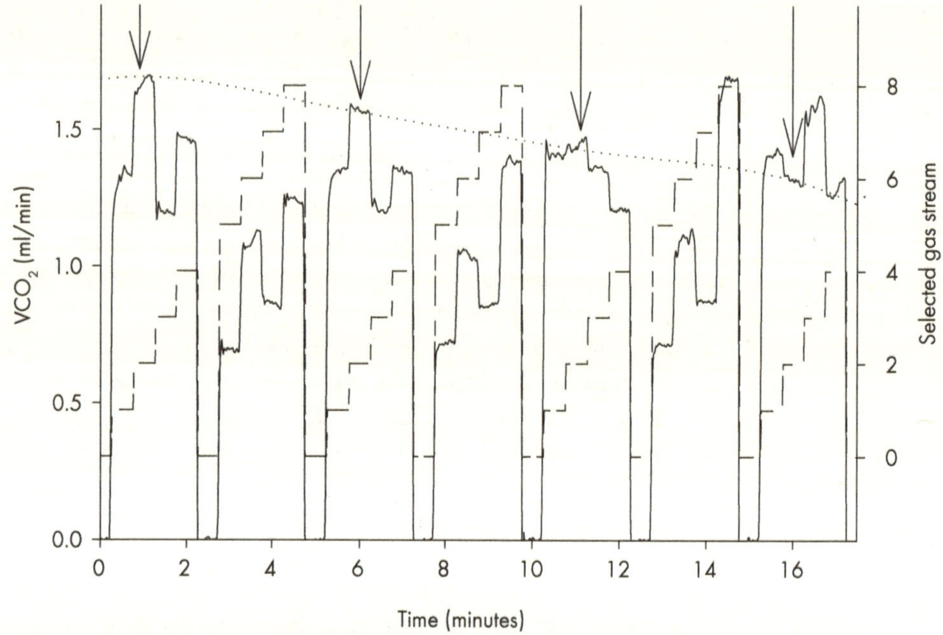

Figure 4. Creation of a pseudo-continuous vector corresponding to a single animal's gas exchange rate (in this case VCO_2 of animal 2 in an 8-cage multiplexed system), from a multiplexed vector consisting of brief measurements from multiple animals in rapid succession. The gas analyser chain is switched from baseline to chambers 1-2-3-4, baseline, chambers 5-6-7-8, repeatedly. Original multiplexed VCO_2 data = solid line, interpolated pseudo-continuous data for animal 2 = dotted line, indicator of currently selected airstream (0 = baseline, 1-8 = chambers 1 to 8; right Y axis) = dash-dot-dot-dashed line. The arrows correspond to the selection of metabolic rate data for animal 2.

Recall that the instantaneous transform calculates the first derivative of the signal, multiplies it by a large number referred to as 'Z', and then adds it back to the original signal (Lighton, 2012; Lighton and Halsey, 2010). Because the first derivative is exquisitely sensitive to noise in the form of sample to sample variation, noise should be minimised prior to engaging it. The instantaneous or 'Z' transform can be applied by utilising the value for its critical parameter, Z, as calculated from chamber volume and flow rate:

$$Z = 1 - e^{-FR(dT/EV)} \tag{12}$$

where FR is flow rate, dT is the interval between successive samples in the same time units, and EV is the 'effective volume' of the system in the same volumetric units as the flow rate, for example l if the flow rate is in l/min.

Alternatively, and preferably, the Z coefficient can be calculated empirically by disturbing the equilibrium of the chamber, for example by injecting pure nitrogen at a rate about fivefold greater than anticipated O_2 consumption rates, and then calculating the value of Z from the changing O_2 concentration as it progresses back towards equilibrium, as described in Bartholomew *et al.* (1981). This is probably a superior technique, because it acts on the empirical response characteristics of the entire system.

When applying a substantial Z coefficient to a VO_2 or VCO_2 data vector, for example when dealing with a signal derived from a chamber with a substantial time constant, any noise derived from the gas analysers or data acquisition system will be remorselessly amplified. A variety of signal-processing techniques, such as Savitsky-Golay smoothing and differentiation (Savitsky and Golay, 1964) can be employed to further facilitate response correction, but ultimately the quality and resolution of gas analysers play a pivotal role in final metabolic data excellence.

The utility of response correction ultimately depends on the temporal resolution of metabolic signals required by the researcher. If separation and resolution of brief episodes of activity are required, for example when quantifying the energy cost of specific activities, or conversely when discarding metabolic data contaminated by activity, response correction is important and may be essential. If, on the other hand, the researcher simply requires an overall mean metabolic rate, the added analytical complexity of response correction is not warranted.

3.8 Summary

Regular baselining, mathematical compensation for the presence of water vapour, multiplexing of gas streams, and response correction are not only compatible, they are synergistic and yield the highest accuracy in metabolic measurement if due care is taken. It is hoped that this chapter will reduce some of the aura of mystery that many researchers experience around these topics. Readers with further questions are encouraged to contact me directly (john@johnlighton.org).

References

Bartholomew, G.A., Vleck, D. and Vleck, C.M., 1981. Instantaneous measurements of oxygen consumption during pre-flight warm-up and post-flight cooling in sphingid and saturniid moths. Journal of Experimental Biology 90: 17-32.

Catmull, E. and Rom, R., 1974. A class of local interpolating splines. In: Barnhill, R. and Risenfeld, R. (eds.) Computer aided geometric design. Academic Press, New York, NY, USA, pp. 317-326.

Lighton, J.R.B., 2008. Measuring metabolic rates: a manual for scientists. Oxford University Press, New York, NY, USA.

Lighton, J.R.B., 2012. 'Instantaneous' metabolic measurement. Journal of Experimental Biology 215: 1605-1606.

Lighton, J.R.B. and Halsey, L.G., 2011. Flow-through respirometry applied to chamber systems: pros and cons, hints and tips. Comparative Biochemistry and Physiology 158: 265-275.

Lipták, B.G., 2003. Instrument engineers' handbook: process control and optimization (4th ed.). CRC Press, Boca Raton, FL, USA.

Melanson, E.L., Ingebrigtsen, J.P., Bergouignan, A., Ohkawara, K., Kohrt, W.M. and Lighton, J.R.B., 2010. A new approach for flow-through respirometry measurements in humans. American Journal of Physiology 298: R1571-R1579.

Prosser, C.L., 1991. Environmental and metabolic animal physiology. Wiley-Liss, New York, NY, USA.

Savitsky, A. and Golay, M.J.E., 1964. Smoothing and differentiation of data by simplified least squares procedures. Analytical Chemistry 36: 1627-1639.

Stephens, B.B., Bakwin, P.S., Tans, P.P., Teclaw, R.M. and Baumann, D.D., 2007. Application of a differential fuel-cell analyzer for measuring atmospheric oxygen variations. Journal of Atmospheric and Oceanic Technology 24: 82-94.

Tohjima, Y., Machida, T., Tomonori, W., Akam, I., Amara, T. and Moriwak, Y., 2005. Preparation of gravimetric standards for measurement of atmospheric oxygen and reevaluation of atmospheric oxygen concentration. Journal of Geophysical Research 110: D11302.

Weir, J.B., 1949. New methods of calculating metabolic rate with special reference to protein metabolism. Journal of Physiology 109: 1-9.

Xu, T., 2007. Characterization of minienvironments in a clean room: design characteristics and environmental performance. Building and Environment 42: 2993-3000.

Chapter 4: Modelling gas exchanges to partition heat production between fasting, thermic effect of feeding and physical activity in farm animals

E. Labussière[1,2], S. Dubois[1,2], J. Van Milgen[1,2] and J. Noblet[1,2]*
[1]INRA, UMR 1348 PEGASE, 35590 Saint-Gilles, France; [2]Agrocampus Ouest, UMR 1348 PEGASE, 35000 Rennes, France; etienne.labussiere@rennes.inra.fr

Abstract

When animals are housed in a respiration chamber, the utilisation of high frequency gas analysers helps in understanding and quantifying variation in heat production (Q) and partitioning of metabolisable energy (ME) intake. The objective of this chapter is to describe a methodology based on mathematical modelling of gas exchanges between animals and the respiration chamber and to highlight some key-results that were obtained about partitioning of ME intake. The model assumes that O_2 consumption and CO_2 production of animals only result from processes associated with fasting heat production (FQ, constant during the day), thermic effect of feeding (TEF, induced by feed intake, digestion and metabolism of nutrients) and physical activity. From these gas exchanges, FQ, TEF and activity-related heat production (AQ) were calculated respectively, in growing pigs, calves, broilers and turkeys. The data obtained from the model were used to estimate the exponent to calculate metabolic body weight, which equalled 0.60 in growing pigs, 0.70 in birds and 0.85 in calves. These exponents range in the same order as specific allometric growth coefficients of viscera in these species, which contribute to a great extent to whole body Q. In *ad libitum* fed animals, FQ accounted for 30% of ME intake in non-ruminants but for 49% in calves. In pigs and calves, it has been shown that FQ depends on the previous level of feeding. When animals are fed a standard diet, TEF was similar between non-ruminants (from 13 to 17% of ME intake) but rather lower in calves, due to the low fibre content of the milk replacer. Except for turkeys (13% of ME intake), AQ averaged 9% of ME intake. Taking the effect of feeding level on FQ into account, the variable components of Q (TEF, AQ) were used to calculate energetic efficiencies of energy retention. In growing pigs, these were estimated at 60 and 80% for energy retained as protein and as fat, respectively.

Keywords: mathematical modelling, fasting heat production, metabolic body weight, growing farm animals

DOI 10.3920/978-90-8686-261-0_4, © Wageningen Academic Publishers 2015

4.1 Introduction

Animals produce heat because of metabolic inefficiencies and use of energy from ATP and related intermediates to support body maintenance and productive functions (e.g. growth, milk, eggs, fur and wool production, pregnancy, incubation), in addition to supplementary heat production (Q) because of possible thermoregulation, physical activity, or inflammatory and immune responses. The measurement of Q in respiration chambers offers the opportunity to evaluate variation both within and among animals depending on their genotype and on environmental conditions. The development of high-frequency measurements of Q allows determining the dynamic patterns of Q within a day. It is commonly accepted that Q is not constant over the day, but varies according to energy metabolism because of changes in basal metabolic rate (BMR), thermic effect of feeding (TEF) and physical activity. Originally, BMR was defined as the rate of metabolism of adult, non-reproductive, and inactive humans within the zone of thermoneutrality and in a post-absorptive state and measured as fasting heat production (FQ; McNab, 1997). In growing animals, the concept of BMR requires distinguishing two components of energy metabolism (BMR and growth), in addition to components related to physical activity and thermoregulation; those related to inflammatory and immune responses are often not considered, or included in BMR. Therefore, FQ is thought to be representative of BMR in growing animals because it excludes the contribution of digestion and tissue deposition in energy metabolism (Baker *et al.*, 1991), whereas it accounts for metabolic efficiencies and ATP use to support body maintenance functions. Nevertheless, the duration of fasting (Q can decrease up to 4 days after feed withdrawal; Close and Mount, 1975) and the contribution of physical activity to whole body Q remains unaccounted for and classical values for FQ can be overestimated by heat production due to physical activity by more than 7% (Van Milgen *et al.*, 1998). The first objective of this chapter is to present a methodology developed in our laboratory (Van Milgen *et al.*, 1997) to partition total Q between FQ, TEF, and physical activity. The second objective of this chapter is to highlight some key results based on this modelling approach of Q in growing animals and partitioning of metabolisable energy (ME) intake.

4.2 Measurements of Q kinetics

The data were obtained during previous experiments conducted in open-circuit respiration chambers located at INRA, Saint-Gilles, France, which are similar to those described by Vermorel *et al.* (1973). Three respiration chambers (either 1.7 or 12 m^3 in volume) were used with ventilation rates ranging from 1 to 12 m^3/h, depending on the body weight (BW) of the animals used. The O_2 and CO_2 concentrations in outgoing air were measured using a paramagnetic differential analyser (Oxymat 6, Siemens, Munich, Germany) and an infrared analyser (Ultramat 6, Siemens, or Unor 600, Sick Mahiak, Houston, TX, USA), respectively. The O_2 and CO_2 concentrations in ingoing air were measured at the beginning and at the end of each measurement period (usually 24 h), and the drift was assumed to be constant during the day. The ventilation rate was measured with a mass flow meter (Teledyne Brown Engineering, Huntsville, AL, USA; Figure 1).

The animals were either housed individually (pigs, calves), or group-housed in a metabolism cage (broilers, turkeys), which was placed on four force sensors (9104A, Kistler, Jonsered, Sweden) that produced an electrical signal proportional to the mechanical force exerted vertically by the animal (Quiniou *et al.*, 2001). Animals spent at least seven days in the respiration chamber. The first day in the respiration chamber was considered as an adaptation day and was not included in the calculations. During the following days, the animals were offered different types of diets, at different feeding levels depending on the protocol of each experiment. The trough was placed on a load cell to determine the time, duration and quantity of feed ingested at each meal. The animals were starved for at least 20 h during the last day in the respiration chamber to estimate FQ.

Gas concentrations, ventilation rate, physical characteristics of the gas in the respiration chamber (i.e. pressure, temperature and relative humidity) and signals from force sensors, infrared beams and load cell placed under the trough were measured at a frequency of 60 times per second, averaged over ten-second intervals and recorded for further calculations. Taking into account the effect of respiratory quotient (RQ, CO_2/O_2) on the difference between inflow and outflow (Vermorel *et al.*, 1973), volumes of O_2 consumption and CO_2 (VO_2 and VCO_2, respectively) and CH_4 production were calculated for 10 s intervals that were cumulated over the day. According to the Brouwer (1965) equation, kinetics of Q were calculated from VO_2 and VCO_2, assuming that the correction factors relative to CH_4 production and nitrogen excreted in urine accounted for less than 1% of Q (Chapter 5, Van den Borne *et al.*, 2015).

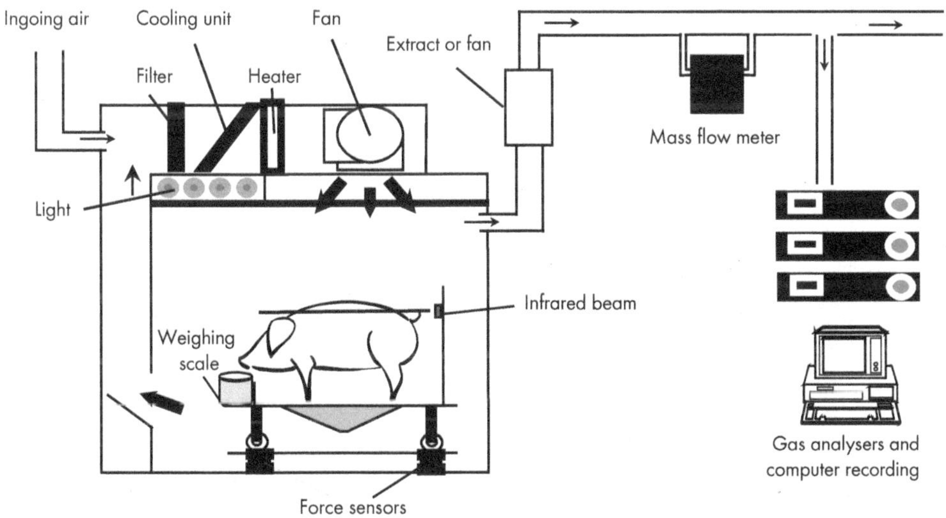

Figure 1. Description of respiration chambers equipped with force sensors under the metabolism cage where the animals are housed, weighing scale under the trough and infrared beams to measure heat production kinetics in farm animals.

4.3 Mathematical modelling of heat production partitioning

Total Q was partitioned between components due to basal metabolism at no physical activity (FQ_0), physical activity (AQ) and TEF (Figure 2) through analysis of the dynamic patterns of O_2 and CO_2 concentrations in the outgoing air of the respiration chamber (Van Milgen *et al.*, 1997). The model assumes that the instantaneous variations in O_2 and CO_2 concentrations are related to VO_2 and $VCO_2$by the animal or group of animals (sub-model 'animal'; Figure 3), in addition to variation induced by ventilation of the respiration chamber and variation of physical characteristics of the gas within the respiration chamber (sub-model 'chamber'). A complete description of the mathematical model is given by Van Milgen *et al.* (1997).

4.3.1 Mathematical modelling of gas exchanges

The structure of the model was similar for VO_2 and VCO_2. When animals were fed, the sub-model 'animal' considered that the instantaneous O_2 consumption or CO_2 production (in standard conditions of temperature and pressure: 0 °C and 1 atm) equaled the sum of O_2 consumption or CO_2 production due to physical activity and short-term TEF (TEF_s), in addition to constant O_2 consumption or CO_2 production associated with resting metabolism (VOFED and VCFED, respectively; Figure 2). It was assumed that the O_2 consumption or CO_2 production due to physical activity was proportional to the electrical signal from force

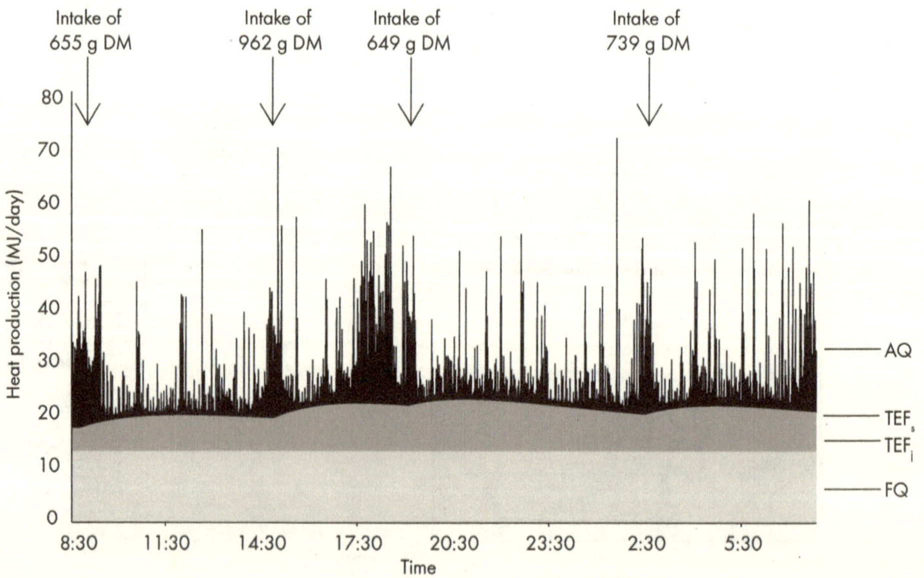

Figure 2. Example of heat production partitioning between components due to basal metabolic rate (fasting heat production, FQ), physical activity (AQ) and thermic effect of feeding (TEF_s, short term and TEF_l, long term) in growing pig (measured on 10 February 2011). DM = dry matter.

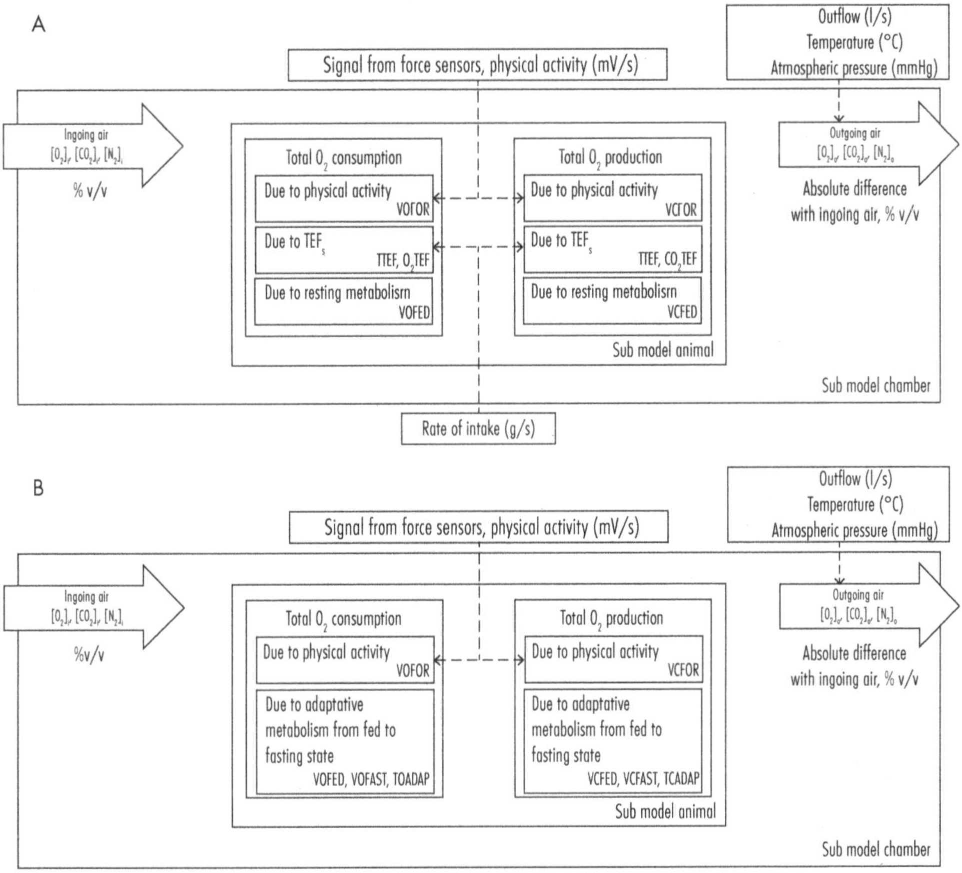

Figure 3. Description of the mathematical model used to partition total heat production from kinetics of O_2 consumption and CO_2 production. (A) Components when animals are in a fed state; (B) components when animals are in a fasted state.

Ingoing air: $[O_2]_i$ 20.95% v/v; *$[CO_2]_i$* 0.03% v/v; *$[N_2]_i$* 79.02% v/v

Parameters of the model:

- *VOFED (VCFED): O_2 consumption (CO_2 production) due to resting metabolism when fed, l/h*
- *VOFAST (VCFAST): O_2 consumption (CO_2 production) due to basal metabolic rate, l/h*
- *TOADAP (TCADAP): rate of decline of O_2 consumption (CO_2 production) due to adaptative metabolism from fed to fasted state, /h*
- *O_2TEF): O_2 consumption (CO_2 production) due to feed intake, l/g of feed*
- *TTEF: mean time between feed intake and its related O_2 consumption and CO_2 production, h*
- *VOFOR (VCFOR): O_2 consumption (CO_2 production) due to physical activity, l/mV from force sensors*

Outputs of the model:

- *$[O_2]_o$: O_2 concentration in outgoing air, % v/v*
- *$[CO_2]_o$: CO_2 concentration in outgoing air, % v/v*
- *$[N_2]_o$: N_2 concentration in outgoing air, % v/v*

sensors with different parameters for O_2 and CO_2 (VOFOR and VCFOR, respectively). The O_2 consumption or CO_2 production due to TEF_s followed a gamma distribution (Van Milgen *et al.*, 1997). This approach was taken because TEF_s is supposed to represent not only the Q during ingestion, but also the Q due to digestion, absorption and (part of) metabolism. The TEF_s was modelled as the output of a two-compartment system, which was filled in the first compartment by feed intake (recorded by the weighing scale placed under the trough) and parameterized by the volume of O_2 consumed or CO_2 produced per g of feed intake (O2TEF and CO2TEF, respectively) and by mean time between feed intake and its related O_2 consumption or CO_2 production (TTEF), representing the average dynamics of Q after ingestion of a meal. Mathematically, the content of each compartment was modelled from its first-order derivative with respect to time and fractional emptying rates were assumed to be identical for both compartments (Van Milgen *et al.*, 1997). In addition to these well-identified contributors to O_2 consumption and CO_2 production, early experiments indicated that modest and time-limited variations in O_2 and CO_2 concentrations in the respiration chamber can occur irrespective of feed intake or physical activity (Van Milgen and Noblet, 2000). Although the contribution of these phenomena to the total VO_2 or VCO_2 is small (<0.5%), they can affect the estimates of parameters of the model when not accounted for. These events were manually included in the model to ensure proper parameter estimation as instantaneous VO_2 and VCO_2. During the day of fasting, VO_2 or VCO_2 during resting (when the contribution of physical activity was removed) were lower than when animals were fed. The decline in VO_2 or VCO_2 was described as a first-order decline between VO_2 or VCO_2 at a fed state (VOFED and VCFED, respectively) to horizontally asymptotic values for VO_2 or VCO_2 during fasting (VOFAST and VCFAST, respectively). It was hypothesised that the rate of decline (TOADAP and TCADAP, respectively) may be different for O_2 and CO_2. Finally, the sub-model 'animal' allowed calculating VO_2 and VCO_2 using feed intake and signals from the force sensors as inputs and seven parameter for the fed days and eight parameters for the fasting day.

The sub-model 'chamber' described the variation in physical characteristics of the gas and considered that the air in the respiration chamber was composed of O_2, CO_2 and N_2. Because only the flow of outgoing air was measured, the inflow was calculated as the flow of air required to fill the physical volume of the respiration chamber when VO_2, VCO_2, and outflow were considered. The physical volume of the respiration chamber was calculated in standard conditions of temperature and pressure (0 °C, 1 atm). The concentration of each gas in the respiration chamber was then calculated from its volume divided by the sum of the volumes of O_2, CO_2 and N_2.

Equations of the model were written in Fortran and compiled in a dynamic linked library that was loaded in R (R Development Core Team, 2010). Package deSolve (Soetaert *et al.*, 2010) was used to solve the ordinary differential equations with an integration step-size of 10 s, after smoothing the data from force sensors, temperature, pressure and outflow to ensure their continuity. Parameters of the model were estimated for each day according to a three-step procedure: parameters directly related to VO_2 or VCO_2 were first estimated separately and then together to minimize the sum of squared differences between predicted and measured O_2 or CO_2 concentrations (Nelder and Mead, 1965).

4.3.2 Calculation of heat production components

Energy expenditure due to fasting metabolism, physical activity, and TEF_s were calculated from VO_2 and VCO_2 according to the Brouwer equation (Brouwer, 1965). The difference between resting heat production when fed (i.e. total Q minus AQ and TEF_s) and FQ was attributed to long-term TEF (TEF_l) and total TEF was calculated as the sum of its short and long term components. Assuming that FQ is representative of the BMR of *ad libitum* fed animals (Baker *et al.*, 1991), the heat increment (HI) was calculated as the sum of AQ and TEF. The efficiency of utilising ME for maintenance and growth (k_{mg}, %) was simply calculated as (1 – HI / ME) × 100. Maintenance ME requirements (ME_m) were calculated as FQ × 100 / k_{mg} (Labussière *et al.*, 2011; Figure 4).

4.4 Determination of the most appropriate exponent to calculate metabolic body size in growing animals

The metabolic body weight used to be defined as the BW raised to a certain exponent. In the past, several values have been proposed for the exponent with the objective to represent the surface of energy exchange between the body and its environment. The FQ, when expressed relative to metabolic body weight, should be constant over a large range of BW for a group of animals. The exponent 0.75 has been widely used for comparing energy metabolism in mature animals across species (Kleiber, 1965), but its use has been questioned for conceptual, methodological and statistical reasons (Da Silva *et al.*, 2006; White and Seymour, 2005). One of the most relevant issues concerns the conditions of fasting that may differ between species. Furthermore, applying a concept of metabolic body weight, intended for interspecies comparison, for intraspecies comparison in growing animals questions the validity of the exponent 0.75. It can be hypothesized that BMR and its associated Q vary according to the variations in weight of metabolically active organs such as the visceral mass (more than 35% of FQ due to liver, heart and kidney; Ferrell *et al.*, 1976) as animals get heavier. According to our methodology for mathematical modelling of Q during fasting, FQ is defined as the horizontal asymptotic value of the decreasing dynamics of Q after the initiation of fasting and at no physical activity. We used this technique for growing pigs, growing calves, growing turkeys, and broilers and we estimated specific values for the exponent for different species (0.60 for pigs – Noblet *et al.*, 1999; 0.85 for calves – Labussière *et al.*, 2008a; 0.70 for turkeys – Rivera-Torres *et al.*, 2010a; 0.70 for broilers – Noblet *et al.*, 2013). These values were estimated over the BW range that corresponds to the production stages (25-100 kg in pigs; 60-265 kg in veal calves; 0.4-24 kg in turkeys and 0.6-2.8 kg in broilers. Except for turkeys, these exponents differed significantly from 0.75 and the classification between species follows the same order as the classification of allometric growth of visceral organs relative to the whole body (from 0.7 in pigs to more than 1.0 in calves; Noblet *et al.*, 1999; Rivera-Torres *et al.*, 2011; Robelin, 1986).

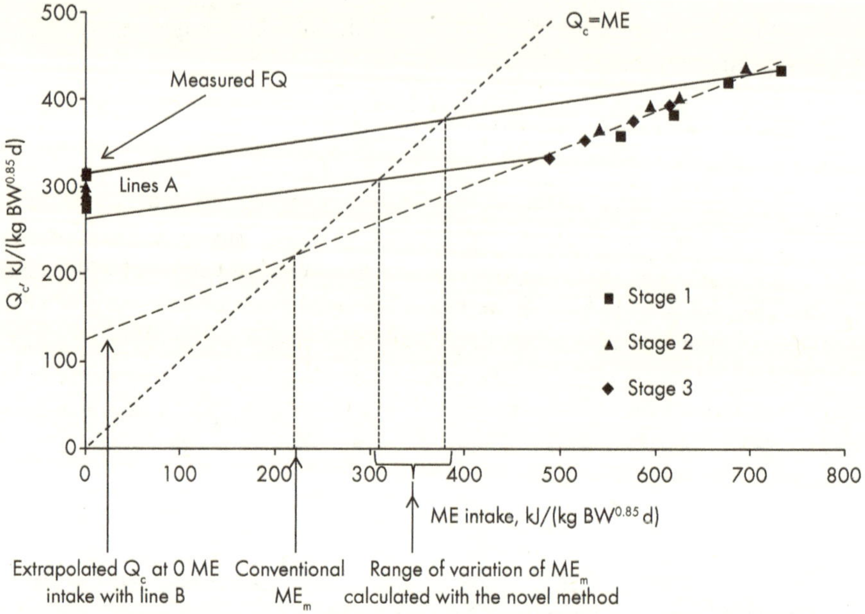

Figure 4. Relationship between heat production (Q_c, corrected for physical activity) or fasting heat production (at zero activity Q) and metabolisable energy (ME) intake in veal calves (Labussière et al., *2009b) for calculating maintenance ME requirements (ME_m). The two continuous parallel ('lines A') originate from the relationship between Q at zero ME intake (i.e. fasting heat production – FQ) and Q at the amount of ME intake by each calf for the lowest and highest ME intake. The slope of lines A equals (Q_c-FQ)/ME and the complement equals the efficiency of utilising ME for maintenance and growth. Line B corresponds to the conventional relationship between Q and ME intake. The complement of the slope equals the efficiency of utilising ME for growth.*

4.5 Utilisation of metabolisable energy in growing animals fed under close to *ad libitum* conditions

The same modelling technique was applied to data obtained in growing pigs (individually housed, castrated), calves (individually housed, male), broilers (group-housed, male) and turkeys (group-housed, male). The animals were housed at thermoneutrality. Because of the effects of diet composition on TEF and Q (Van Milgen *et al.*, 2001), the animals were fed standard diets formulated to meet their requirements, that consisted:

- for growing pigs, broilers and turkeys in a cereals and soybean meal based diet with low content in crude fibre;
- for calves in milk replacers with low content in starch.

When animals were fed under close to *ad libitum* conditions, between 54 and 65% of ME intake was dissipated as Q but this proportion varied between species (Table 1) and age of the animals (Labussière *et al.*, 2009b; Rivera-Torres *et al.*, 2010b). Among Q components, the largest is due to FQ, which accounted for about 30% of ME intake in monogastric growing animals, whereas FQ increased up to 49% of ME intake in veal calves, likely because of the higher amount of viscera in calves (20% of empty BW; Diaz *et al.*, 2001) than in monogastric animals (e.g. in pigs: 12% of empty BW; Noblet *et al.*, 1999) and the important contribution of visceral mass to whole body Q (Johnson *et al.*, 1990; Ortigues *et al.*, 1995). Similarly, the proportion of ME intake which was dissipated as TEF was similar in growing pigs, broilers and turkeys (between 13 and 18% of ME intake), whereas it accounted for less than 8% of ME intake in veal calves, which is probably related to the high digestibility, high fat and low fibre content of the milk replacer diet. Finally, the contribution of physical activity to total HP was rather constant across species (about 9% of ME intake), but slightly higher in turkeys (Table 1), because of increased social interactions between animals as they reached maturity (Rivera-Torres *et al.*, 2010b). It should be noted that most AQ estimates were obtained with animals housed in cages, which allowed limited activity (standing, sitting, walking some meters); physical activity will likely be higher under free living conditions. Consequently, estimates of energetic costs of physical activity from indirect calorimetry are probably in the lower range of values for energy expenditure for physical activity under more practical conditions.

Table 1. Interspecies comparison of losses of metabolisable energy (ME) intake as heat production (Q) and its components: fasting heat production (FQ), thermic effect of feeding (TEF) and activity-related heat production (AQ) in growing pigs, veal calves, broilers and turkeys fed ad libitum *a standard diet.*[1]

	Pigs[2]	Calves[3]	Broilers[4]	Turkeys[5]
MBW[1]	$BW^{0.60}$	$BW^{0.85}$	$BW^{0.70}$	$BW^{0.70}$
BW, kg	72 (26)	156 (67)	1.47 (0.06)	9.3 (7.8)
ME intake, kJ/kg MBW per day	2.59 (0.25)	0.62 (0.07)	1.65 (0.21)	1.35 (0.23)
ME intake, kJ/kg $BW^{0.75}$ per day	1.38 (0.17)	1.02 (0.08)	1.62 (0.21)	1.25 (0.28)
Utilisation of ME intake, % of ME				
Q	55.0 (2.1)	65.3 (3.5)	54.1 (6.5)	60.6 (4.8)
FQ	29.7 (1.2)	48.9 (2.9)	26.5 (2.8)	34.0 (5.4)
TEF	16.9 (1.9)	7.9 (1.0)	17.9 (2.4)	13.4 (5.7)
AQ	8.4 (1.1)	8.4 (0.9)	9.7 (2.1)	13.2 (5.9)

[1] Values are means and their standard deviations in brackets; MBW = metabolic body weight; BW = body weight.

[2] Barea *et al.*, 2010; De Lange *et al.*, 2006; Labussière *et al.*, 2013; Le Bellego *et al.*, 2001; Le Goff *et al.*, 2002.

[3] Labussière *et al.*, 2008b, 2009a,b.

[4] Noblet *et al.*, 2007, 2009.

[5] Rivera-Torres *et al.*, 2010a,b.

4.6 Variations of fasting heat production in growing animals

Within species, variations in FQ of different genotype (including gender and breed) may occur, but these are likely explained by differences in body composition, assuming that animals with a high visceral mass also have a higher FQ (Noblet *et al.*, 1999; Tess *et al.*, 1984a; Van Milgen *et al.*, 1998). Since BMR was originally defined for adult humans eating close to their maintenance requirements, the possible effects of feed intake on BMR were not considered in the definition of BMR, and more generally in the estimation of ME_m where ME intake is completely dissipated as Q; the latter is of special concern when ME_m is calculated as FQ/k_{mg}. In growing animals fed at different levels, variations in FQ may occur because the weight of visceral organs – that contribute to a great extent to FQ (Ferrell *et al.*, 1976) – varies according to feeding level (Ferrell *et al.*, 1986; Koong *et al.*, 1985). We conducted two experiments using growing pigs and calves that were fed at different levels for at least three weeks before estimating FQ according to the modelling approach (De Lange *et al.*, 2006; Labussière *et al.*, 2009b). The estimated FQ changed by 0.14 kJ in pigs and 0.22 kJ in calves per kJ ME intake before fasting (Table 2; Labussière *et al.*, 2011), which agrees with previous results in calves (Holmes and Davey, 1976). Consequently, the classical method used for determining ME_m in growing animals by extrapolating the relationship between energy retention and ME intake (Baldwin, 1995; Kielanowski, 1965) at zero energy retention may not be adequate, unless animals are allowed sufficient time to adjust Q to altered feeding levels. The adaptation of the animals to the lower feeding levels will result in a lower FQ, yielding low estimates of ME_m calculated as FQ/k_{mg}. In growing animals, it is not possible to directly distinguish FQ from the Q associated with growth. Therefore, care should be taken when considering the relationship between ME intake and Q in growing animals, because estimates of FQ (and ME_m) and the marginal efficiency of using ME intake for energy retention are confounded. Considering the theoretical relationship between Q and its associated FQ (i.e. for the same ME intake; Figure 4), it is possible to determine different ME_m values for different feeding levels. Using this

Table 2. Effect of body weight (BW) and previous metabolisable energy (ME) intake on fasting heat production (FQ) of growing pigs and calves (Labussière et al., *2011).*

Species	MBW[1]	BW kg	Mean ME intake[2] (range) kJ/kg MBW per day	Mean FQ[2] (range) kJ/kg MBW per day	ΔFQ/ΔME[1]
Pigs (n=18)	$BW^{0.60}$	55.4	2,084[a] (1,578-2,598)	687 (566-757)	0.14
		70.4	2,082[a] (1,562-2,606)	697 (568-766)	
		86.7	2,003[b] (1,485-2,528)	631 (561-761)	
Calves (n=47)	$BW^{0.85}$	72.8	647[a] (525-769)	298[a] (270-328)	0.23
		150.6	619[b] (533-716)	301[a] (272-343)	
		236.8	554[c] (467-644)	280[b] (255-319)	

[1] MBW = metabolic BW; ΔFQ/ΔME = slope of the relationship between FQ and ME intake.

[2] LS-means within each species with different superscript letters are significantly different ($P<0.05$).

technique, ME_m varies by 0.17 kJ in pigs and 0.27 kJ in calves per additional kJ ME intake (Labussière *et al.*, 2011). From a practical point of view, the classical estimation of energy requirements in growing animals as the sum of a constant maintenance component and a variable component due to BW gain and its composition is useful for calculating targeted feed intake required to reach the objectives for growth performance. Our findings imply that part of the increase in energy intake is used for an increased ME_m and will not be available for increased growth performance.

4.7 Calculation of energetic efficiencies for energy retention

The classic approach to calculate energy efficiency for energy retention (either total, or as protein and fat) in growing animals is to compute the first-order derivative of the relationship between energy retention and ME intake (Kielanowski, 1965). In this relationship, ME intake at zero energy retention (when ME intake equals Q) is defined as ME_m. When using this regression method, no direct estimate of ME_m is obtained. Moreover, estimates of the energetic efficiency for energy retention require observations at multiple levels of energy intake and are confounded with estimates obtained for ME_m. From this method, estimates of energy efficiencies for protein and lipid deposition have been proposed, but with a wide range of variation (from 36 to 77% and from 58 to 99%, respectively; Tess *et al.*, 1984b). The combined estimation of Q and FQ in the same animal using modelling techniques allows calculating energy efficiencies for each animal and to determine to what extent dietary treatments affect these efficiencies. Additionally, variable part of the inefficiency can be attributed to changes in physical activity, which, in the modelling approach has been standardised to avoid bias in estimates of energy efficiencies (Labussière *et al.*, 2013). In the latter approach, it was assumed that a fixed amount of ME intake should be dissipated as AQ (AQ_c) to standardize Q between animals for difference in physical activity. When AQ was higher than AQ_c, the difference between AQ and AQ_c resulted in a positive variation of ME available for other metabolic pathways, which was dissipated as TEF or retained as fat. The amount which was dissipated as supplementary TEF can be calculated as $(AQ - AQ_c) \times TEF / (ME - FQ - AQ)$ and added to TEF to calculate a corrected TEF (TEF_c) that contributed to a corrected Q (Q_c).

When the variations in FQ were accounted for, marginal energy efficiencies no longer depend on feeding level (Labussière *et al.*, 2011) and estimates of the energy efficiency for protein deposition was close to 60% and the estimate for lipid deposition was close to 80% (Noblet *et al.*, 1999; Van Milgen and Noblet, 1999). However, the latter depends on the nutrients used for lipid synthesis. It varied from 52%, when protein was the source of energy, to 88%, when lipid was the source of energy. Furthermore, the energy efficiency for lipid synthesis also depends also depends on the nutrients used to generate ATP to support lipid and protein deposition (Van Milgen *et al.*, 2001).

4.8 Conclusions

Refinement in systems for measuring energy metabolism and quantifying physical activity of growing animals improves the interpretation of data obtained from respiration chambers. Mathematical modelling of daily dynamics of Q can be used to better understand the partitioning of energy expenditure between maintenance, growth functions and physical activity. Based on this technique, it appears that specific parameters must be used for calculating metabolic BW of different growing animals species, likely because of the difference in allometric growth of viscera. Moreover, measurements of FQ may be combined with observations on total Q, to assess variation in energetic efficiencies between individual animals. From a practical point of view, improved understanding of energy partitioning will help nutritionists to better estimate energy requirements and optimise feeding and management strategies.

References

Baker, J.F., Buckley, B.A., Dickerson, G.E. and Nienaber, J.A., 1991. Body composition and fasting heat production from birth to 14 months of age for three biological types of beef heifers. Journal of Animal Science 69: 4406-4418.

Baldwin, R.L., 1995. Energy requirements for maintenance and production. In: Baldwin, R.L. (ed.) Modeling ruminant digestion and metabolism. Chapman & Hall, London, UK, pp. 148-188.

Barea, R., Dubois, S., Gilbert, H., Sellier, P., Van Milgen, J. and Noblet, J., 2010. Energy utilization in pigs selected for high and low residual feed intake. Journal of Animal Science 88: 2062-2072.

Brouwer, E., 1965. Report of sub-comittee on constants and factors. In: Blaxter, K.L. (ed.) Energy metabolism, EAAP Publication no. 11. EAAP, Rome, Italy, pp. 441-443.

Close, W.H. and Mount, L.E., 1975. The rate of heat loss during fasting in the growing pig. British Journal of Nutrition 34: 279-290.

Da Silva, J.K.L., Garcia, G.J.M. and Barbosa, L.A., 2006. Allometric scaling laws of metabolism. Physics of Life Reviews 3: 229-261.

De Lange, K., Van Milgen, J., Noblet, J., Dubois, S. and Birkett, S., 2006. Previous feeding level influences plateau heat production following a 24 h fast in growing pigs. British Journal of Nutrition 95: 1082-1087.

Diaz, M.C., Van Amburgh, M.E., Smith, J.M., Kelsey, J.M. and Hutten, E.L., 2001. Composition of growth of Holstein calves fed milk replacer from birth to 105-kilogram body weight. Journal of Dairy Science 84: 830-842.

Ferrell, C.L., Garrett, W.N., Hinman, N. and Grichting, G., 1976. Energy utilization by pregnant and non-pregnant heifers. Journal of Animal Science 42: 937-950.

Ferrell, C.L., Koong, L. and Nienaber, J.A., 1986. Effect of previous nutrition on body composition and maintenance energy costs of growing lambs. British Journal of Nutrition 56: 595-605.

Holmes, C.W. and Davey, A.W.F., 1976. The energy metabolism of young jersey and friesian calves fed fresh milk. Animal Production 23: 43-53.

Johnson, D.E., Johnson, K.A. and Baldwin, R.L., 1990. Changes in liver and gastrointestinal tract energy demands in response to physiological workload in ruminants. Journal of Nutrition 120: 649-655.

Kielanowski, J., 1965. Estimates of the energy cost of protein depostion in growing animals. In: Blaxter, K.L. (Ed.) Energy metabolism, pp. 13-20.

Kleiber, M., 1965. Metabolic body size. In: Blaxter, K.L. (Ed.) Energy metabolism, pp. 427-435.

Koong, L.J., Ferrell, C.L. and Nienaber, J.A., 1985. Assessment of interrelationships among levels of intake and production, organ size and fasting heat production in growing animals. Journal of Nutrition 115: 1383-1390.

Labussière, E., Dubois, S., Van Milgen, J. and Noblet, J., 2013. Partitioning of heat production in growing pigs as a tool to improve the determination of efficiency of energy utilization. Frontiers in Physiology 4: 146.

Labussière, E., Dubois, S., Van Milgen, J., Bertrand, G. and Noblet, J., 2008a. Fasting heat production and energy cost of standing activity in veal calves. British Journal of Nutrition 100: 1315-1324.

Labussière, E., Dubois, S., Van Milgen, J., Bertrand, G. and Noblet, J., 2008b. Effects of dietary crude protein on protein and fat deposition in milk-fed veal calves. Journal of Dairy Science 91: 4741-4754.

Labussière, E., Dubois, S., Van Milgen, J., Bertrand, G. and Noblet, J., 2009a. Effect of solid feed on energy and protein utilization in milk-fed veal calves. Journal of Animal Science 87: 1106-1119.

Labussière, E., Maxin, G., Dubois, S., Van Milgen, J., Bertrand, G. and Noblet, J., 2009b. Effect of feed intake on heat production and protein and fat deposition in milk-fed veal calves. Animal 3: 557-567.

Labussière, E., Van Milgen, J., De Lange, C.F.M. and Noblet, J., 2011. Maintenance energy requirements of growing pigs and calves are influenced by feeding level. Journal of Nutrition 141: 1855-1861.

Le Bellego, L., Van Milgen, J., Dubois, S. and Noblet, J., 2001. Energy utilization of low-protein diets in growing pigs. Journal of Animal Science 79: 1259-1271.

Le Goff, G., Dubois, S., Van Milgen, J. and Noblet, J., 2002. Influence of dietary fiber level on digestive and metabolic utilization of energy in growing and finishing pigs. Animal Research 51: 245-259.

McNab, B.K., 1997. On the utility of uniformity in the definition of basal rate of metabolism. Physiological Zoology 70: 718-720.

Nelder, J.A. and Mead, R., 1965. A simplex method for function minimization. The Computer Journal 7: 308-313.

Noblet, J., Dubois, S., Van Milgen, J., Warpechowski, M., Le Bellego, L. and Carré, B., 2007. Influence de la teneur en protéines de l'aliment sur l'utilisation métabolique de l'énergie chez le poulet en croissance. In: 7th Journées de la Recherche Avicole, pp. 141-144.

Noblet, J., Karege, C., Dubois, S. and Van Milgen, J., 1999. Metabolic utilization of energy and maintenance requirements in growing pigs: effects of sex and genotype. Journal of Animal Science 77: 1208-1216.

Noblet, J., Labussière, E., Dubois, S., De Lange, C.F.M., Barea, R., Lasnier, J., Rivera-Torres, V., Warpechowski, M. and Van Milgen, J., 2013. Fasting heat production and metabolic body size in non-ruminant growing farm animals. In: Oltjen, J.W., Kebreab, E. and Lapierre, H. (eds.) Energy and protein metabolism and nutrition in sustainable animal production, EAAP publication No. 134. Wageningen Academic Publishers, Wageningen, the Netherlands, pp. 313-314.

Noblet, J., Warpechowski, M., Dubois, S., Van Milgen, J. and Carré, B., 2009. Influence de la teneur en matières grasses de l'aliment sur l'utilisation métabolique de l'énergie chez le poulet. In: 8th Journées de la Recherche Avicole, pp. 164-168.

Ortigues, I., Martin, C., Durand, D. and Vermorel, M., 1995. Circadian changes in energy-expenditure in the preruminant calf – whole animal and tissue-level. Journal of Animal Science 73: 552-564.

Quiniou, N., Noblet, J., Van Milgen, J. and Dubois, S., 2001. Modelling heat production and energy balance in group-housed growing pigs exposed to low or high ambient temperatures. British Journal of Nutrition 85: 97-106.

R Development Core Team 2010. R: a language and environment for statistical computing. R foundation for Statistical Computing, Vienna, Austria.

Rivera-Torres, V., Noblet, J. and Van Milgen, J., 2011. Changes in chemical composition in male turkeys during growth. Poultry Science 90: 68-74.

Rivera-Torres, V., Noblet, J., Dubois, S. and Van Milgen, J., 2010a. Energy partitioning in male growing turkeys. Poultry Science 89: 530-538.

Rivera-Torres, V., Noblet, J., Dubois, S. and Van Milgen, J., 2010b. Dynamics of energy utilization in male and female turkeys during growth. Animal 5: 202-210.

Robelin, J., 1986. Composition corporelle des bovins: évolution au cours du développement et différences entre races. PhD Thesis. Université de Clermont-Ferrand II, Clermont-Ferrand, France.

Soetaert, K., Petzoldt, T. and Setzer, R.W., 2010. Solving differential equations in R: package deSolve. Journal of Statistical Software 33: 1-25.

Tess, M.W., Dickerson, G.E., Nienaber, J.A. and Ferrell, C.L., 1984a. The effects of body composition on fasting heat production in pigs. Journal of Animal Science 58: 99-110.

Tess, M.W., Dickerson, G.E., Nienaber, J.A., Yen, J.T. and Ferrell, C.L., 1984b. Energy costs of protein and fat deposition in pigs fed *ad libitum*. Journal of Animal Science 58: 111-122.

Van den Borne, J.J.G.C., Heetkamp, M.J.W., Alferink, S.J.J. and Gerrits, W.J.J., 2015. Moving from a complete energy balance towards substrate oxidation: use of stable isotopes. In: Gerrits, W.J.J. and Labussière, E. (eds.) Indirect calorimetry. Techniques, computations and applications. Wageningen Academic Publishers, Wageningen, the Netherlands, pp. 87-114.

Van Milgen, J. and Noblet, J., 1999. Energy partitioning in growing pigs: the use of a multivariate model as an alternative for the factorial analysis. Journal of Animal Science 77: 2154-2162.

Van Milgen, J. and Noblet, J., 2000. Modelling energy expenditure in pigs. In: McNamara, J.P., France, J. and Beever, D.E. (eds.) Modelling nutrient utilization in farm animals. CAB International, Oxon, UK, pp. 103-114.

Van Milgen, J., Bernier, J.F., Lecozler, Y., Dubois, S. and Noblet, J., 1998. Major determinants of fasting heat production and energetic cost of activity in growing pigs of different body weight and breed/castration combination. British Journal of Nutrition 79: 509-517.

Van Milgen, J., Noblet, J. and Dubois, S., 2001. Energetic efficiency of starch, protein and lipid utilization in growing pigs. Journal of Nutrition 131: 1309-1318.

Van Milgen, J., Noblet, J., Dubois, S. and Bernier, J.F., 1997. Dynamic aspects of oxygen consumption and carbon dioxide production in swine. British Journal of Nutrition 78: 397-410.

Vermorel, M., Bouvier, J.C., Bonnet, Y. and Fauconneau, G., 1973. Construction et fonctionnement de deux chambres respiratoires du type 'circuit ouvert' pour jeunes bovins. Annales De Biologie Animale Biochimie Biophysique 13: 659-681.

White, C.R. and Seymour, R.S., 2005. Allometric scaling of mammalian metabolism. Journal of Experimental Biology 208: 1611-1619.

Chapter 5: Moving from a complete energy balance towards substrate oxidation: use of stable isotopes

J.J.G.C. van den Borne[1,3], M.J.W. Heetkamp[2], S.J.J. Alferink[1] and W.J.J. Gerrits[1]*
[1]Animal Nutrition Group, Wageningen University, P.O. Box 338, 6700 AH Wageningen, the Netherlands; [2]Adaptation Physiology Group, Wageningen University, P.O. Box 338, 6700 AH Wageningen, the Netherlands; [3]Current address: FrieslandCampina, Stationsplein 4, 3818 LE Amersfoort, the Netherlands; joost.vandenborne@frieslandcampina.com

Abstract

Substrate oxidation can be estimated from gas exchange in combination with urinary nitrogen excretion measurements. The estimates are net rates of oxidation, because several metabolic processes, such as lipogenesis and gluconeogenesis, may affect the respiratory quotient, resulting in under- or overestimation of absolute rates of substrate oxidation. If these limitations, as well as accumulation of ketone bodies, lactate and urea, are taken into consideration, indirect calorimetry can be an effective method to quantify substrate oxidation in animals and humans. Methane should be excluded from the equations for calculating oxidation rates. Using ^{13}C labelled substrates provides additional possibilities in oxidation studies. Whole body nutrient oxidation can be measured when tracer amounts of the nutrient of interest are continuously infused and when $^{13}CO_2$ expiration and ^{13}C enrichment of the precursor for oxidation are measured in blood plasma. Oxidation of exogenous substrates can be distinguished from that of endogenous substrates by using ^{13}C enriched (or depleted) nutrients in the diet. In addition, labelling of individual monosaccharides, fatty acids or amino acids can provide insight in the metabolic fate of the tracee irrespective of availability of other substrates from the same macronutrient class (i.e. carbohydrate, fat, protein). Finally, selection of appropriate isotopomers may allow quantification of pre-identified metabolic processes (e.g. lipogenesis). Fluctuations in background enrichment, isotope administration strategy, sequestration of CO_2 and isotopic gradient in the body are among the factors that should be considered when designing and interpreting oxidation studies with labelled nutrients. Modelling approaches should be in place or be developed to maximise the learnings from these studies. This appears especially challenging when animals are studied in a non-steady physiological state. In conclusion, using ^{13}C labelling in combination with indirect calorimetry provides unique opportunities to quantify the contribution of individual substrates (endogenous as well as exogenous) to whole body energy metabolism.

Keywords: ^{13}C, energy metabolism, gas exchange, respiratory quotient, tracers

DOI 10.3920/978-90-8686-261-0_5,

5.1 Estimating substrate oxidation from indirect calorimetry

With indirect calorimetry, the Fick principle is used to estimate heat production (Q): air flows through the chamber at a known rate and the amounts of gas consumed (VO_2) and produced (VCO_2) are estimated from this flow rate and their concentrations in ingoing and outgoing air as well as gas accumulation inside the chamber. As described in Chapter 1 of this book (Gerrits *et al.*, 2015), formulae based on stoichiometry (e.g. Brouwer, 1965; Weir, 1949) are used to calculate Q from gas exchange. In addition, these gas exchange data, in combination with measurement of urinary nitrogen excretion, can be used to estimate rates of oxidation of each macronutrient. Almost 100 years ago, measurements of pulmonary gas exchange were already used to measure the contribution of carbohydrate and fat to Q at rest and during exercise (Krogh and Lindhard, 1920). Macronutrients differ in their chemical composition and in the amounts of oxygen needed per unit of carbon for their complete oxidation. For animal calorimetry, the constants of Brouwer (1965; Table 1) are often used. Minor variations in these constants were adopted earlier by others, including Zuntz (1897), Rubner (1901), Magnus-Levy (1907), Benedict and Talbot (1914), Benedict and Tompkins (1916) and Lusk (1917, 1918), as reviewed by Brockway (1987). These differences mainly originate from assuming different monomer composition of proteins, fats and carbohydrates, as described in Chapter 1 of this book (Gerrits *et al.*, 2015). The energy equivalent of O_2 and CO_2 depends on the type of nutrient oxidised, and can be calculated by dividing the enthalpy (-ΔHc; heat of combustion) by the VO_2 or VCO_2. The respiratory quotient (RQ) is VCO_2 divided by VO_2. When energy expenditure is calculated from VO_2 only (e.g. exercise studies) or from VCO_2 only (e.g. doubly labelled water), RQ is assumed to be constant based on an average RQ in the target population or based on the food quotient which can be calculated from the macronutrient composition of the diet. For example, 1 l of oxygen consumed is often associated with 20 kJ of energy expenditure (McArdle *et al.*, 2007). The validity of such assumptions should be taken into account, especially when feeding very specific diets (e.g. high fat or high carbohydrate) or under particular physiological conditions (e.g. fasting, high intensity exercise).

At whole body level, RQ reflects the metabolism of a mixture of carbohydrate, fat and protein. The amount of protein, or more specifically amino acids, oxidized can be estimated from

Table 1. Oxygen consumption (VO_2), carbon dioxide production (VCO_2), heat of combustion ($-\Delta H_c$), and respiratory quotient (RQ) for carbohydrate (i.e. starch), fat and protein (Brouwer, 1965).

Oxidized substrate	VO_2 (l/g)	VCO_2 (l/g)	$-\Delta H_c$ (kJ/g)	RQ
Carbohydrate (starch)	0.829	0.829	17.57	1.00
Fat	2.013	1.431	39.75	0.71
Protein	0.957	0.774	18.41	0.81

the amount of N excreted in urine (as N × 6.25). Oxidation of amino acids begins with deamination, and the ammonia produced is excreted with urine as urea or other N-containing compounds, such as uric acid, allantoin, ammonia and hippuric acid. If protein oxidation is estimated from urinary N excretion then carbohydrate and fat oxidation can be calculated based on the ratio of CO_2 produced relative to O_2 consumed. Carbohydrate and fat oxidation can be calculated in short-term time intervals from gas exchange data, whereas protein oxidation is commonly assumed to be constant over the day (i.e. based on 24-h urine N). The overall stochiometry of these calculations for glucose (Equation 1) is displayed below. For fat oxidation, palmitate (Equation 2a) or palmitoyl-stearoyl-oleoyl-glycerol (Equation 2b) is often taken as a representative of the average composition of triglycerides.

$$C_6H_{12}O_6 + 6\,O_2 \rightarrow 6\,CO_2 + 6\,H_2O \tag{1}$$

$$C_{16}H_{32}O_2 + 23\,O_2 \rightarrow 16\,CO_2 + 16\,H_2O \tag{2a}$$

$$C_{55}H_{104}O_6 + 78\,O_2 \rightarrow 55\,CO_2 + 52\,H_2O \tag{2b}$$

Based on the equations for protein, carbohydrate and fat and three unknown variables (contribution of each of the substrates to oxidation) a system with 3 equations can be solved, and formulae for oxidation of substrates have been developed (Table 2). A correction for methane is often not applied in human studies, but has been suggested to be especially relevant when studying ruminants or monogastric animals fed relatively high levels of fibres (Brouwer, 1958). If not relevant, the same formula can be used but without the methane correction (Brouwer, 1957).

The contribution of protein oxidation to Q is relatively small, usually between 10 and 15% (and 11% in the example in Table 3). Mathematically, urinary nitrogen excretion does not

Table 2. Calculation of the net rates of substrate oxidation (in g) from gas exchange (in l) and urinary nitrogen excretion (in g).

Substrate	Equation	Reference
Carbohydrate	$-2.965 \times VO_2 + 4.170 \times VCO_2 - 2.438 \times N\ (-1.759 \times VCH_4)$	Brouwer, 1957,
Fat	$1.718 \times VO_2 - 1.718 \times VCO_2 - 1.969 \times N\ (+1.718 \times VCH_4)$	1958
Protein	$6.25 \times N$	
Carbohydrate	$-3.311 \times VO_2 + 4.650 \times VCO_2 - 3.581 \times N$	Livesey and Elia,
Fat	$1.720 \times VO_2 - 1.720 \times VCO_2 - 1.776 \times N$	1988
Protein	$6.25 \times N$	
Carbohydrate	$-3.21 \times VO_2 + 4.55 \times VCO_2 - 2.87 \times N$	Frayn, 1983
Fat	$1.67 \times VO_2 - 1.67 \times VCO_2 - 1.92 \times N$	
Protein	$6.25 \times N$	

Table 3. Example of calculating heat production and oxidation of carbohydrates, fat and protein from gas exchange and urinary nitrogen excretion (human subject, 24 h). Calculations based on Brouwer (1958; see Gerrits et al., *2015).*

Measurement	Value	Heat (kJ)	Contribution to substrate oxidation		
			Carbohydrate (g)	Fat (g)	Protein (g)
VO_2, l	600	9,708	-1,781	1,031	-
VCO_2, l	525	2,636	2,191	-902	-
N, g	12	-72	-29	-24	75
VCH_4, l	8	-17	-14	14	-
Total, kJ or g	-	12,254	367	119	75

substantially contribute to Q (0.6% = 72/12,254; Table 3), but its contribution to estimated rates of carbohydrate and fat oxidation is much more pronounced (-8% = -29/367, and -20%, respectively; Table 3). Apart from measuring gas exchange, complete collection of urine over a representative period is therefore important for accurately estimating substrate oxidation. In addition, the period of urine collection should be representative for the period of measuring gas exchange. Due to the irregular pattern of urination and the delay in urinary N excretion when compared to respiration, short-term measurements in combination with a non-steady physiological state can possibly result in erroneous estimates, even when urine and gas are measured over the same time period.

Substrate oxidation from the abovementioned equations assumes that all substrates of endogenous and exogenous origin are completely oxidized to CO_2 and H_2O and that the main nitrogenous end-product is urea. In addition, it is assumed that temporary storage of metabolites derived from these macronutrients is negligible. The following topics will be discussed in this chapter: (1) limitations of indirect calorimetry to measure substrate oxidation; (2) using stable isotope tracers to measure substrate oxidation in combination with indirect calorimetry; (3) other applications combining stable isotopes and gas exchange measurements.

5.2 Limitations of indirect calorimetry to measure substrate oxidation

Indirect calorimetry measures the net disappearance rate of a substrate regardless of the metabolic interconversions that the substrate may undergo before its disappearance from its metabolic pool. Under most circumstances, direct oxidation represents the major route by which a substrate disappears from its metabolic pool. However, under conditions when rates of gluconeogenesis, ketogenesis, or lipogenesis are elevated, the presumed equivalence between oxidation and disappearance may no longer apply, even though the actual measurements derived from indirect calorimetry remain valid. As a consequence, these metabolic processes can affect the RQ and may therefore impact the contribution of carbohydrates and fat to total Q. In addition, ethanol (RQ = 0.667) should not contribute substantially to Q, as this will underestimate the contribution of carbohydrates and overestimate that of fat.

5.2.1 Influence of other metabolic processes

Lipogenesis from carbohydrates results in a high RQ. The overall stoichiometry of the synthesis of triglycerides from glucose is associated with an RQ of 5.6 (168/30):

$$83 \text{ glucose} + 30 \ O_2 \rightarrow 6 \text{ triglycerides} + 168 \ CO_2 \qquad (3)$$

This process is stimulated by a high level of energy intake from carbohydrates, especially when animals are fed at high levels with a rapidly digestibly source of carbohydrates such as in growing pigs (e.g. Gerrits *et al.*, 2012; Figure 1). Substrate oxidation rates, as estimated from indirect calorimetry, are net oxidation rates at whole body level. For example, glucose may be used to fill hepatic glycogen stores and (partly) be converted to body fat. This will not necessarily be reflected by an RQ>1 because at the same time fatty acids may be oxidized in other tissues.

Gluconeogenesis from amino acids, but not from carbohydrate precursors (e.g. lactate or propionate) has an RQ lower than 0.7. The actual value depends on the amino acid that is

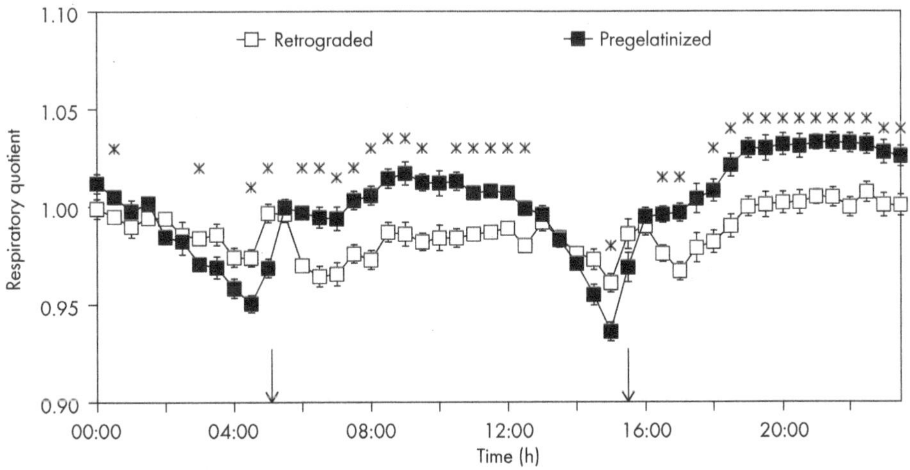

Figure 1. Effect of feeding pregelatinised vs resistant starch (45% inclusion in diet) in two meals (arrows) to growing pigs on the diurnal pattern of the respiratory quotient. Asterisk indicates difference (P<0.05) between treatments (Gerrits et al.*, 2012; Figure reused with permission from the Journal of Nutrition).*

used as a precursor, but commonly alanine is assumed to be a representative glucogenic amino acid. Gluconeogenesis from alanine results in a RQ of 0.13 (3/23):

$$\text{palmitate} + 26\ \text{alanine} + 23\ O_2 \rightarrow 13\ \text{glucose} + 3\ CO_2 + 13\ \text{urea} \qquad (4)$$

Under most circumstances, however, the rate of glucose oxidation exceeds that of gluconeogenesis, so that the results obtained from gas exchange measurement indicate net glucose oxidation. In highly gluconeogenic states, such as fasting, exercise or lactation, this may result in substantial underestimation of carbohydrate oxidation and overestimation of fat oxidation. Indeed, the non-protein RQ decreased below 0.7 in pigs when fasted for more than 1 day (Chwalibog *et al.*, 2005). If the endogenously produced glucose is oxidised, substrate oxidation will be ascribed to protein oxidation, as the overall RQ is independent of intermediate metabolic processes.

Also ketogenesis from incompletely oxidised fatty acids may affect RQ and therefore estimates of carbohydrate and fat oxidation (for details: Frayn, 1983). Synthesis of ketone bodies from acetyl-CoA requires O_2 but no CO_2 is produced. If the ketone bodies are retained in the body or excreted in urine or breath, this will result in an RQ less than 0.70. If the ketone bodies are oxidized, then the RQ will be similar to that of fatty acid oxidation. High rates of ketone body production may also result in changes in the bicarbonate pool to compensate for the metabolic acidosis associated with ketogenesis and therefore affect VCO_2.

5.2.2 Changes in bicarbonate and urea pools

When calculating substrate oxidation, we assume that the body pools that could influence the measurements (i.e. O_2, CO_2 and urea) remain stable. This assumption will hold for O_2, because O_2 pools in the body are negligible compared with VO_2. Although O_2 also binds to myoglobin in muscle, blood haemoglobin is generally considered the body's main O_2 pool. At whole body level, approximately 1 l of O_2 is bound to haemoglobin (assuming 5 l blood; 1.39 ml O_2/g haemoglobin and 150 g haemoglobin/l blood; ICSH, 1967) and the ratio of O_2 pool size (mmol) to consumption rate (mmol/min) is about 2.4:1.

The situation is more critical for CO_2 due to the large size of the endogenous CO_2 (bicarbonate) pool and the fact that this pool is sensitive to changes in acid-base balance (Tappy and Schneiter, 1997). Bicarbonate partitions over multiple compartments in the body, but the ratio between the pool size of freely exchangeable bicarbonate (mol) and CO_2 production (mol/min) is approximately 149:1 (Irving *et al.*, 1983). A period of hyperventilation, for example, results in depletion of the endogenous CO_2 pool and overestimation of CO_2 production. Although this could affect estimates of substrate oxidation over a short period of time, it will be compensated for by a period of hypoventilation and carbon dioxide retention. More prolonged challenges to the acid-base balance may be a more serious problem. Metabolic alkalosis will result in urinary excretion of bicarbonate in substantial amounts, resulting in underestimation of VCO_2 (Burnier *et al.*, 1993). High rates of lactate production, such as during high-intensity exercise, may induce metabolic acidosis and depletion of the body's bicarbonate pool to buffer lactic acid. In the latter circumstance, estimates of carbohydrate oxidation based on indirect calorimetry measurements will be erroneously high (Jeukendrup and Wallis, 2005).

Decreases or increases in the urea pool during a study will respectively over- or underestimate protein oxidation. This can be corrected by measuring blood urea nitrogen at the start and at the end of the study, and assuming a fixed urea distribution volume (usually 60% of body weight, corresponding with extra- and intracellular water volume). For urea, it should be noted that the pool size is relatively large compared to the production rate. Early estimates for human subjects indicated a ratio of urea pool size (mmol) to production rate (mmol/min) of 750:1 (Matthews and Downey, 1984), but urea production was estimated from urinary urea excretion. If urea recycling is taken into account, this ratio would be approximately 560:1 which was confirmed by Jahoor and Wolfe (1987).

5.2.3 Accuracy of gas exchange measurements

The assessment of Q and substrate oxidation by indirect calorimetry is generally valid when measured over a prolonged period of time, but relatively small over- or underestimations in either VO_2 or VCO_2 gas exchange may substantially affect estimated rates of carbohydrate and fat oxidation calculated over a short time period (Arch *et al.*, 2006; Schutz, 1997). Accurate measurements of VO_2 is very important for calculating Q, because an error in VO_2 affects the ultimate estimate three times more than an error in VCO_2 (Brouwer, 1957). For estimating substrate oxidation, however, errors in VO_2 measurements are not more important than those in VCO_2. For example, VO_2 and VCO_2 contribute in equal amounts to the calculated fat oxidation, and VCO_2 contributes 40% more to the calculated carbohydrate oxidation than VO_2 (Table 4).

5.2.4 Including the methane correction

In the example in Table 3, methane significantly contributes to the net rates of carbohydrate (-4%) and – especially – fat (+12%) oxidation. The sum of heat produced from oxidation

Table 4. Calculated consequences of errors in VO_2 and VCO_2 measurements for estimates of Q and substrate oxidation using the formula of Brouwer (1965).

Variable	True value	Overestimation of VCO_2 by 5%	Overestimation of VO_2 by 5%
Heat production, kJ/min	5.02	5.07 (+1%)	5.22 (+4%)
Carbohydrate oxidation, kJ/min	1.335	2.144 (+59%)	0.633 (-47%)
Fat oxidation, kJ/min	2.822	2.147 (-24%)	3.657 (+30%)
Ratio carbohydrate-to-fat oxidation	0.47	1.00	0.17

of macronutrients, however, exceeds Q calculated from the Brouwer equation, and this inconsistency will become more pronounced with increasing methane production (e.g. dairy cows). This is a direct consequence of adding methane with a negative coefficient to the Q equation. Although defendable for the total energy expended – when assuming complete aerobic oxidation of methane (see Chapter 1 by Gerrits *et al.*, 2015) – the impact on net substrate oxidation rates have not been evaluated in much detail. As the formulae for substrate oxidation have been derived from VO_2, VCO_2 and the heat of combustion of carbohydrates, fat and protein (Table 1), and a methane correction was applied afterwards (Brouwer, 1958), this may inappropriately affect true substrate oxidation rates, especially in ruminants. For example, a dairy cow (VO_2: 5,212 l/d; VCO_2: 5,758 l/d; VCH_4: 640 l/d; urinary N: 150 g/d; data from baseline measurements in cows, Burgos *et al.*, 2001) produces 111 MJ of heat according to the Brouwer equation. The sum of carbohydrate oxidation (124 MJ/d), fat oxidation (-5 MJ/d) and protein oxidation (17 MJ/d) is 136 MJ/d, and exceeds this value by 23%. This is mainly due to the large contribution of methane to the calculated oxidation rates of carbohydrate (-20 MJ/d) and fat (44 MJ/d). We therefore suggest to exclude methane from the equations for substrate oxidation. In order to improve these equations for ruminants, and to include methane, more quantitative insight in the interdependencies between VO_2, VCO_2 and VCH_4 and their relationship with whole body substrate oxidation is required.

In summary, glucose and fat oxidation can be calculated from indirect calorimetry when the non-protein RQ is less than 1, when ethanol does not contribute substantially to total Q, when pool sizes of intermediate and final products are steady, and when the measurement period is sufficiently long. When RQ exceeds 1, the net rate of lipogenesis can be estimated, and rates of carbohydrate and fat oxidation are apparent values (Chwalibog *et al.*, 1992). The (current) methane correction coefficients should not be used for predicting substrate oxidation rates.

5.3 Using ^{13}C-labelled substrates for measuring substrate oxidation

5.3.1 Stable isotopes

The above mentioned limitations can be (partly) overcome by using stable isotopes. Isotope tracers can be used to quantify the rates of oxidation of specific substrates by non-invasive or minimally invasive methods. Isotopes are atoms of the same element (i.e. same number of protons and electrons) but different numbers of neutrons. Isotopes are radioactive when there is a large excess or deficit of neutrons relative to protons, resulting in respectively β minus and β plus decay. For stable isotopes, the numbers of protons and neutrons are more similar, with the less abundant isotopes usually being heavier (i.e. one or two extra neutrons).

Stable isotopes can be used as a tracer; a compound that is chemically and functionally identical to the naturally occurring compound of interest (the tracee), but is distinct in some way that enables detection (i.e. by mass difference). By assessing the fate of a tracer, the digestion and metabolism of its tracee can be derived. The general methodology for using stable isotopes in digestive and metabolic research has been described extensively (Beylot, 2006; Bier, 1987; Jeukendrup and Wallis, 2005; Koletzko *et al.*, 1998; Rennie, 1999; Schoeller, 2002; Wolfe and Chinkes, 2005; Young and Ajami, 1999).

Any stable isotope that is given in an amount that causes a pharmacological effect cannot be considered a tracer dose. Limitations of a tracer dose are that the tracer should not (1) produce a pharmacological effect (e.g. insulin release after glucose or arginine administration), and (2) be given in significant amounts relative to the daily intake of a nutrient (e.g. leucine supply with low-protein diets). A rule of thumb is that the rate of tracer administration should not exceed 10% of the flux (rate of appearance) of the metabolite being measured (D.E. Matthews, personal communication). This criterion means that the enrichment of the metabolite should not exceed 10 mol%, in excess over background (MPE), whereas enrichments of <1 MPE are more common in tracer studies.

One may distinguish stable isotope dilution and stable isotope incorporation studies. Dilution studies can provide insight in the disappearance (e.g. irreversible loss rates of amino acids) or appearance of tracee (e.g. glucose entry rates). Incorporation studies can indicate the rates of utilization of nutrients into various 'end-products', such as muscle protein synthesis from ^{13}C-labelled phenylalanine or $^{13}CO_2$ production from oxidation of ^{13}C-labelled fatty acids. Whole body oxidation from ^{13}C-labelled substrates requires a combination of accurately measuring gas exchange (indirect calorimetry) and ^{13}C enrichment in expired CO_2. Several units exist to express ^{13}C enrichment:

- Tracer-to-tracee ratio (TTR). A TTR of 0.05 indicates that the amount of tracer is 5% of the amount of tracee.
- Mole percent (mol%) or atom percent (at%), which refers to the molar ratio of the tracer molecule or atom and the sum of the tracer and tracee molecules or atoms. A mol% or at% of 0.05 indicates that the amount of tracer is 5% of the amount of tracer plus tracee.
- Mole percent excess (MPE) or atom percent excess (APE) is similar to respectively mol% and at%, but then corrected by subtracting the background enrichment (also expressed in mol% or at%) which is usually obtained from measurements before starting the tracer study or from measurements during a similar time period but without tracer administration.
- Delta value relative to an international standard (PDB). The PDB-standard is a limestone originating from the Pee Dee Belemnite formation (South Carolina, USA). The ^{13}C enrichment of this limestone standard is relatively high, resulting in a negative δ value for most natural occurring samples.

$$\delta\ ^{13}C\ (‰) = [\frac{(TTR_s - TTR_{PDB})}{TTR_{PDB}}] \times 1000 \qquad (5)$$

where $TTR_s = {}^{13}C/{}^{12}C$ of the sample, $TTR_{PDB} = {}^{13}C/{}^{12}C$ of the standard PDB (= 0.0112372).

Approximately 1.1% of all carbon on earth is ^{13}C. There is, however, a difference between substrates obtained from plants with different photosynthetic pathways. For example, C4 plants (incl. maize, millet, sugar cane, pineapple) have a higher natural ^{13}C enrichment than C3 plants. These plants differ in the molecule into which CO_2 is fixed during photosynthesis; the 3 carbon intermediate pathway is known as the Calvin-Benson route (Calvin, 1962), and the 4 carbon pathway as the Hatch-Slack route (Hatch and Slack, 1970). More than 95% of earth's plant species can be characterised as C3 plants. In addition, some plants use Crassulacean Acid Metabolism (CAM) and possess both the C3 and C4 cycles. However, unlike C4 plants, both cycles operate together in the same cells, but at different times of the day in these plants. CAM plants are not commonly found, except in deserts where stomata are open at night, when transpiration rates would be lowest and CO_2 is then stored as C4 acid, but during day time after decarboxylation refixation occurs via the C3 cycle. The isotopic composition of animals or humans reflects that of their diet. Greater consumption of maize and maize-derived products in the United States than in Europe, for example, impacts ^{13}C enrichment in the body as well as expired air. This also holds for animal products, such as eggs, milk or meat. A 50-kg human subject, for example, contains approximately 225 g of the heavier stable isotopes, i.e. ^{2}H, ^{13}C, ^{15}N, ^{17}O and ^{18}O (Figure 2). This increases to 315 g for a 70-kg person and to 405 g for a 90-kg person. The isotopes ^{13}C (61%) and ^{18}O (31%) are the main contributors.

5.3.2 Measurement of ^{13}C enrichment in expired CO_2

Measuring ^{13}C enrichment in CO_2 can be done by collecting air samples and analysis by isotope ratio mass spectrometry (IRMS) or by non-dispersive infrared spectroscopy (NDIR).

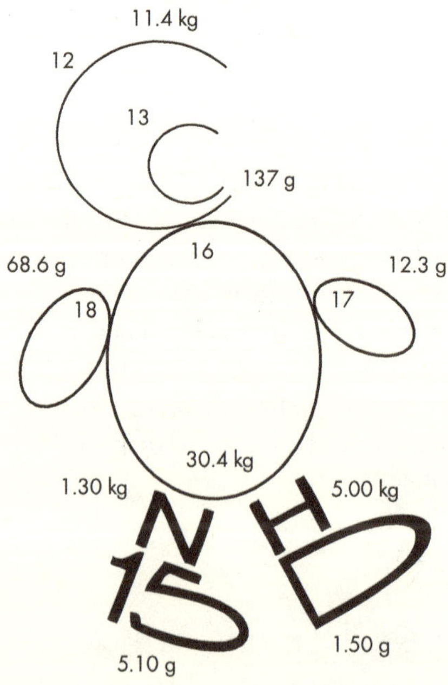

Figure 2. Isotopic body composition as a reflection of isotopic composition of the diet; the contribution of stable isotopes to body weight in a 50 kg human subject (Wada, 2009; Figure reused with permission from the Proceedings of the Japan Academy, Series B).

These are highly sensitive analytical methods that allow detection of very small changes in isotopic enrichment with a precision of ~0.0002% (Schierbeek et al., 2012). Although IRMS is the golden standard to measure ^{13}C enrichment in air samples (Ferrannini, 1988; Meier-Augenstein, 1999; Schierbeek et al., 2012), NDIR has the advantage that the analyser can be integrated in series with analysers for O_2, CO_2 and CH_4 for indirect calorimetry, can be linked to existing data acquisition systems, and shows good correlation with IRMS (Alferink et al., 2003; Barth et al., 1998; Schadewaldt et al., 1997). In our laboratory we use an Advance Optima Uras-14 NDIR analyser system (ABB automation analytical division, Frankfurt a. M., Germany) for separate analysis of total CO_2 and $^{13}CO_2$. This analyser is coupled to existing open circuit indirect calorimetry facilities and integrated in the automated data-acquisition system where incoming and outgoing air of sets of respiration chambers are analysed (Chapter 2; Heetkamp *et al.*, 2015). This analyser system was put in series with existing O_2 and CH_4 analysers. Raw data from NDIR are corrected based on a 3-step procedure including calibration gasses (calibration is performed daily): (1) CO_2 containing (1.08%) and CO_2 depleted (0.00%) calibration gas mixtures are used for linear zero and span point corrections for all gases, including $^{13}CO_2$; (2) a gas mixture with $^{12}CO_2$ but no $^{13}CO_2$ (^{13}C depleted) is used to correct step 1 for overlapping infrared absorption spectra of $^{12}CO_2$ and $^{13}CO_2$; (3) correction of $^{13}CO_2$ as measured in step 1 based on the (linear) correction factor obtained in step 2 (Alferink *et al.*, 2003).

5.3.3 Whole body substrate oxidation

Oxidation of a substrate of interest can be measured by continuous intravenous infusion of that ^{13}C enriched substrate and measuring rates of $^{13}CO_2$ expiration. At isotopic equilibrium, the rate of irreversible loss of ^{13}C is equal to the rate of tracer infusion. The time required for reaching an isotopic equilibrium can be markedly shortened by an appropriate priming dose to enrich the tracee and bicarbonate pools (Wolfe and Chinkes, 2005). The rate of uptake and appearance of tracee are also equal when the tracee concentration is in steady state. It is assumed that the tracee and tracer are not distinguished, hence the same percentages of each that irreversibly leave the substrate pool will be oxidised to CO_2 (Wolfe and Chinkes, 2005). The rate of uptake of the tracer equals the rate of tracer infusion (F), and the percentage oxidized can then be calculated:

$$\%\text{ uptake oxidised} = \frac{^{13}CO_2\text{ excretion }(\mu mol/h)}{F\,(\mu mol/h) \times c \times 100\%} = \frac{E_{CO_2} \times V_{CO_2}}{F \times c \times 100\%} \tag{6}$$

Where E_{CO_2} is the ^{13}C enrichment in expired CO_2 and V_{CO_2} is the rate of CO_2 production, c is a correction factor for incomplete bicarbonate recovery which is explained in more detail below.

The absolute rate of substrate oxidation can be calculated by multiplying the rate of uptake (i.e. the rate of disappearance R_d, F/E_p, which equals the rate of appearance R_a at isotopic steady state) and the percentage of uptake oxidised:

$$Rate\ of\ oxidation\ (\mu mol/h) = \frac{F / E_p \times {}^{13}CO_2\ excretion}{F \times c} = \frac{{}^{13}CO_2\ excretion\ (\mu mol/h)}{E_p \times c} \quad (7)$$

E_p is the ^{13}C enrichment in the precursor pool, ideally measured close to the site of oxidation. For practical reasons, the plasma enrichment of the compound of interest is usually considered representative of true precursor enrichment. In some cases, enrichment of a derived metabolite may be used. For example, enrichment of the keto-acid of leucine (keto-isocaproic acid) in plasma is more similar to the true intracellular precursor enrichment than that of plasma leucine. For glucose and fatty acids, precursor enrichment is usually determined in plasma and proxies for intracellular enrichment do not exist.

Over a non-protein RQ range from 0.76 to 0.90, the contribution of glucose oxidation to Q was similar when calculated from gas exchange (i.e. using coefficients like in Table 2) compared to when calculated from rate of oxidation of labelled glucose (Glamour *et al.*, 1995; Figure 3). Glucose oxidation by the tracer method was quantified to be higher than that calculated from gas exchange when the non-protein RQ was <0.76; it was lower than the respiratory calorimetry estimate when the non-protein RQ was >0.90. This can be explained by the contribution of gluconeogenesis to gas exchange at a low RQ and that of lipogenesis at a high RQ.

Another example includes continuous intravenous infusion of a ^{13}C-labelled fatty acid for estimating plasma free fatty acid oxidation. It is assumed that the labelled carbon appears in breath as $^{13}CO_2$ when the fatty acid tracer is oxidised (Klein and Wolfe, 1987). In humans, total parenteral nutrition often results in an RQ>1 but when combined with intravenous

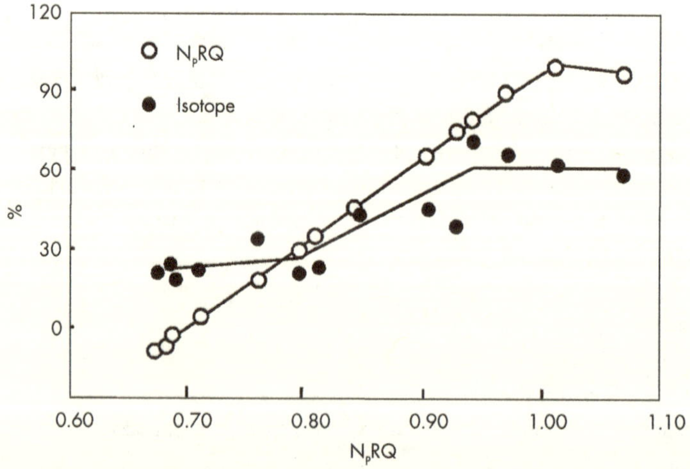

Figure 3. Comparing two methods ('Isotope', [U-^{13}C]glucose vs 'N_pRQ', non-protein respiratory quotient) for estimating whole body glucose oxidation, expressed as a percentage of total heat production, in human subjects over a broad RQ range. This range was created by intravenous infusion of (unlabelled) glucose (Glamour et al.*, 1995; Figure reused with permission from the American Journal of Physiology).*

infusion of ^{13}C-palmitate $^{13}CO_2$ has been detected in breath (Goodenough and Wolfe, 1984). This indicates that true whole body fatty acid oxidation can be underestimated with indirect calorimetry when the RQ exceeds unity.

Such substrate oxidation approaches yield valuable insights in processes at whole-body level, but mask the large differences that exist between tissues and between individual carbohydrates, fats and proteins. The combination of ^{13}C-labelled substrates and $^{13}CO_2$ production can also be used to quantify substrate oxidation by specific organs. This requires robust measurements of $^{13}CO_2$ in arterial and venous blood and accurate detection of blood flow. Such studies need sophisticated and well-validated methods but the underlying principles are very similar to those for whole body oxidation (Beaufort-Krol *et al.*, 1998; Wolfe and Chinkes, 2005).

5.3.4 Oxidation of exogenous substrates

Oxidation of dietary substrates can be measured by including ^{13}C-enriched substrates in the diet and measuring $^{13}CO_2$ production (i.e. ^{13}C enrichment in CO_2 as well as total CO_2 production). For example, exogenous glucose oxidation has been measured by including [U-^{13}C]glucose in the diet and measuring ^{13}C recovery in breath CO_2 in several species, including human (Caron *et al.*, 2004; Jeukendrup *et al.*, 1997; Lifschitz *et al.*, 1988; Moseley *et al.*, 2005), mice (Ishihara *et al.*, 2002), pigs (Bruininx *et al.*, 2011), poultry (Buyse *et al.*, 2004) and calves (Van den Borne *et al.*, 2007). Oxidation of dietary glucose can be affected by the metabolic status of the experimental subjects. Factors such as physical activity, insulin sensitivity, age, level of energy intake, and the rate of digestion have been shown to affect oxidation of exogenous glucose.

Using ^{13}C-labelled substrates may also allow estimation of whole body oxidation of that substrate (i.e. exogenous plus endogenous), but this requires measurements of ^{13}C enrichment in precursor pools (e.g. plasma glucose). Due to the non-steady metabolic and isotopic state, frequent sampling is usually required in order to obtain reliable oxidation estimates, and care should be taken in calculating whole body oxidation rates from samples with relatively low enrichments. Also dual tracer approaches, such as an oral bolus of ^{13}C-glucose (or ^{13}C-phenylalanine) combined with a continuous intravenous infusion of ^{2}H-glucose (or ^{2}H-phenylalanine), exist and can be used to quantify rates of dietary glucose or amino acid appearance in the systemic circulation. This represents the rate of absorption from the intestinal lumen minus first-pass metabolism across the splanchnic tissues.

Similar oxidation studies have been done by supplementing the diet with tracers for fatty acids and triglycerides (Bruininx *et al.*, 2011; Demmelmair *et al.*, 1998; Jones *et al.*, 1998; Kalivianakis *et al.*, 1997; Metges and Wolfram, 1991; Murphy *et al.*, 1995; Vermunt *et al.*, 2001) or amino acids (Labayen *et al.*, 2004; Metges *et al.*, 2000; Millward *et al.*, 2002; Millward *et al.*, 2000; Schadewaldt *et al.*, 1998; Van den Borne *et al.*, 2012). When such studies are combined with measurements of precursor enrichment in plasma, absolute rates of substrate oxidation – rather than recovery of ^{13}C as $^{13}CO_2$ can also be calculated (e.g. Van den Borne *et al.*, 2012). Such measurements should be taken in short time intervals due to the non-steady state induced by feeding.

When using dietary labelled substrates and measuring $^{13}CO_2$ production, exogenous substrate oxidation can be discriminated from total substrate oxidation, and oxidation of endogenous substrates (including glycogen, protein, fat) can be calculated (e.g. Chu *et al.*, 2011; Folch *et al.*, 2001). Also glycogen formation from exogenous carbohydrates can be estimated when labelled starch is fed (Folch *et al.*, 2001). This is done by subtracting the difference between the total amount of glucose provided by the meal and the amount of exogenous glucose oxidized (calculated from ^{13}C recovery in breath), taking into account the amount of glucose converted into fat, if any, and assuming that all the glucose released from the starch had been absorbed. This is not possible with indirect calorimetry only.

5.3.5 Quantifying oxidation of individual carbohydrates, fats and proteins

Indirect calorimetry can be adequately used to estimate oxidation of carbohydrates, fats and protein, but these are representing classes of nutrients that contain many different types of monomers, dimers and polymers which may vary in their digestive, absorptive and metabolic route. Using ^{13}C-enriched substrates allows tracing the metabolic fate of individual components, such as specific monosaccharides from the diet (Blondin *et al.*, 2012; Jentjens *et al.*, 2003a; Jentjens *et al.*, 2003b). For example, fructose is absorbed from the intestine by GLUT-5 transporters and intestinal glucose transport occurs via a sodium-dependent glucose transporter (Ferraris and Diamond, 1997). Galactose has been suggested to be more rapidly absorbed than glucose (Groen, 1937; Holdsworth and Dawson, 1964). To study such differences, individual monosaccharides can be enriched with ^{13}C and tracer appearance in blood as well as expired CO_2 can be measured (e.g. Jentjens *et al.*, 2003a; Leijssen *et al.*, 1995). Similarly, effects of individual dietary fatty acids on their oxidative utilisation have been assessed (reviewed by Emken, 2001) which has yielded insight in the effect of nutritional and physiological conditions on fatty acid utilisation (Vermunt *et al.*, 1999), but also allowed comparison of fatty acids with different chain length or degree of saturation (Bruininx *et al.*, 2011; Jones *et al.*, 1999; McCloy *et al.*, 2004; Schmidt *et al.*, 1999).

An important advantage of using ^{13}C-enriched nutrients is that their oxidation can be measured despite the presence of other nutrients from the same class of macronutrients. For example, fructose oxidation can be traced when also glucose and galactose are available, or stearic acid oxidation can be measured when also other fatty acids are available. Such information cannot be obtained by calculating substrate oxidation from gas exchange only.

5.3.6 Type of ^{13}C-enriched substrates and background enrichment

A good tracer molecule reflects the digestion, absorption and metabolism of its tracee. Using free [U-^{13}C]glucose as a tracer for starch, for example, may not be appropriate, because the glucose tracer can be expected to become more rapidly metabolically available and – as a consequence – be metabolised differently compared to its tracee (i.e. starch-derived glucose). In order to obtain representative tracer products the use of intrinsic labels in dietary ingredients may be desirable. This can be achieved by selecting plant-derived ingredients with a naturally high ^{13}C enrichment (e.g. Peronnet *et al.*, 1990; Petzke and Klaus, 2008; Schoeller *et al.*, 1980) or animal products from animals fed C4 plants (Metges *et al.*, 1992; Vonk *et al.*, 2001) and using these in nutritional studies. Alternatively, ingredients that are depleted in ^{13}C could be used, such as protein from bacteria that have been grown on methane gas.

A more substantial increase in intrinsic ^{13}C enrichment can be obtained by providing enriched precursors during the process of protein, fat or carbohydrate synthesis. For example, intravenous infusion of ^{13}C-labelled amino acids in dairy cows has resulted in enriched cow's milk protein (Boirie *et al.*, 1995; Pennings *et al.*, 2011) and meat (Pennings *et al.*, 2013), and feeding L[1-^{13}C]leucine also increased its enrichment in egg protein (Evenepoel *et al.*, 1997). Increasing ^{13}C enrichment in atmospheric CO_2 during plant growth results in enriched plants, such as wheat (Eelderink *et al.*, 2012) and grass (Pellikaan *et al.*, 2013) from which various fractions can be extracted for oxidation studies. Importantly, the molecules in these plants will be uniformly labelled (i.e. at all C-atoms). Therefore, it is not possible to label, for instance, exclusively the carboxyl atom of specific amino acids when using an atmosphere rich in $^{13}CO_2$.

The choice of tracer is generally driven by the research questions, the analytical possibilities and budgets available (for purchasing or producing tracer molecules). As a consequence, one may opt for intrinsic vs extrinsic labels, natural abundance vs synthetic enrichment, and uniform vs specific labels. The use of ^{13}C-enriched substrates in oxidation studies requires accurate knowledge of the starting ^{13}C abundance in expired air. This baseline value marks the point of departure from where oxidation is determined. Typically, >80% of the $^{13}CO_2$ production originates from non-enriched substrates (i.e. background enrichment, ~1.08% ^{13}C) and the remainder is introduced by oxidation of tracer. Baseline ^{13}C enrichment in expired CO_2, and its fluctuation during the day, may vary depending on the differences in natural enrichment between dietary ingredients and between differences in endogenous (body fat and protein, glycogen) and exogenous substrates for oxidation during fed and fasted states. Also sequential use of stable isotopes may increase background enrichment. Appropriate corrections should be applied to correct for these changes during energy metabolism studies.

5.3.7 Isotopomers

An isotopomer is a combination of isomers having the same number of each isotopic atom but differing in their positions (e.g. [1-^{13}C]leucine and [2-^{13}C]leucine). In general, analysing the ^{13}C-mass isotopomer distribution in various plasma metabolites after administration of a deliberately chosen tracer molecule (e.g. [U-^{13}C]glucose) can provide insight in partitioning of (atoms from) nutrients over metabolic pathways (e.g. Bequette *et al.*, 2006; Hellerstein and Neese, 1992). With the choice of the right precursor labelling, chemical derivative, ionisation mode and measured fragment ions, one can determine both the location and the quantity of the label involved in the downstream metabolic product. Development of methods to use isotopomer information requires validation studies with respect to the accuracy of assumptions (e.g. exchanges at intermediate steps) and possible isotopic effects (e.g. incompleteness of isomerase reactions). Isotopomer analysis in metabolites is commonly done by gas chromatography-mass spectrometry and not by gas chromatography-combustion IRMS as isotopomer information is destroyed when the derivative enters the combustion furnace.

In whole body oxidation studies, isotopomers may also be useful to trace nutrient partitioning. For example, the metabolic fate of dietary glucose in growing animals can be measured by using glucose isotopomers. Differences in oxidation of [3-^{13}C]glucose and [2-^{13}C]glucose would reflect, indirectly, fatty acid synthesis from glucose, with a greater reduction in the amount of $^{13}CO_2$ released from [2-^{13}C]glucose if pyruvate is converted to acetate, by the action of pyruvate dehydrogenase, and this is used for fatty acid synthesis. The C_2-atom from glucose is not released if acetate is formed and used for fatty acid synthesis, whereas the C_3-atom is released as CO_2. Similarly, [U-^{13}C]glucose may be used instead of the (expensive) tracer [3-^{13}C]glucose (e.g. Van den Borne *et al.*, 2007).

Importantly, the labelling (i.e. number and position of ^{13}C atoms) also affects flux calculations and interpretation of studies on substrate oxidation. Due to sequestration of ^{13}C in the body, labels from oxidised substrates are not completely recovered in expired air during the course of a metabolic study. Hence, corrections may be required to take these apparent losses into account.

5.3.8 Correcting for bicarbonate and acetate sequestration

CO_2 which is produced by metabolic processes in the body is not for 100% recovered in expired breath, as demonstrated by measurements of ^{13}C-recovery after intravenous infusion of ^{13}C-labelled bicarbonate. The bicarbonate correction factor can be used to correct the oxidation rates for the proportion of labelled CO_2 that is produced via oxidation (i.e. from decarboxylation and tricarboxylic acid cycle) but not secreted in expired air (Owen *et al.*, 2002; Van Hall, 1999). CO_2 retention mainly occurs in the body stores of bicarbonate (e.g. bone, intracellular reserves, metabolic intermediates) and can be determined by intravenous infusion of ^{13}C-labelled bicarbonate (as a bolus or continuous infusion, in accordance with the experimental protocol) and measuring label recovery in breath (Irving *et al.*, 1983). As reviewed by Leijssen and Elia (1996), recovery of ^{13}C from bicarbonate in expired air is approximately 80% in resting human subjects, but higher during feeding (84%) than during fasting (74%), and higher after continuous infusion (84%) than after a bolus (69%). Sequestration of bicarbonate in the body commonly decreases with increasing metabolic rate, and label recovery in expired air is close to 100% during exercise. In addition, bicarbonate kinetics are very sensitive to changes in acid-base status (Van Hall, 1999).

Apart from ^{13}C sequestered in CO_2, labelled C from the metabolised substrate can also be temporarily fixed in other metabolites, mainly in the tricarboxylic acid cycle (Van Hall, 1999) and especially via conversion of α-ketoglutarate to glutamate and glutamine. This is due to isotopic exchange reactions in which labelled and unlabelled C-atoms are shunted. Although the substrate is completely oxidised, its label does not appear exclusively as CO_2 in breath, hence substrate oxidation rates are underestimated. Ultimately, the labelled intermediates will re-enter the oxidative pathways but, as these pools are so large, this will occur at a later point in time. This label sequestration can be determined by an intravenous infusion of labelled acetate and measurement of label recovery in breath (i.e. acetate correction factor). Acetate is converted to acetyl-CoA and thereafter it behaves like acetyl-CoA originating from fatty acids or glucose. The likelihood of label being retained is higher for labelled C-2 than for labelled C-1 (Schrauwen *et al.*, 2000; Van Hall, 1999; Van Loon *et al.*, 2003). For example, 81% of infused [1-^{13}C]acetate but only 53% of [2-^{13}C]acetate was recovered after a 4-h infusion period (Wolfe and Jahoor, 1990). If universally labelled substrates are used in the study, the [1,2-^{13}C] acetate recovery factor should be considered, for [1-^{13}C] and [2-^{13}C] labelled substrates the [1-^{13}C] and [2-^{13}C] forms of acetate, respectively, seem more appropriate. The accuracy – and relevance – of the acetate correction factor depends on two major assumptions: (1) the acetate infused is fully oxidised via the tricarboxylic acid cycle and not used for lipogenesis and ketogenesis; (2) acetate is metabolised in the same tissue to an extent proportional to the substrate. The first assumption does not hold, for example, in growing pigs and broilers in which acetate is an important precursor for *de novo* fatty acid synthesis. Recovery of [1,2-^{13}C] acetate in growing pigs, fed at a rate of 2.5 times the energy requirements for maintenance, averaged 50%, which was similar to that of [U-^{13}C]glucose (unpublished data). The second assumption may not hold for animals at low levels of physical activity, as acetate is mainly oxidised in liver at rest whereas substrate oxidation mainly occurs in peripheral tissues. For most studies in growing or producing (milk, eggs) animals, the acetate correction factor is therefore rarely used, but this may differ when studying for example nutrient oxidation in companion animals.

5.4 Other applications combining ^{13}C-labelling and indirect calorimetry

The combined use of ^{13}C-tracers and indirect calorimetry is also applied for purposes other than measuring substrate oxidation for energy metabolism. Two examples are briefly discussed.

5.4.1 Breath tests for clinical purposes

The combined use of gas exchange measurements and stable isotopes can be used to study aspects related to health and digestion in a non-invasive way. Breath tests are designed to monitor the activity of a specific enzyme or physiological system *in vivo*. Tests have been developed to measure specific metabolic pathways of interest, such as gastric ulcers, liver function, gastric emptying or pancreatic function and are used in research and as a clinical diagnostic tool in practice. The simple principle of the breath test is ingestion of a ^{13}C substrate as a marker for the limiting step of metabolic pathway of interest, the functional ^{13}C labelled group is cleaved (e.g. enzymatically, by a well-defined gastrointestinal function or by bacterial metabolism) and $^{13}CO_2$ is measured in collected breath samples. For medical applications, most breath tests provide qualitative or semi-quantitative results, because total CO_2 production is not measured (by indirect calorimetry). Nonetheless, the breath test has several advantages over the conventional methods; the test is not invasive, can be done in a short time, is relatively simple to perform and can be executed at low costs (Ghoos *et al.*, 2002). Conditions for a successful breath test include: (1) there should be a steady baseline label production; (2) of the pathway that is studied one process is the rate limiting step; (3) the amount of label should be large enough to produce a significant signal, but should be small enough not to interfere with the studied (patho)-physiological process; (4) CO_2 production should be steady or should be measured during the test; and (5) there should be no recycling of the label or loss of label by other pathways. The ^{13}C-breath test is used for diagnostic and research applications (Table 5).

The most known breath test is the use of [^{13}C]urea to detect stomach ulcers. The C-atom from urea is converted to CO_2 by the obligate presence of urease from Helicobacter pylori which causes the ulceration. Another example is the use of ^{13}C-labelled nutrients to determine the rate of gastric emptying by measuring recovery in expired CO_2; ^{13}C octanoic acid is used as a hydrophobic label (i.e. solid phase) and ^{13}C acetic acid as a hydrophilic label (i.e. liquid phase).

5.4.2 Indirect amino acid oxidation

Amino acid requirements have been studied by measuring $^{13}CO_2$ production from a non-limiting ^{13}C-labelled indispensable amino acid in response to graded levels of the limiting indispensable amino acid. The rationale of this indicator amino acid oxidation (IAAO) technique is that $^{13}CO_2$ production from unbalanced amino acids decreases with increasing supplementation of the limiting amino acid until the balance between amino acids is restored and therefore a plateau (i.e. the requirement) is reached (Elango *et al.*, 2012; Pencharz and Ball, 2003). The body conserves the labelled amino acid when it is supplied at levels below the minimum requirement, but oxidises increasing amounts as the amino acid intake is increased above the minimum requirement (Zello *et al.*, 1995). Many IAAO studies have been conducted during the past two decades in several species (including man, pigs, chickens). A physiological as well as an isotopic steady state is important for accurate measurements and therefore intravenous tracer (usually L[1-^{13}C]phenylalanine) infusion and frequent meals are commonly applied. However, less-invasive approaches have been developed in which the tracer is provided with the diet and sampling occurs from urine instead of from blood (Bross *et al.*, 1998; Kriengsinyos *et al.*, 2002). Providing the tracer as a bolus rather than via continuous infusion appears also possible (Gooding *et al.*, 2013).

In addition, a modified version of the IAAO technique can be used to assess the metabolic availability of amino acids from dietary proteins (Elango *et al.*, 2009; Moehn *et al.*, 2005; Moehn *et al.*, 2007). This method can be used to determine effects of different dietary protein sources, processing and interaction with other nutrients on availability in the systemic circulation by

Table 5. Examples of diagnostic and research applications of the ^{13}C *breath test (Braden, 2009; Braden* et al.*, 2007; Fischer and Wetzel, 2002; Ghoos* et al.*, 2002; Koletzko* et al.*, 1998; Rating and Langhans, 1997; Weaver, 1998).*

Function tested	Tracer
Infection	
Helicobacter pylori	[^{13}C]urea
Gastrointestinal passage rate	
Gastric emptying	Solid phase: [^{13}C]octanoic acid
	Liquid phase: [^{13}C]acetic acid, [^{13}C]glycine
Orocecal transit	Lactose [^{13}C]ureide
Exocrine pancreatic function[1]	
Lipase activity[2]	1,3 distearyl 2-[^{13}C]carboxyl-octanoyl glycerol [^{13}C]trioctanoin, [^{13}C] palmitin, [^{13}C]hiolein, [^{13}C]triolein
Trypsin activity	[^{13}C]protein (e.g. chicken egg white)
Amylase activity	[^{13}C]starch
Lactase deficiency	[^{13}C]lactose
Cholesterol esterase activity	[^{13}C]cholesterol-octanoate
Liver function	
Microsomal enzyme function	[^{13}C]aminopyrine, [^{13}C]methacetin, [^{13}C]caffeine, [^{13}C]phenylalanine
Cytosolic enzyme function	[^{13}C]galactose
Mitochondrial enzyme function	[^{13}C]keto-isocaproic acid, [^{13}C]methionine, [^{13}C]octanoic acid
Bacterial overgrowth	[^{13}C]xylose, [^{13}C]glycocholate

[1] Combinations of ^{13}C-breath tests are often applied in order to separate maldigestion (by ^{13}C-labelled polymers) from malabsorption (by ^{13}C-labelled monomers).

[2] Lipase activity is considered as first-limiting step for $^{13}CO_2$ production from these tracers, but consider various phases involved in fat digestion (lipolysis, solubilization, mucosal uptake) when selecting appropriate tracers (e.g. Weaver, 1998).

using titration with a synthetic amino acid (e.g. lysine) as a reference for which complete metabolic availability is assumed (Table 6).

In conclusion, several limitations of assessing whole body nutrient oxidation by indirect calorimetry can be circumvented when using ^{13}C-tracer methodology in combination with indirect calorimetry. This approach also allows discrimination of oxidation of exogenous vs endogenous substrates. Labelling of individual monosaccharides, fatty acids or amino acids can provide insight in the metabolic fate of the tracee irrespective of availability of other substrates from the same macronutrient class (i.e. carbohydrate, fat, protein). Importantly, technical considerations (e.g. fluctuations in background enrichment, isotope administration strategy, sequestration of CO_2 and isotopic gradient in the body) should be considered a

Table 6. Example of assessing metabolic availability of amino acids from peas and heated peas in pigs by using the IAAO technique. Metabolic availability of synthetic lysine was set at 100% and the oxidation response of other proteins was expressed relative to that of synthetic lysine (Moehn et al., *2005).*

	Lysine intake in excess of basal diet (g/d)	Indicator amino acid oxidation (% of dose)	Oxidation response (% of dose/g Lys)	Metabolic availability (%)
Lys – 50%	0.00	17.38		
Lys – 60%	0.77	16.94		
Lys – 70%	1.63	14.66		
Lys – 50%	2.12	10.50	-3.16	100.0
Peas	2.28	12.56	-2.81	88.8
Heated peas	2.42	14.46	-1.73	54.8

priori and physiologically relevant conditions – which are often non-steady state – require appropriate mathematical models. Finally, facilities combining indirect calorimetry and tracer methodology could also use these techniques for, amongst others, ^{13}C-breath tests or amino acid oxidation studies.

References

Alferink, S.J.J., Van den Borne, J.J.G.C., Gerrits, W.J.J., Lammers-Wienhoven, S.C.W. and Heetkamp, M.J.W., 2003. On-line, continuous determination of $^{13}CO_2/^{12}CO_2$ ratios by non-dispersive infrared absorption in indirect calorimetry facilities. In: Souffrant, W.B. and Metges, C.C. (eds.) Progress in Research on Energy and Protein Metabolism, Rostock-Warnemünde, Germany. EAAP Publication No. 109. Wageningen Academic Publishers, Wageningen, the Netherlands, pp. 465-468.

Arch, J.R.S., Hislop, D., Wang, S.J.Y. and Speakman, J.R., 2006. Some mathematical and technical issues in the measurement and interpretation of open-circuit indirect calorimetry in small animals. International Journal of Obesity 30: 1322-1331.

Barth, E., Tugtekin, I., Weidenbach, H., Wachter, U., Vogt, J., Radermacher, P., Adler, G. and Georgieff, M., 1998. Determination of $^{13}CO_2/^{12}CO_2$ ratio by IRMS and NDIRS. Isotopes in Environment and Health Studies 34: 209-213.

Beaufort-Krol, G.C.M., Takens, J., Molenkamp, M.C., Smid, G.B., Zijlstra, W.G. and Kuipers, J.R.G., 1998. Determination of organ substrate oxidation *in vivo* by measurement of $^{13}CO_2$ concentration in blood. Journal of Mass Spectrometry 33: 328-222.

Bequette, B.J., Sunny, N.E., El-Kadi, S.W. and Owens, S.L., 2006. Application of stable isotopes and mass isotopomer distribution analysis to the study of intermediary metabolism of nutrients. Journal of Animal Science 84 (E. Suppl.): E50-E59.

Beylot, M., 2006. Use of stable isotopes to evaluate the functional effects of nutrients. Current Opinion in Clinical Nutrition and Metabolic Care 9: 734-739.

Bier, D.M., 1987. The use of stable isotopes in metabolic investigation. Bailliere's Clinical Endocrinology and Metabolism 1: 817-836.

Blondin, D.P., Péronnet, F. and Haman, F., 2012. Coingesting glucose and fructose in the cold potentiates exogenous CHO oxidation. Medicine and Science in Sports and Exercise 44: 1706-1714.

Boirie, Y., Fauquant, J., Rulquin, H., Maubois, J.L. and Beaufrère, B., 1995. Production of large amounts of ^{13}C-leucine-enriched milk proteins by lactating cows. Journal of Nutrition 125: 92 98.

Braden, B., 2009. Methods and functions: Breath tests. Best Practice & Research Clinical Gastroenterology 23: 337-352.

Braden, B., Lembcke, B., Kuker, W. and Caspary, W.F., 2007. ^{13}C-breath tests: Current state of the art and future directions. Digestive and Liver Disease 39: 795-805.

Brockway, J.M., 1987. Derivation of formulae used to calculate energy expenditure in man. Human Nutrition: Clinical Nutrition 41C: 463-471.

Bross, R., Ball, R.O. and Pencharz, P.B., 1998. Development of a minimally invasive protocol for the determination of phenylalanine and lysine kinetics in humans during the fed state. Journal of Nutrition 128: 1913-1919.

Brouwer, E., 1957. On simple formulae for calculating the heat expenditure and the quantities of carbohydrate and fat oxidized in metabolism of men and animals, from gaseous exchange (oxygen intake and carbonic acid output) and urine-N. Acta Physiologica Neerlandica 6: 795-802.

Brouwer, E., 1958. On simple formulae for calculating the heat expenditure and the quantities of carbohydrate and fat metabolized in ruminants, from data on gaseous exchange and urine-N. In: First Symposium on Energy Metabolism, Copenhagen. EAAP, Rome, Italy, pp. 182-194.

Brouwer, E., 1965. Report of sub-committee on constants and factors. In: Blaxter, K.L. (ed.) Energy metabolism. Academic Press, London, UK, pp. 441-443.

Bruininx, E., Van den Borne, J., Van Heugten, E., Van Milgen, J., Verstegen, M. and Gerrits, W., 2011. Oxidation of dietary stearic, oleic, and linoleic acids in growing pigs follows a biphasic pattern. Journal of Nutrition 141: 1657-1663.

Burgos, M.S., Senn, M., Sutter, F., Kreuzer, M. and Langhans, W., 2001. Effect of water restriction on feeding and metabolism in dairy cows. American Journal of Physiology 280: R418-R427

Burnier, P., Tappy, L., Jéquier, E., Schneeberger, D. and Chioléro, R., 1993. Metabolic and respiratory effects of infused sodium acetate in healthy human subjects. American Journal of Physiology 263: 1271-1276.

Buyse, J., Geypens, B., Malheiros, R.D., Moraes, V.M., Swennen, Q. and Decuypere, E., 2004. Assessment of age-related glucose oxidation rates of broiler chickens by using stable isotopes. Life Sciences 75: 2245-2255.

Calvin, M., 1962. The path of carbon in photosynthesis. Science 135: 879-889.

Caron, A., Lavoie, C., Péronnet, F., Hillaire-Marcel, C. and Massicotte, D., 2004. Oxidation of [^{13}C] glucose ingested before and/or during prolonged exercise. European Journal of Applied Physiology 91: 217-223.

Chu, L., Riddell, M.C., Takken, T. and Timmons, B.W., 2011. Carbohydrate intake reduces fat oxidation during exercise in obese boys. European Journal of Applied Physiology 111: 3135-3141.

Chwalibog, A., Jakobsen, K., Henckel, S. and Thorbek, G., 1992. Estimation of quantitative oxidation and fat retention from carbohydrate, protein and fat in growing pigs. Journal of Animal Physiology and Animal Nutrition 68: 123-135.

Chwalibog, A., Jakobsen, K., Tauson, A.-H. and Thorbek, G., 2005. Energy metabolism and nutrient oxidation in young pigs and rats during feeding, starvation and re-feeding. Comparative Biochemistry and Physiology Part A 140: 299-307.

Demmelmair, H., Baumheuer, M., Koletzko, B., Dokoupil, K. and Kratl, G., 1998. Metabolism of U-^{13}C-labeled linoleic acid in lactating women. Journal of Lipid Research 39: 1389-1396.

Eelderink, C., Moerdijk-Poortvliet, T.C., Wang, H., Schepers, M., Preston, T., Boer, T., Vonk, R.J., Schierbeek, H. and Priebe, M.G., 2012. The glycemic response does not reflect the *in vivo* starch digestibility of fiber-rich wheat products in healthy men. Journal of Nutrition 142: 258-263.

Elango, R., Ball, R.O. and Pencharz, P.B., 2009. Amino acid requirements in humans: with a special emphasis on the metabolic availability of amino acids. Amino Acids 37: 19-27.

Elango, R., Ball, R.O. and Pencharz, P.B., 2012. Recent advances in determining protein and amino acid requirements in humans. British Journal of Nutrition 108: S22-S30.

Emken, E.A., 2001. Stable isotope approaches, applications, and issues related to polyunsaturated fatty acid metabolism studies. Lipids 36: 965-973.

Evenepoel, P., Hiele, M., Luypaerts, A., Geypens, B., Buyse, J., Decuypere, E., Rutgeerts, P. and Ghoos, Y., 1997. Production of egg proteins, enriched with L-leucine-13C-1, for the study of protein assimilation in humans using the breath test technique. Journal of Nutrition 127: 327-331.

Ferrannini, E., 1988. The theoretical bases of indirect calorimetry: A review. Metabolism: Clinical and Experimental 37: 287-301.

Ferraris, R.P. and Diamond, J., 1997. Regulation of intestinal sugar transport. Physiology Reviews 77: 257-302.

Fischer, H. and Wetzel, K., 2002. The future of ^{13}C-breath tests. Food and Nutrition Bulletin 23: 53-56.

Folch, N., Péronnet, F., Massicotte, D., Duclos, M., Lavoie, C. and Hillaire-Marcel, C., 2001. Metabolic response to small and large ^{13}C-labelled pasta meals following rest or exercise in man. British Journal of Nutrition 85: 671-680.

Frayn, K.N., 1983. Calculation of substrate oxidation rates *in vivo* from gaseous exchange. Journal of Applied Physiology 55: 628-634.

Gerrits, W.J.J., Bosch, M.W. and Van den Borne, J.J.G.C., 2012. Quantifying resistant starch using novel, *in vivo* methodology and the energetic utilization of fermented starch in pigs. Journal of Nutrition 142: 238-244.

Gerrits, W.J.J., Van den Borne, J.J.G.C. and Labussière, E., 2015. Deriving heat production from gaseous exchange: validity of the approach. In: Gerrits, W.J.J. and Labussière, E. (eds.) Indirect calorimetry. Techniques, computations and applications. Wageningen Academic Publishers, Wageningen, the Netherlands, pp. 19-34.

Ghoos, Y., Geypens, B. and Rutgeerts, P., 2002. Stable isotopes and $^{13}CO_2$ breath tests for investigating gastrointestinal functions. Food and Nutrition Bulletin 23: 166-168.

Glamour, T.S., McCullough, A.J., Sauer, P.J.J. and Kalhan, S.C., 1995. Quantification of carbohydrate oxidation by respiratory gas exchange and isotopic tracers. American Journal of Physiology 268: E789-E796.

Goodenough, R.D. and Wolfe, R.R., 1984. Effect of total parenteral nutrition on free fatty acid metabolism in burned patients. Journal of Parenteral and Enteral Nutrition 8: 357-360.

Gooding, M., Cant, J., Pencharz, P., Davenport, G., Atkinson, J. and Shoveller, A., 2013. Oral and intravenous l-[1-^{13}C] phenylalanine delivery measure similar rates of elimination when gastric emptying and splanchnic extraction are accounted for in adult mixed hounds. Journal of Animal Physiology and Animal Nutrition 97: 181-189.

Groen, J., 1937. Absorption of hexoses from the upper part of the small intestine in man. Journal of Clinical Investigation 16: 245-255.

Hatch, M.D. and Slack, C.R., 1970. Photosynthetic CO_2-fixation pathways. Annual Review of Plant Physiology 21: 141-162.

Heetkamp, M.J.W., Alferink, S.J.J., Zandstra, T., Hendriks, P., Van den Brand, H. and Gerrits, W.J.J., 2015. Design of climate respiration chambers, adjustable to the metabolic mass of subjects. In: Gerrits, W.J.J. and Labussière, E. (eds.) Indirect calorimetry. Techniques, computations and applications. Wageningen Academic Publishers, Wageningen, the Netherlands, pp. 35-56.

Hellerstein, M.K. and Neese, R.A., 1992. Mass isotopomer distribution analysis: A technique for measuring biosynthesis and turnover of polymers. American Journal of Physiology 263: E988-E1001.

Holdsworth, C.D. and Dawson, A.M., 1964. The absorption of monosaccharides in man. Clinical Science London 27: 371-379.

International Committee for Standardization in Haematology (ICSH), 1967. Recommendations for haemoglobinometry in human blood. British Journal of Haematology 13 (Suppl.): 71-76.

Irving, C.S., Wong, W.W., Shulman, R.J., O'Brian Smith, E. and Klein, P.D., 1983. [^{13}C]bicarbonate kinetics in humans: intra- vs. interindividual variations. American Journal of Physiology 245: R190-R202.

Ishihara, K., Oyaizu, S., Mizunoya, W., Fukuchi, Y., Yasumoto, K. and Fushiki, T., 2002. Use of ^{13}C-labeled glucose for measuring exogenous glucose oxidation in mice. Bioscience, Biotechnology, and Biochemistry 66: 426-429.

Jahoor, F. and Wolfe, R.R., 1987. Reassessment of primed constant-infusion tracer method to measure urea kinetics. American Journal of Physiology 252: E557-E564.

Jentjens, R.L.P.G., Moseley, L., Waring, R.H., Harding, L.K. and Jeukendrup, A.E., 2003a. Oxidation of combined ingestion of glucose and fructose. Journal of Applied Physiology 96: 1285-1291.

Jentjens, R.L.P.G., Venables, M.C. and Jeukendrup, A.E., 2003b. Oxidation of exogenous glucose, sucrose, and maltose during prolonged cycling exercise. Journal of Applied Physiology 96: 1285-1291.

Jeukendrup, A.E., Mensink, M., Saris, W.H.M. and Wagenmakers, A.J.M., 1997. Exogenous glucose oxidation during exercise in endurance-trained and untrained subjects. Journal of Applied Physiology 82: 835-840.

Jeukendrup, A.E. and Wallis, G.A., 2005. Measurement of substrate oxidation during exercise by means of gas exchange measurements. International Journal of Sports Medicine, Supplement 26: S28-S37.

Jones, A.E., Murphy, J.L., Stolinski, M. and Wootton, S.A., 1998. The effect of age and gender on the metabolic disposal of [1-^{13}C] palmitic acid. European Journal of Clinical Nutrition 52: 22-28.

Jones, A.E., Stolinski, M., Smith, R.D., Murphy, J.L. and Wootton, S.A., 1999. Effect of fatty acid chain length and saturation on the gastrointestinal handling and metabolic disposal of dietary fatty acids in women. British Journal of Nutrition 81: 37-43.

Kalivianakis, M., Verkade, H.J., Stellaard, F., Van Der Werf, M., Elzinga, H. and Vonk, R.J., 1997. The ^{13}C-mixed triglyceride breath test in healthy adults: determinants of the $^{13}CO_2$ response. European Journal of Clinical Investigation 27: 434-442.

Klein, S. and Wolfe, R.R., 1987. The use of isotopic tracers in studying lipid metabolism in human subjects. Bailliere's Clinical Endocrinology and Metabolism 1: 797-816.

Koletzko, B., Demmelmair, H., Hartl, W., Kindermann, A., Koletzko, S., Sauerwald, T. and Szitanyi, P., 1998. The use of stable isotope techniques for nutritional and metabolic research in paediatrics. Early Human Development 53: S77-S97.

Kriengsinyos, W., Wykes, L.J., Ball, R.O. and Pencharz, P.B., 2002. Oral and intravenous tracer protocols of the indicator amino acid oxidation method provide the same estimate of the lysine requirement in healthy men. Journal of Nutrition 132: 2251-2257.

Krogh, A. and Lindhard, J., 1920. The relative value of fat and carbohydrate as sources of muscular energy. Biochemical Journal 14: 290-363.

Labayen, I., Díez, N., Parra, D., González, A. and Martínez, J.A., 2004. Basal and postprandial substrate oxidation rates in obese women receiving two test meals with different protein content. Clinical Nutrition 23: 571-578.

Leijssen, D.P.C. and Elia, M., 1996. Recovery of $^{13}CO_2$ and $^{14}CO_2$ in human bicarbonate studies: a critical review with original data. Clinical Science 91: 665-677.

Leijssen, D.P.C., Saris, W.H.M., Jeukendrup, A.E. and Wagenmakers, A.J.M., 1995. Oxidation of exogenous [^{13}C]galactose and [^{13}C]glucose during exercise. Journal of Applied Physiology 79: 720-725.

Lifschitz, C.H., Boutton, T.W., Carrazza, F., Beyreiss, K., Schmitz, J., Ricour, C., Shulman, R. and Nichols, B.L., 1988. A carbon-13 breath test to characterize glucose absorption and utilization in children. Journal of Pediatric Gastroenterology and Nutrition 7: 842-847.

Livesey, G. and Elia, M., 1988. Estimation of energy expenditure, net carbohydrate utilization, and net fat oxidation and synthesis by indirect calorimetry: evaluation of errors with special reference to the detailed composition of fuels. American Journal of Clinical Nutrition 47: 608-628.

Matthews, D.E. and Downey, R.S., 1984. Measurement of urea kinetics in humans: a validation of stable isotope tracer methods. American Journal of Physiology 246: E519-E527.

McArdle, W.D., Katch, F.I. and Katch, L., 2007. Exercise Physiology; energy, nutrition & human performance. Lippincott Williams & Wilkins, New York, NY, USA.

McCloy, U., Ryan, M.A., Pencharz, P.B., Ross, R.J. and Cunnane, S.C., 2004. A comparison of the metabolism of eighteen-carbon 13C-unsaturated fatty acids in healthy women. Journal of Lipid Research 45: 474-485.

Meier-Augenstein, W., 1999. Use of gas chromatography-combustion-isotope ratio mass spectrometry in nutrition and metabolic research. Current Opinion in Clinical Nutrition and Metabolic Care 2: 465-470.

Metges, C.C. and Wolfram, G., 1991. Medium- and long-chain triglycerides labeled with ^{13}C: A comparison of oxidation after oral or parental administration in humans. Journal of Nutrition 121: 31-36.

Metges, C.C., El-Khoury, A.E., Selvaraj, A.B., Tsay, R.H., Atkinson, A., Regan, M.M., Bequette, B.J. and Young, V.R., 2000. Kinetics of L-1-^{13}C-Leucine when ingested with free amino acids, unlabeled or intrinsically labelled casein. American Journal of Physiology 278: E1000-E1009.

Metges, C.C., Schmidt, H.L. and Eichinger, H., 1992. Use of naturally enriched mixed food in 13C breath tests applied in young sucking calves. British Journal of Nutrition 67: 43-55.

Millward, D.J., Fereday, A., Gibson, N.R., Cox, M. and Pacy, P.J., 2002. Efficiency of utilization of wheat and milk protein in healthy adults and apparent lysine requirements determined by a single-meal [1-^{13}C]leucine balance protocol. American Journal of Clinical Nutrition 76: 1326-1334.

Millward, D.J., Fereday, A., Gibson, N.R. and Pacy, P.J., 2000. Human adult amino acid requirements: 1-13C-leucine balance evaluation of the efficiency of utilization and apparent requirements for wheat protein and lysine compared with those for milk protein in healthy adults. American Journal of Clinical Nutrition 72: 112-121.

Moehn, S., Bertolo, R.F., Pencharz, P.B. and Ball, R.O., 2005. Development of the indicator amino acid oxidation technique to determine the availability of amino acids from dietary protein in pigs. Journal of Nutrition 135: 2866-2870.

Moehn, S., Martinazzo-Dallagnol, É., Bertolo, R.F., Pencharz, P.B. and Ball, R.O., 2007. Metabolic availability of lysine in feedstuffs determined using oral isotope delivery. Livestock Science 109: 24-26.

Moseley, L., Jentjens, R.L.P.G., Waring, R.H., Harris, R.M., Harding, L.K. and Jeukendrup, A.E., 2005. Measurement of exogenous carbohydrate oxidation: a comparison of [U-^{14}C]glucose and [U-^{13}C] glucose tracers. American Journal of Physiology 289: 206-211.

Murphy, J.L., Jones, A., Brookes, S. and Wootton, S.A., 1995. The gastrointestinal handling and metabolism of (1-13C)palmitic acid in healthy women. Lipids 30: 291-298.

Owen, O.E., Kalhan, S.C. and Hanson, R.W., 2002. The key role of anaplerosis and cataplerosis for citric acid cycle function. Journal of Biological Chemistry 277: 30409-30412.

Pellikaan, W., Verstegen, M., Tamminga, S., Dijkstra, J. and Hendriks, W., 2013. δ ^{13}C as a marker to study digesta passage kinetics in ruminants: a combined *in vivo* and *in vitro* study. Animal 7: 754-767.

Pencharz, P.B. and Ball, R.O., 2003. Different approaches to define individual amino acid requirements. Annual Reviews of Nutrition 23: 101-116.

Pennings, B., Groen, B.B., van Dijk, J.-W., de Lange, A., Kiskini, A., Kuklinski, M., Senden, J.M. and van Loon, L.J., 2013. Minced beef is more rapidly digested and absorbed than beef steak, resulting in greater postprandial protein retention in older men. American Journal of Clinical Nutrition 98: 121-128.

Pennings, B., Pellikaan, W.F., Senden, J.M., Van Vuuren, A.M., Sikkema, J. and Van Loon, L.J., 2011. The production of intrinsically labeled milk and meat protein is feasible and provides functional tools for human nutrition research. Journal of Dairy Science 94: 4366-4373.

Peronnet, F., Massicotte, D., Brisson, G. and Hillaire-Marcel, C., 1990. Use of ^{13}C substrates for metabolic studies in exercise: methodological considerations. Journal of Applied Physiology 69: 1047-1052.

Petzke, K.J. and Klaus, S., 2008. Reduced postprandial energy expenditure and increased exogenous fat oxidation in young woman after ingestion of test meals with a low protein content. Nutrition and Metabolism 5: 25.

Rating, D. and Langhans, C.D., 1997. Breath tests: concepts, applications and limitations. European Journal of Pediatrics 156: S18-S23.

Rennie, M.J., 1999. An introduction to the use of tracers in nutrition and metabolism. Proceedings of the Nutrition Society 58: 935-944.

Schadewaldt, P., Bodner, A., Brosicke, H., Hammen, H.W. and Wendel, U., 1998. Assessment of whole body L-leucine oxidation by noninvasive L-[1-^{13}C]leucine breath tests: a reappraisal in patients with maple syrup urine disease, obligate heterozygotes, and healthy subjects. Pediatric Research 43: 592-600.

Schadewaldt, P., Schommartz, B., Wienrich, G., Brösicke, H., Piolot, R. and Ziegler, D., 1997. Application of isotope-selective nondispersive infrared spectrometry (IRIS) for evaluation of [^{13}C]octanoic acid gastric-emptying breath tests: comparison with isotope ratio-mass spectrometry. Clinical Chemistry 43: 518-522.

Schierbeek, H., Van den Akker, C.H.P., Fay, L.B. and Van Goudoever, J.B., 2012. High-precision mass spectrometric analysis using stable isotopes in studies of children. Mass Spectrometry Reviews 31: 312-330.

Schmidt, D.E., Allred, J.B. and Kien, C.L., 1999. Fractional oxidation of chylomicron-derived oleate is greater than that of palmitate in healthy adults fed frequent small meals. Journal of Lipid Research 40: 2322-2332.

Schoeller, D.A., 2002. Uses of stable isotopes in the assessment of nutrient status and metabolism. Food and Nutrition Bulletin 23: 17-20.

Schoeller, D.A., Klein, P.D., Watkins, J.B., Heim, T. and MacLean, W.C., 1980. ^{13}C abundances of nutrients and the effect of variations in ^{13}C isotopic abundances of test meals formulated for $^{13}CO_2$ breath tests. American Journal of Clinical Nutrition 33: 2375-2385.

Schrauwen, P., Blaak, E.E., Van Aggel-Leijssen, D.P.C., Borghouts, L.B. and Wagenmakers, A.J.M., 2000. Determinants of the acetate correction factor: implications for estimation of [^{13}C]substrate oxidation. Clinical Science 98: 587-592.

Schutz, Y., 1997. On problems of calculating energy expenditure and substrate utilization from respiratory exchange data. Zeitschrift für Ernährungswissenschaft 36: 255-262.

Shulman, R.J., Wong, W.W., Irving, C.S., Nichols, B.L. and Klein, P.D., 1983. Utilization of dietary cereal by young infants. Journal of Pediatrics 103: 23-28.

Tappy, L. and Schneiter, P., 1997. Measurement of substrate oxidation in man. Diabetes and Metabolism 23: 435-442.

Van den Borne, J.J.G.C., Alferink, S.J.J., Heetkamp, M.J.W., Jacobs, A.A.A., Verstegen, M.W.A. and Gerrits, W.J.J., 2012. Asynchronous supply of indispensable amino acids reduces protein deposition in milk-fed calves. Journal of Nutrition 142: 2075-2082.

Van den Borne, J.J.G.C., Lobley, G.E., Verstegen, M.W.A., Muijlaert, J.-M., Alferink, S.J.J. and Gerrits, W.J.J., 2007. Body fat deposition does not originate from carbohydrates in milk-fed calves. Journal of Nutrition 137: 2234-2241.

Van Hall, G., 1999. Correction factors for ^{13}C-labelled substrate oxidation at whole body and muscle level. Proceedings of the Nutrition Society 58: 979-986.

Van Loon, L.J.C., Koopman, R., Schrauwen, P., Stegen, J. and Wagenmakers, A.J.M., 2003. The use of the [1,2-^{13}C]acetate recovery factor in metabolic research. European Journal of Applied Physiology 89: 377-383.

Vermunt, S.H.F., Mensink, R.P., Simonis, A.M.G. and Hornstra, G., 1999. Effects of age and dietary n-3 fatty acids on the metabolism of [^{13}C]-alpha-linolenic acid. Lipids 34 Suppl.: S127.

Vermunt, S.H.F., Mensink, R.P., Simonis, M.M.G., Wagenmakers, A.J.M. and Hornstra, G., 2001. The metabolism of linoleic acid in healthy subjects after intake of a single dose of ^{13}C-linoleic acid. European Journal of Clinical Nutrition 55: 321-326.

Vonk, R.J., Stellaard, F., Priebe, M.G., Koetse, H.A., Hagedoorn, R.E., De Bruijn, S., Elzinga, H., Lenoir-Wijnkoop, I. and Antoine, J.-M., 2001. The ^{13}C/^{2}H-glucose test for determination of small intestinal lactase activity. European Journal of Clinical Investigation 31: 226-233.

Wada, E., 2009. Stable δ^{15}N and δ^{13}C isotope ratios in aquatic ecosystems. Proceedings of the Japan Academy, Series B 85: 98-107.

Weaver, L.T., 1998. Stable isotope breath tests. Nutrition 14: 826-829.

Weir, J.B., 1949. New methods of calculating metabolic rate with special reference to protein metabolism. Journal of Physiology109: 1-9.

Wolfe, R.R. and Chinkes, D.L., 2005. Isotope tracers in metabolic research – Principles and practice of kinetic analysis. John Wiley and Sons, Hobroken, NJ, USA.

Wolfe, R.R. and Jahoor, F., 1990. Recovery of labeled CO_2 during the infusion of C-1 vs C-2-labeled acetate: implication for tracer studies of substrate oxidation. American Journal of Clinical Nutrition 51: 248-252.

Young, V.R. and Ajami, A.M., 1999. Isotopes in nutrition research. Proceedings of the Nutrition Society 58: 15-32.

Zello, G.A., Wykes, L.J., Ball, R.O. and Pencharz, P.B., 1995. Recent advances in methods of assessing dietary amino acid requirements for adult humans. Journal of Nutrition 125: 2907-2915.

Chapter 6: Indirect calorimetry for elucidating dynamics in energy metabolism of farm animals

*B. Kuhla**, *M. Derno and C.C. Metges*
Institute of Nutritional Physiology 'Oskar Kellner', Leibniz Institute for Farm Animal Biology (FBN), Wilhelm-Stahl-Allee 2, 18196 Dummerstorf, Germany; b.kuhla@fbn-dummerstorf.de

Abstract

Respiration chambers equipped with O_2, CO_2 and CH_4 gas analysers allow the short-term measurement of gas exchange as a prerequisite to study dynamics in energy metabolism of farm animals. Starting from those technical aspects and limitations of dynamic gas exchange recordings, this chapter covers the mathematical evaluation of data obtained from short-term gas exchange measurements, circadian variation of fat and carbohydrate oxidation and the dynamics of methane production in relation to feed intake. The final section of this chapter deals with the identification of specific components of energy expenditure from dynamic records of gas exchange such as the thermic effect of feeding, heat production due to physical activity, basal metabolic rate, resting metabolic rate, and fasting heat production.

Keywords: short-term energy expenditure, macronutrient oxidation, multi curve evaluation

6.1 Introduction

In the middle of the 20th century open-circuit respiration chambers for farm animals were used all over the world as a technical and experimental platform for evaluating the energetic value of feed for animal production. By using this technique, C balance was determined from 24-h measurements involving the analysis of C intake and C excretion into faeces, urine, CO_2 and CH_4. The additional measurement of N balance allowed the calculation of energy retention as fat and protein in the body. Decades of scientific work performed, for example at the 'Oskar Kellner' Institute in Rostock, Germany, at the University of California, Davis, USA, at the Institut National de la Recherche Agronomique (INRA), France, at Wageningen University, the Netherlands, and at many more places at which respiration chambers were used, has resulted in the establishment of Net Energy-based feed evaluation systems (Beyer *et al.*, 2003; NRC, 1981; Schiemann *et al.*, 1971; Van Es, 1978). The retention of energy in the body, e.g. deposition of body tissue, or in animal products such as eggs and milk, can be predicted from data provided by these systems. In the last third of the 20th century, the measurement of heat production (Q) by indirect calorimetry, and its relationship with thermoregulation and physical activity, gained more and more importance for the calculation of energy requirements of animals. Thus, feed energetics and animal energetics are two sides of the same coin which

DOI 10.3920/978-90-8686-261-0_6, © Wageningen Academic Publishers 2015

both rely on 24-h measurements of gas exchange in respiration chambers over a few days. To this end a cumulative air sample was obtained by continuously withdrawing air from the chamber over 24 h, well mixed and subsequently pumped through a gas analyser to measure gas concentration. Thus, only one measurement per chamber and day was available. Respiration chambers were technically inappropriate for short-term measurements that require more frequent sampling. However, determination of short-term responses of energy metabolism to e.g. feed intake, feed withdrawal, acute climatic change or oestrus is crucial to understand physiological mechanisms regulating feed intake, circadian rhythm, or the adaptation to challenges such as restricted energy supply. Respiration chambers later developed were equipped with gas analysers allowing the measurement of gas concentrations at shorter intervals. Short intervals between measurements, however, may lead to noisy data and can pose a problem when Q is calculated (Brown *et al.* 1984; McDonald *et al.* 1988; McLean and Watts, 1976). Van Milgen *et al.* (1997) proposed a method to circumvent this issue by modelling the observed concentration of gases by a series of differential equations rather than using direct calculations of Q. In the meantime further refinements towards continuous gas exchange measurement were performed. This combined with continuous recordings of feed and water intake, physical activity, body temperature and heart rate provides an opportunity for a comprehensive monitoring of variables related to energy metabolism. Additional frequent blood sampling to evaluate metabolic status in concert with e.g. continuous rumen pH monitoring while gas exchange is measured in the respiration chamber provides a large methodological repertoire to investigate physiological characteristics of farm animals *in vivo* (Derno *et al.*, 2009, 2013; Van Milgen *et al.*, 1997).

6.2 Technical aspects and limitations of dynamic gas exchange recordings

For short-term measurements, it is presumed that the animal in the chamber produces a certain volume of CO_2 and CH_4 and consumes a certain volume of O_2 in the same time interval as the data are recorded at the gas analysers. However, due to the large volume of the chamber, the default air flow rate and the distance between the chamber and the gas analyser, instantaneous changes in O_2 consumption or CO_2/CH_4 production by the animal in the chamber are detected at the gas analyser with some delay, thus slowing the response time after a unit step change (Brown *et al.*, 1984). When for example a single CO_2 bolus is injected into the chamber (e.g. reflecting the CO_2 increase in response to exhaling), the gas concentration (c_{gas}, l/m^3) detected at the analyser first increases in an exponential manner according to Equation 1a whereupon it decreases after time (*t*, h) according to Equation 1b,

$$c_{gas} = c_0 + a \times (1 - \exp(-t \times F/V) \quad (1a)$$

$$c_{gas} = c_0 + a \times \exp(-t \times F/V) \quad (1b)$$

where V is the volume of the chamber (m^3), F is the air flow rate (m^3/h), a is a chamber-specific coefficient (l/m^3). In Equation 1a, c_0 is the gas concentration in the chamber immediately after a virtual CO_2 bolus application, while in Equation 1b, c_0 is the gas concentration of the incoming air (l/m^3). The area under the curve formed by these two equations (Figure 1) equals the total CO_2 from the single injection event. Gas production by an animal can be considered as a series of many single injections forming a series of enrichment and decay curves (Figure 1). The sum of the gas concentrations determined from curve 1, 2, 3, etc. at time t is the concentration measured at the gas analyser (measuring concentration; Figure 1) reflecting the real gas exchange of the animal but with some lag time.

The length of the lag time depends primarily on the construction of the chamber, e.g. the distance between the chamber and the gas analyser but also on the flow rate. This delay needs to be taken into account when data from gas exchange measurements are compared or correlated with data recorded in real time (e.g. feed intake, physical activity, or heart rate). Technically, the time delay is determined by the tubing length between the chamber outlet and the gas analyser and the gas flow. However, the lag time between measurement at the analyser and production by the animal is not always constant. Some physiological factors of the animal may influence the time between gas production/consumption, its exhalation and its detection at the gas analyser. For example, CO_2 eructed from the rumen or exhaled from the lungs is in part inhaled, reabsorbed and metabolized in the carboanhydrase reaction (Section 6.3) by the host. Conditions influencing the carboanhydrase metabolism such as acidosis or alkalosis but also the reabsorption of CH_4 may affect the lag time described above.

Another issue of dynamic gas exchange recordings is the time between two consecutive measurement points (Figure 1). A long time interval between two measurements results in a poorer resolution and loss of dynamic information. Apart from predefining a short measurement interval, high resolution of gas concentration changes can only be achieved in an 'enriched atmosphere', which will establish when an animal is brought into the chamber several hours before data recording. It is also important to avoid or at least to minimise openings of the respiration chamber to the outer environment. The latter issue can be minimised by using an airlock, rinsed by chamber air. This airlock serves for short-term entry into the chamber for milking, feeding, cleaning, etc. (Derno *et al.*, 2009).

Data originating from short-term gas exchange measurements (minutes to hours) can be used for the calculation of short-term energy expenditure or substrate oxidation (Section 6.3) because CO_2 and CH_4 production and O_2 consumption of an animal are continuously occurring. In contrast, nitrogen excretion via urine and faeces is rather a long-term (several hours), discontinuous process and the concentration of nitrogen in urine may vary over time to a large extent. Therefore, the measurement of short term protein oxidation processes is not possible despite their contribution to short-term energy expenditure. Anyway, the contribution of protein oxidation as compared with the contribution from carbohydrates and fat to total energy expenditure is usually small and accounts only for up to 10% in ruminants and pigs (Simonson and DeFronzo, 1990).

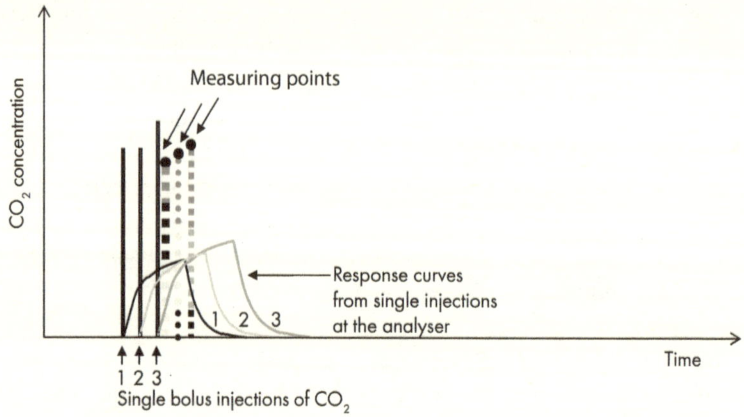

Figure 1. Relationship between single bolus injection of CO_2 into the chamber and the CO_2 concentration detected at the analyser. A CO_2 bolus injection is detected by the CO_2 analyser as a temporary increase followed by a decay both forming a response curve. The sum of the CO_2 concentrations determined from curve 1, 2, 3 ... etc. at time t is the concentration measured at the gas analyser (measuring point). The delay between the single injection 1 and measuring point 1 is the lag time.

6.3 Mathematical data evaluation

The total CO_2 production (VCO_2) measured from ruminants originates from two sites; CO_2 which is produced during microbial fermentation primarily in the rumen and the hindgut, and CO_2 which is produced during oxidative degradation of macronutrients in the intermediary metabolism of the host. Separate detection of fermentative and metabolic CO_2 is technically not feasible in a respiration chamber. Even a separate detection of CO_2 from the rumen and CO_2 from the lungs would not allow a clear conclusion on the origin of fermentative or metabolic CO_2 because of the carboanhydrase reaction in the rumen wall. Carboanhydrase catalyses the rapid conversion of CO_2 and water to bicarbonate and protons or *vice versa*, and thus enables the exchange of either CO_2 or bicarbonate between the host and the rumen. Nevertheless, CO_2 produced from fermentation can be calculated from the stoichiometric equations expressing fermentation of carbohydrates (Wolin, 1960) or estimated from *in vitro* experiments. Blümmel *et al.* (1999) found that *in vitro* fermentation of a variety of diets with different compositions results in a surprisingly constant ratio between CO_2 from fermentation and CH_4 of 1.4 to 1.7. Based on these *in vitro* studies, Chwalibog *et al.* (1996) suggested that for commonly used diets for lactating cows, the ratio between CO_2 from fermentation and CH_4 is close to 1.7 and therefore, fermentative CO_2 production ($VCO_{2\ ferm}$) can be estimated as follows:

$$VCO_{2\ ferm}\ (l) = 1.7 \times VCH_4\ (l) \qquad (2)$$

Accordingly, metabolic CO_2 production ($VCO_{2\ metabol}$) can be calculated from the total CO_2 production (VCO_2) measured in the respiration chamber as:

$$VCO_{2\ metabol}\ (l) = VCO_2\ (l) - VCO_{2\ ferm}\ (l) \tag{3}$$

The differentiation between VCO_2 and $VCO_{2\ metabol}$ thus allows the calculation of parameters related to metabolism of the host, e.g. the respiratory quotient $RQ_{\ metabol}$:

$$RQ_{\ metabol} = \frac{VCO_{2\ metabol}\ (l)}{VO_2\ (l)} \tag{4}$$

which differs from the respiratory quotient calculated from total CO_2 production measured in the chamber:

$$RQ = \frac{VCO_2\ (l)}{VO_2\ (l)} \tag{5}$$

Total oxygen consumption (VO_2) and $VCO_{2\ metab}$ can be used to calculate the net 'disappearance rate' of macronutrient substrates (mostly carbohydrates and lipids) from their respective metabolic pools according to Simonson and DeFronzo (1990). A ruminant diet is primarily composed of carbohydrates. Carbohydrate substrates are mostly converted to intermediates such as volatile fatty acids (VFA) and lactate in the rumen, but gas exchange measurements in a respiration chamber detect only the net loss of carbohydrates by oxidation (COX_{net}) to CO_2 and H_2O, regardless of any formation of intermediates, exchanges or cycling in metabolic pathways (Ferrannini, 1992). These aspects are described in details in Chapter 1 (Gerrits *et al.*, 2015) and Chapter 5 (Van den Borne *et al.*, 2015).

The CO_2 and CH_4 production and O_2 consumption of an animal in a respiration chamber is measured typically at successive time points (t=1, 2, 3, ...) equally spaced at uniform time intervals (usually several minutes) generating a time series of data points. Due to the natural one-way ordering of time, time series data reflect a dynamic, in case of a 24 h-measurement, circadian changes of CO_2 and CH_4 production and O_2 consumption, respectively. When, for example, the Brouwer equation (1965) is applied to each gas production data point, a time series with dynamic changes of metabolic Q will be obtained. Methods for evaluating the time-domain may be divided into two classes: the analysis of a single time series and the analysis of two independent time series. For the evaluation of a single time series, interpolation between discrete measured data points may be achieved by curve fitting or non-linear regression analysis. Although regression analysis focuses more on statistical questions such as how much uncertainty is present in a curve that is fitted to data observed with random errors, it also provides a mathematical function (Figure 2).

A temporal increase or decay in a time series of gas data points can be best regressed with an exponential function which provides various coefficients depending on the number of parameters chosen for the exponential regression. If a 3-parameter exponential function is applied for regression, such as shown in Figure 2, the coefficients y_0, a and b provide information on the kinetics (Figure 12A and B) by which an animal adapts to a certain

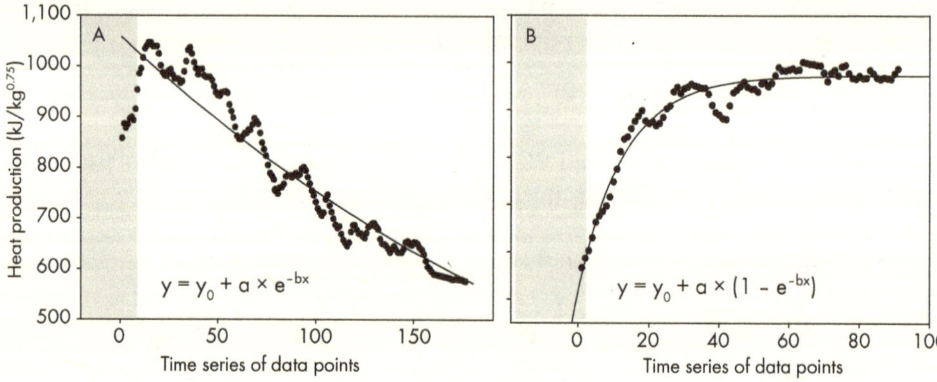

Figure 2. (A) Time series data points of metabolic heat production (HP, kJ/kg$^{0.75}$) of an individual dairy cow during feeding (grey shaded box) and in response to feed deprivation. The decay of heat production after feed deprivation was regressed using the 3-parameter exponential function $y = y_0 + a \cdot e^{-bx}$ (solid line). In this case, y_0 indicates fasting heat production (FQ=HP$_{0,}$ kJ/kg$^{0.75}$). (B) Time series data points of metabolic heat production of a dairy cow in response to feed consumption after fasting (grey shaded box). The rise of heat production after feed consumption was regressed using the 3-parameter exponential function $y = y_0 + a \cdot (1\text{-}e^{(-bx)})$ (solid line). In this case, y_0 indicates thermic effect of feeding (TEF=HP$_{0,}$ kJ/kg$^{0.75}$). The coefficients a equals to HP (kJ/kg$^{0.75}$) at the beginning of fasting (a) or at the beginning of feeding after fasting (B) at t = 0, respectively. The coefficient b (min^{-1}, or d^{-1}) is an animal individual time constant.

threshold (y_0). For example, fasting heat production during feed deprivation or the thermic effect of feeding after fasting can be determined as a threshold y_0 = HP_0 (kJ/kg$^{0.75}$) using simple exponential regression (Figure 2). Furthermore, the coefficient a equals to HP (kJ/kg$^{0.75}$) at the beginning of feed deprivation (Figure 2A) or at the beginning of feeding after fasting (Figure 2B), respectively. The coefficient b (min^{-1}, or d^{-1}) is an animal individual time constant.

However, multiple feeding and non-feeding events lead to a response in HP, which often occurs in wave form. Those wave forms can be evaluated by curve fitting which constructs a mathematical function or curve with the best fit to a series of data points. The curve fit is mostly based on a polynomial or sine functions. However, large time series data require complex polynomial functions yielding numerous mathematical coefficients that are often hard to interpret in a physiological or biochemical sense. However, fitted curves assist in data visualization, simplification of curve progressing, or reduction of noise (Figure 3A).

For precise quantification of dynamic data originating from gas exchange measurements, calculation of the area under the curve (AUC) from measured data is the method of choice (Figure 3B). The calculation of AUC, curve fitting or regression analysis can be performed by various software programs such as SigmaPlot (SYSTAT Software Inc., Chicago, IL, USA), TableCurve 2D (SYSTAT Software), or SAS (SAS Institute Inc., Cary, NC, USA).

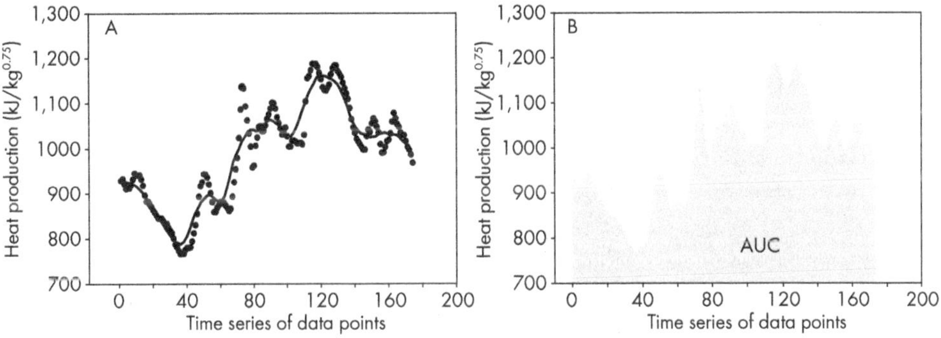

Figure 3. (A) Time series data points of metabolic heat production (HP, kJ/kg$^{0.75}$) of an individual dairy cow during ad libitum *feeding. The curve fitting involves a 'smooth' function (solid line) which approximately fits the time series data. (B) The area under the curve (AUC) shaded in grey for the time series data points displayed in (A).*

For the evaluation of the time-domain of two greater time series, cross correlation analysis may be applied when data resembles a wave form. Cross correlation is a measure of similarity between two waveforms as a function of a time-lag applied to one of the waves. It identifies the time-lag τ between two waves, which is valid for each data point of the wave but it can be best seen as the time difference between a maximum/minimum of one wave and a maximum/minimum of the second wave (Figure 4A and 4B). Cross correlation analysis does not require the same resolution in both time series data sets (Figure 4A). However, low resolution in a

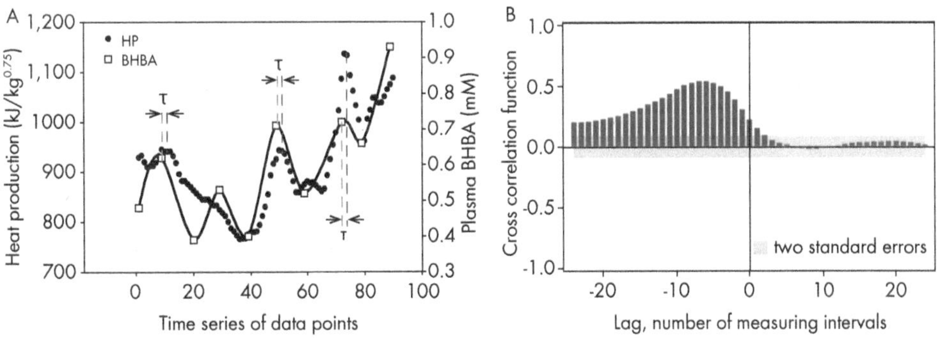

Figure 4. (A) Time series data points of metabolic heat production (HP, kJ/kg$^{0.75}$) and time series data points of plasma beta-hydroxybutyrate (BHBA, mM) concentration of an individual dairy cow during ad libitum *feeding. The two time series data sets represent different time resolution. The time-lag τ (number of measuring intervals) identified from cross correlation analysis indicates the offset between HP and BHBA. (B) The cross correlation function (CCF) between the two waves identifies the number of measuring intervals and thus the time-lag τ at the maximum. One bar represents the measuring interval between two data points of a time series.*

time series bears not only the risk of missing the true maxima and minima of the wave, it may also results in multiple maxima in the cross correlation function (Figure 4B), small cross correlation coefficients and high time-lag values.

The calculation of cross correlation requires software programs such as Proc Timeseries of SAS or LabVIEW (Laboratory Virtual Instrument Engineering Workbench; National Instruments, Austin, TX, USA). A time-lag of $\tau = 0$ point to the (nearly) simultaneous changes of the two physiological parameters, while $\tau > 0$ indicates the time delay by which one parameter responses to changes of the second parameter. In this regard a clear cause-effect relationship can be deduced.

Simultaneous nonlinear multicurve analysis may also be performed as fitting results (see above) are subjected to a Monte Carlo analysis. Herman and Lee (2012) compared a steady-state reaction kinetic complemented with calorimetry data for rigorous assessment of confidence intervals and parameter correlations but this approach will not be further described herein.

6.4 Dynamics of fat and carbohydrate oxidation in relation to feed intake

If voluntary feed intake of farm animals is too low, then milk or meat production may be reduced, whereas if it is too high then excessive fat deposition may occur. Therefore, understanding the regulation of feed intake in farm animals is crucial for ensuring optimal health and performance. The concept of metabolic oxidation of ingested carbohydrates and fats as processes affecting subsequent feed and energy intake has been developed a couple of years ago for humans (Stubbs and O'Reilly, 2000). This concept proposes that among others metabolic processing of macronutrients induces a number of physiological changes, which affect neuroendocrine signalling pathways that connect peripheral signals to the feeding centres in the brain. Changes in metabolic oxidation of macronutrients appear to be an integral component of these peripheral signals (Stubbs and O'Reilly, 2000). Examination of the relationship between metabolic oxidation of macronutrients and feed intake in ruminants necessarily requires real-time observations of feeding behaviour and metabolic changes because ruminants are fed *ad libitum* and have feed available almost all the time. For experimental dissection of this relationship, indirect calorimetry together with on-line continuous measurement of feed intake has become an indispensable tool. In mid-lactating dairy cows, we found that every feed intake event is strongly associated with COX_{net} in such a way that a temporal minimum of COX_{net} ($COX_{net\ (min)}$) coincides with the beginning of feed intake, and that COX_{net} increases during feed intake (Derno *et al.*, 2013). The postprandial COX_{net} peak appears with a time-lag τ ranging between 30 and 40 min after each feed intake event, the period characterised by satiety. The time-lag τ provides information on the time necessary for microbial fermentation before VFA are absorbed and metabolized by the host. The amount of feed ingested per event is positively and linearly correlated with the postprandial COX_{net} rate, but this correlation varies between individual animals (R^2 = 0.3-0.9; Derno *et al.*,

2013). Thus, while carbohydrate ingestion and oxidation appear to correlate with subsequent satiety, a number of exogenous and endogenous factors, including the energy balance status of the animal may affect these (cross) correlations.

Between two consecutive feed intake events, the COX_{net} rate decreased after the COX_{net} maximum with an exponential rate $COX_{net} = COX_{net\ (min)} + a \cdot e^{-bx}$ as illustrated in Figure 2. When the COX_{net} rate approached $COX_{net\ (min)}$, the next feed intake is initiated and prandial plus postprandial COX_{net} started to increase according to $COX_{net} = COX_{net\ (min)} + a \cdot (1-e^{(-b \cdot x)})$. Based on the fact that carbohydrates are the main substrate for oxidation in a ruminant diet, the inverse correlation between metabolic carbohydrate oxidation and hunger suggest an important process underlying metabolic satiety signals.

A RQ value of about 1 is predictive of prevailing carbohydrate oxidation while a RQ value above 1 is related to fat deposition and RQ values close to 0.7 reflects prevailing fat oxidation. Accordingly, the cross-correlation between $RQ_{metabol}$ (Equation 4) and COX_{net} is equal to zero and thus the time-lag τ between feed intake and RQ is the same as between feed intake and COX_{net} (Figure 5).

Ad libitum fed mid-lactating dairy cows possess a negative net fat oxidation (FOX_{net}), indicating a net lipogenesis (Figure 6). The cross-correlation analysis between feed intake and

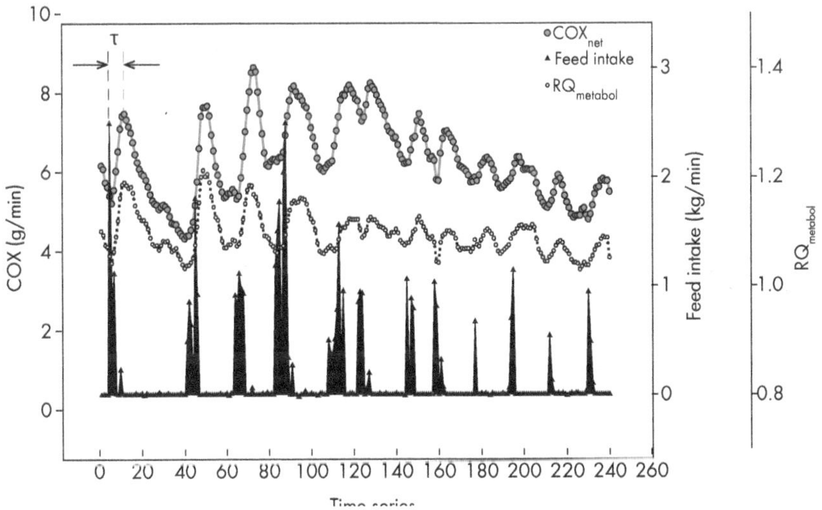

Figure 5. Association between feed intake (kg/min) and COX_{net} of an individual dairy cow during a 24 h period. The time-lag τ (number of measuring intervals) identified from cross correlation indicates the offset between feed intake and COX. The vertical dashed black and grey lines indicate the time t between the maximum of a feed intake event and the subsequent COX_{net} maximum. Due to the parallel course of COX_{net} and $RQ_{metabol}$, the time-lag between feed intake and $RQ_{metabol}$ is the same as determined for COX_{net} and feed intake.

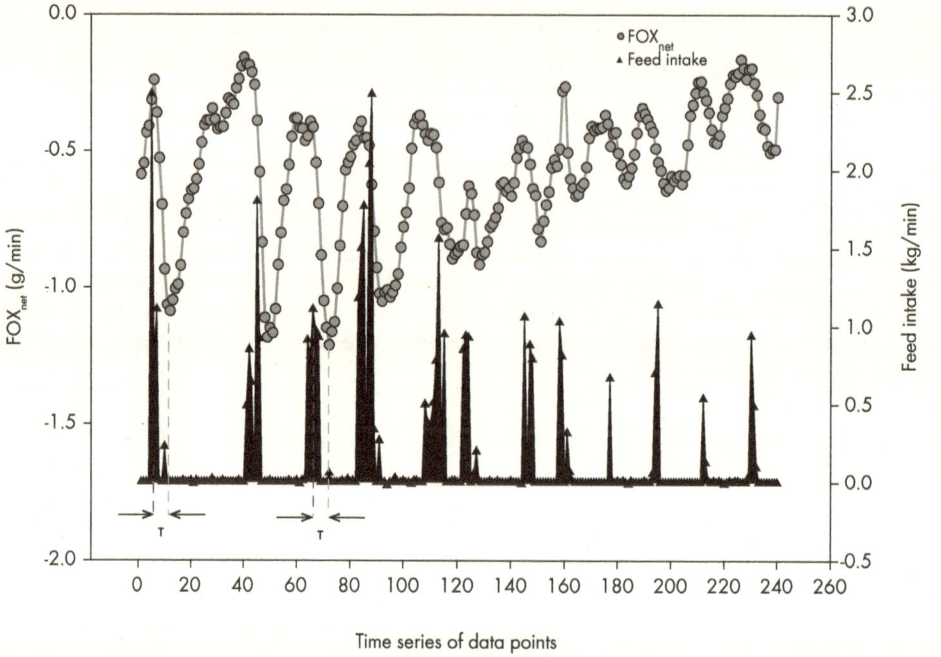

Figure 6. Association between feed intake (kg/min) and FOX_{net} (g/min) of an individual dairy cow during a 24 h period. The time-lag τ (number of measuring intervals) identified from cross correlation indicates the offset between feed intake and FOX. Herein, the time-lag between the maximum of a feed intake event and the subsequent FOX_{net} minimum is indicated by vertical dashed lines and arrows. Data for the feed intake time series originate from the same animal shown in Figure 5.

FOX_{net} revealed that each maximum of a feed intake event was followed by a FOX_{net} minimum with an offset of τ ranging between 30 and 42 min (Derno *et al.*, 2013). During between-feed intake intervals (satiety), FOX_{net} re-increased and obeys the exponential function $FOX_{net} = FOX_{net\ (min)} + a \cdot (1 - e^{(-b \cdot x)})$. When FOX_{net} reached a temporal maximum of -0.7 to -0.2 g/min, the next feed intake event started. Thus, it seems that cows in mid-lactation start to ingest feed in order to keep the FOX_{net} rate from becoming too positive. Langhans and Scharrer (1992) suggested that there may be postprandial elevation of hepatic fat oxidation which contributes to satiety. In line with this idea, the mitochondrial oxidation of various metabolites may lead to the formation of ATP which drives the ATP pump, increases the membrane potential and decreases the discharge rate in afferent nerve fibres projecting to the satiety centres in the brain (Allen and Piantoni, 2013; Langhans and Scharrer, 1992).

However, since ruminant diets usually contain not very much fat, the contribution of FOX_{net} to satiety of mid-lactating cows is certainly much lower than the effect of COX_{net} on satiety. However, whether this relationship holds true for early lactating cows with increased lipid mobilization and fat oxidation needs to be examined.

As it can easily be seen from Figures 5 and 6, where substrate oxidation is inferred from respiration exchange data, COX_{net} bears reciprocal relationship to FOX_{net}. Thus, early rising oxidation of carbohydrates followed by a later onset of increasing fat oxidation indicates that both COX_{net} and FOX_{net} interact to signal satiety. This also agrees to the idea that the minute to minute pattern of macronutrient oxidation rather than the amount of oxidation over longer periods of time (hours or days) – the latter is relatively constant – affect feed intake behaviour (Allen and Piantoni, 2013).

A further example how feed intake can affect fuel selection is the response of COX and FOX to feed restriction (FR). Growing pigs were subjected to FR over 3 weeks at a level of ~60% of the control group fed *ad libitum* (Krüger *et al.*, 2014). Whole body fat and carbohydrate oxidation were measured on day 4 of FR, and on days 4 and 25 of *ad libitum* re-feeding (Figure 7). After overnight feed withdrawal it took 20 min after initiation of meal consumption for FOX to

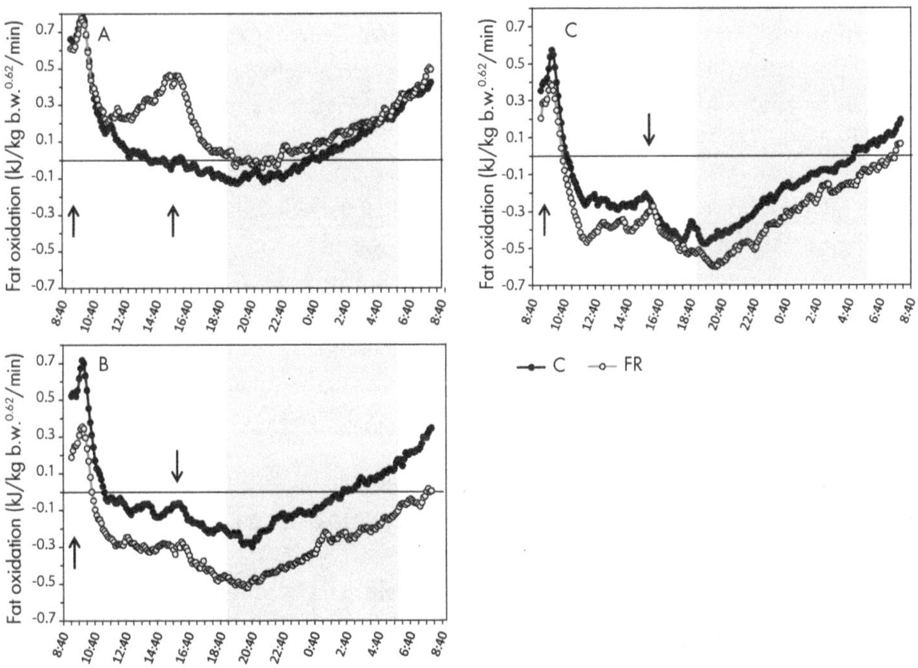

Figure 7. Diurnal course of fat oxidation in growing female pigs (kJ/kg body weight$^{0.62}$/min) at the age of 83 (panel A), 104 (panel B) and 125 d (panel C) either fed ad libitum *(Control, C; n=9-11) or feed restricted (FR; n=10-12) to ~60% of the* ad libitum *group between age 80 and 100 d. Average body weights for C and FR pigs were 32, 47.2, 67.6 kg, and 31.8, 43.1, 64.5 kg at 83, 104, 125 d of age, respectively. Negative values indicate lipogenesis. During measurement respiration chambers were kept at 22 °C and 65% relative humidity. For the sake of clarity error bars are omitted. Pigs were meal fed twice a day with 50% of the daily allowance at t=09:30 and 16:00 h (arrows). Feed was withheld from t=18:00 until 08:00 h. Light was off between t=19:00 and 06:00 h (shaded area).*

decline. In the control group, FOX kept decreasing after the meal until the next meal was consumed (6.5 h later). Shortly thereafter FOX became slightly negative (i.e. lipogenesis) and remained so until 23:00 h. In contrast, in FR pigs FOX started to re-increase at 2 h after the morning meal (9:30 h) to reach a 2.5-fold higher value until the next meal was consumed (16:00 h) whereupon FOX decreased again to reach zero fat oxidation 10-12 h after the morning meal. On the 4th day of re-feeding following 3 weeks of FR, FOX became negative already at 1 h after the morning meal and remained negative throughout the next 20 h of the measurement period (Figure 7b). The control group showed also negative FOX between 11:00 h and 02:00 h, but the decrease was less pronounced than for the FR group indicating less lipogenesis. After 25 days of re-feeding the response of FOX to feeding did not differ among groups. This finding suggests overcompensation of lipogenesis and replenishment of fat stores within the first few days after previous moderate-term energy deficiency, and increasingly higher lipogenesis (i.e. decreasing FOX) with progressing age in pigs.

In pigs carbohydrates are by far the largest dietary energy supplier while fat contents are only approximately 5% of DM. This is reflected in the contribution of COX to Q (Figure 8), which is generally higher than that of FOX (Figure 7). However, negative FOX (i.e. lipogenesis) was balanced by relatively higher COX, as can be seen in Figure 8 and previously reported by Chwalibog *et al.* (2004).

During FR, pigs showed a reduction of COX as a result of the reduced carbohydrate intake from feed (Figure 8a). On the 4th day of re-feeding COX was higher in previously feed-restricted animals than in control pigs. Higher COX involves higher production of pyruvate and thus acetyl CoA via the pyruvate dehydrogenase complex, thus providing building blocks for fatty acid synthesis, reflecting increased utilization of glucose for lipogenesis (Van Milgen, 2002). Accordingly, RQ of previously feed-restricted pigs was higher than in age-matched control pigs fed *ad libitum* throughout the experiment (1.07 vs 0.98; Krüger *et al.*, 2014).

6.5 Dynamics of methane production in relation to feed intake

The conversion of feed to VFA in the rumen is closely associated with methane production, both involving several microbial species. The final step of methane production is carried out by methanogenic archaea, which convert H_2 and CO_2 to CH_4. Therefore, methane production depends on the presence of H_2 and CO_2 which in turn are primarily determined by the amount of feed consumed by the host, the feed composition and the rate of feed fermentation. The more feed is ingested the more methane is usually produced, but the amount of methane per kg dry matter intake (DMI) decreases with increasing feed intake because the retention time of feed in the rumen is reduced and passage rate is increased (Jentsch *et al.*, 2007). Moreover, the fibre fraction, as compared to the starch or sugar fraction of feed, is fermented much slower, has a greater retention time in the rumen, and exerts greater impact on CH_4 production. These different characteristics of feed components do not much affect methane production when expressed as amount per day, but clearly influence methane production in the short-

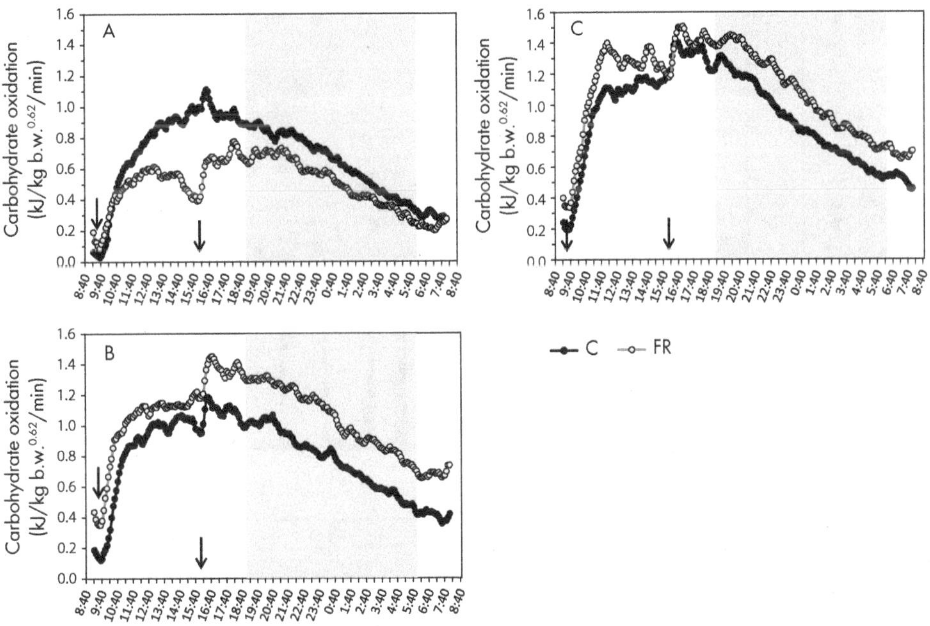

Figure 8. Diurnal course of carbohydrate oxidation in growing female pigs (kJ/kg body weight$^{0.62}$/ min) at the age of 83 (panel A), 104 (panel B) and 125 d (panel C) either fed ad libitum *(Control, C; n=9-11) or feed restricted (FR; n=10-12) to ~60% of the* ad libitum *group between age 80 and 100 d. Average body weights for C and FR pigs were 32, 47.2, 67.6 kg, and 31.8, 43.1, 64.5 kg at 83, 104, 125 d of age, respectively. During measurement respiration chambers were kept at 22 °C and 65% relative humidity. For the sake of clarity error bars are omitted. Pigs were meal fed twice a day with 50% of the daily allowance at t=09:30 and 16:00 h (arrows). Feed was withheld from t=18:00 until 08:00 h. Light was off between t=19:00 and 06:00 h (shaded area).*

term (postprandially), particularly when feed components such as concentrate and forage are fed separately and not as a total mixed ration. Depending on the feed intake behaviour of the animal, prandial and postprandial methane production may not only be influenced by the corresponding feed intake event (Figure 9) but also by the time and the amount of feed ingested far before methane measurement.

These data argue against those approaches that attempt to estimate average daily methane production from measurements lasting for only one hour or less (Robinson *et al.*, 2014). Although the time-lag between a feed intake event and the peak in postprandial methane production is relatively constant for a dairy cow fed total mixed ration (Figure 9), it may highly vary between individuals.

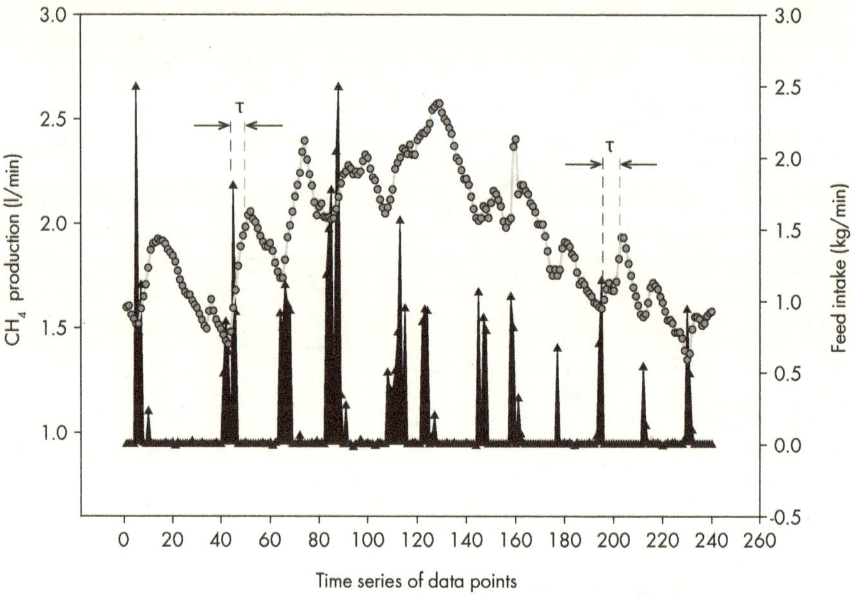

Figure 9. Association between feed intake (kg/min) and methane production (l/min) during a 24 h period. The time-lag τ (number of measuring intervals) identified from cross correlation indicates the offset between feed intake and methane production, which herein, is illustrated as the time between the maximum of a feed intake event and the subsequent methane maximum. Data for the feed intake time series originate from the same animal shown in Figure 5 and 6

6.6 Identifying components of energy expenditure from dynamic records

As shown in Figure 3, Q is not constant and may vary considerably within a day. The thermic effect of feeding (TEF) and heat production due to physical activity (AEE) are thought be the major factors affecting the daily dynamics of total Q, while basal heat production is a constant term without distinguishable diurnal kinetics (Van Milgen *et al.*, 2000). The TEF following a single feed intake event may last for several hours and can be modelled by different exponential functions described herein (Figure 2B) or elsewhere (Reed and Hill, 1996; Van Milgen *et al.*, 1997; Chapter 4, Labussière *et al.*, 2015).

The basal metabolism and the closely related resting metabolism is by far the largest component of total energy expenditure, and is associated with maintaining basal life processes, body temperature and functioning of all organs. In humans, the basal metabolic rate (BMR) is typically determined in a lying and resting subject in a dark room upon waking after sleeping and after 12 hours overnight fast to ensure that the digestive system is almost inactive. However, these conditions cannot be realised in farm animals. The measurement of resting metabolic rate (RMR) requires less restricted conditions than BMR measurement, because the animal does not need to spend the night sleeping in the respiration chamber. However, the animal has to be in a complete rest, in a post-absorptive state, and kept at thermo-neutral environmental temperature to exclude the contribution of energy expenditure due to physical activity, feed intake, and absorption of nutrients. In practice, RMR can be estimated in a post-absorptive state, i.e. after several hours of feed withdrawal in recumbent position, as measured automatically by a photoelectric switch (Derno *et al.*, 2009). For example, averaging the five lowest Q values during longer periods of lying (i.e. >1 h) in the nocturnal period (Figure 10), RMR can be estimated (Derno *et al.*, 2005).

Temporary feed restriction results in a decreased RMR as shown in pigs feed restricted at ~60% of *ad libitum* intake (Krüger *et al.*, 2014; Figure 11). RMR was derived from the 10 lowest HP values over the day and were averaged and normalized to 24 h. The decrease is considered a mechanism to survive deficient energy supply, and it is known that post-absorptive Q directly correlates to the mass of viscera (Koong *et al.*, 1982, 1985). We found that RMR was reduced only after 17 days of FR indicating that it required about 2 weeks for the visceral organ mass to adjust to the lower feeding level.

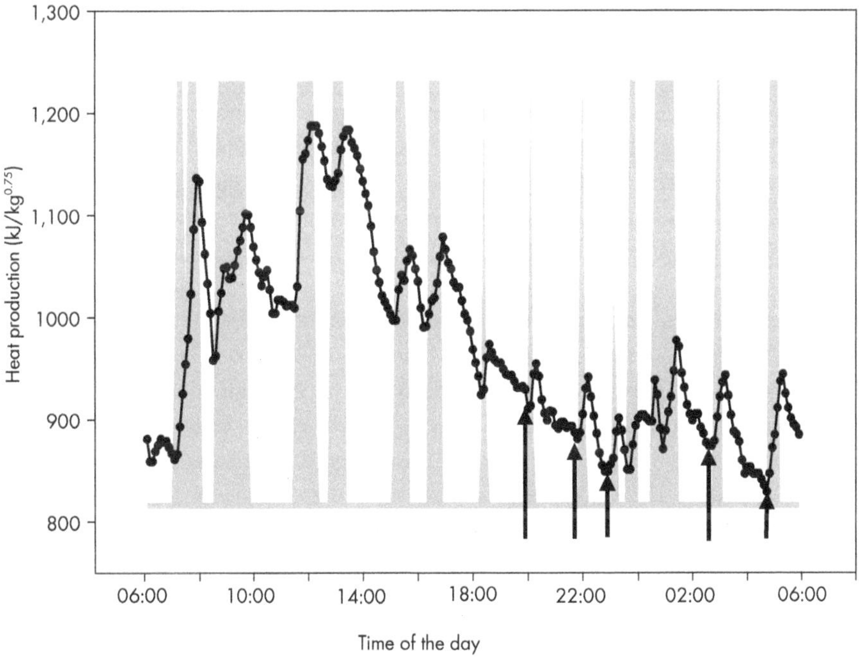

Figure 10. Diurnal heat production (HP, kJ/kg$^{0.75}$) of an individual dairy cow. Grey bars illustrate the standing position (h). Black arrows indicate four lowest heat production values during lying in the nocturnal period (dark period between 19:00 and 06:00). Means of theses lowest HP values yield the resting metabolic rate (RMR, kJ/kg$^{0.75}$).

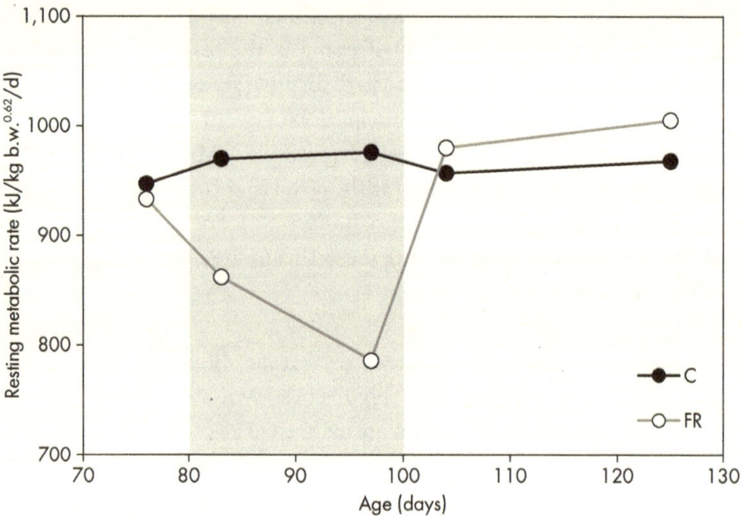

Figure 11. Resting metabolic rate (kJ/kg body weight$^{0.62}$/d) in 9-12 growing female pigs per group and point in time at the age of 76, 83, 97, 104 and 125 days either fed ad libitum *(Control, C) or feed restricted (FR) to ~60% of the* ad libitum *group between age 80 and 100 days (shaded area). Data from Krüger* et al. *(2014).*

Apart from a photoelectric switch that determines only standing and lying positions, physical activity can also be measured by a motion sensor or a force sensor, which are both very sensitive for recording any additional movement including ear shaking and panting. From these recordings AEE can be estimated but with a relatively large variation depending on the detector used (Van Milgen *et al.*, 2000).

In many respiration chamber protocols, animals are fasted at the end of the *ad libitum* period (Derno *et al.*, 2013; Van Milgen *et al.*, 1997) for 24 h or longer. Heat production usually declines with progressing time reaching a plateau (Figure 2A) which estimates fasting heat production (FQ) by regression analysis (Figure 2). However, determination of FQ may not only depend on the length of the fasting period but also on the amount of feed that the animal ingested before fasting. The FQ represents a catabolic state whereas BMR is determined in an anabolic state (see above). The long-term thermic effect of feeding (TEF_{LT}) was defined as the difference between RMR and FQ and reflects the Q during long-term anabolic relative to catabolic processes (Van Milgen *et al.*, 2000).

In conclusion, continuous measurements of O_2, CO_2 and CH_4 concentrations combined with advanced feeding and fasting protocols are important technical tools allowing the study of dynamic energy metabolism in farm animals. Besides simple regression analyses, more complex mathematical models such as cross-correlation analysis provide useful mathematical tools for evaluating time series data. Technical and mathematical tools are (1) a prerequisite to identify different components of heat production such as FQ, RMR, TEF, and AEE and contribute (2) to the understanding how dynamic changes in nutrient metabolism, specifically COX and FOX integrate into the control of short-term feed intake and satiety, and (3) how feed intake and postprandial methane production are related.

References

Allen, M.S. and Piantoni, P., 2013. Metabolic control of feed intake: implications for metabolic disease of fresh cows. Veterinary Clinics of North America: Food Animal Practice 29: 279-297.

Beyer, M., Chudy, A., Hoffmann, L., Jentsch, W., Laube, W., Nehring, K. and Schiemann, R., 2003. Rostock feed evaluation system. Plexus verlag, Amorbach, Germany.

Blümmel, M., Aiple, K.P., Steingaβ, H. and Becker, K., 1999. A note on the stoichiometrical relationship of short chain fatty acid production and gas formation *in vitro* in feedstuffs of widely differing quality. Journal of Animal Physiology and Animal Nutrition 81: 157-167.

Brouwer, E., 1965. Report of sub-committee on constants and factors. In: Blaxter, K.L. (ed.) Energy metabolism of farm animals. Academic Press, London, UK, pp. 441-443.

Brown, D., Cole, T.J., Dauncey, M.J., Marrs, R.W. and Murgatroyd, P.R., 1984. Analysis of gaseous exchange in open-circuit indirect calorimetry. Medical and Biological Engineering and Computing 22: 333-338.

Chwalibog, A., Jensen, K. and Thorbek, G., 1996. Oxidation of nutrients in bull calves treated with β-adrenergic agonists. Archives of Animal Nutrition 49: 255-261.

Chwalibog, A., Tauson, A.-H. and Thorbek, G. 2004. Diurnal rhythm in heat production and oxidation of carbohydrate and fat in pigs during feeding, starvation and re-feeding. Journal of Animal Physiology and Animal Nutrition 88: 266-274.

Derno, M., Elsner, H.G., Paetow, E.A., Scholze, H. and Schweigel, M., 2009. Technical note: a new facility for continuous respiration measurements in lactating cows. Journal of Dairy Science 92: 2804-2808.

Derno, M., Jentsch, W., Schweigel, M., Kuhla, S., Metges, C.C. and Matthes, H.D., 2005. Measurements of heat production for estimation of maintenance energy requirements of Hereford steers. Journal of Animal Science 83: 2590-2597.

Derno, M., Nürnberg, G., Schön, P., Schwarm, A., Röntgen, M., Hammon, H.M., Metges, C.C., Bruckmaier, R.M. and Kuhla, B., 2013. Short-term feed intake is regulated by macronutrient oxidation in lactating Holstein cows. Journal of Dairy Science 96: 971-980.

Ferrannini, E., 1992. Equations and assumptions of indirect calorimetry: some special problems. In: Kinney, J.M. and Tucker, H.N. (eds.) Energy metabolism: tissue determinants and cellular corollaries. Raven Press, Ltd., New York, NY, USA, pp. 1-17.

Gerrits, W.J.J., Van den Borne, J.J.G.C. and Labussière, E., 2015. Deriving heat production from gaseous exchange: validity of the approach. In: Gerrits, W.J.J. and Labussière, E. (eds.) Indirect calorimetry. Techniques, computations and applications. Wageningen Academic Publishers, Wageningen, the Netherlands, pp. 19-34.

Herman, P. and Lee, J.C., 2012. The advantage of global fitting of data involving complex linked reactions. Methods in Molecular Biology 796: 399-421.

Jentsch, W., Schweigel, M., Weissbach, F., Scholze, H., Pitroff, W. and Derno, M., 2007. Methane production in cattle calculated by the nutrient composition of the diet. Archives of Animal Nutrition 61: 10-19.

Koong, L.J., Ferrell, C.L. and Nienaber, J.A., 1985. Assessment of interrelationships among levels of intake and production, organ size and fasting heat production in growing animals. Journal of Nutrition 115: 1383-1390.

Koong, L.J., Nienaber, J.A., Pekas, J.C. and Yen, J.T., 1982. Effects of plane of nutrition on organ size and fasting heat production in pigs. Journal of Nutrition 112: 1638-1642.

Krüger, R., Derno, M., Goers, S., Metzler-Zebeli, B.U., Nuernberg, G., Martens, K., Pfuhl, R., Nebendahl, C., Zeyner, A., Hammon, H.M. and Metges, C.C., 2014. Higher body fatness in intrauterine growth retarded juvenile pigs is associated with lower fat and higher carbohydrate oxidation during *ad libitum* and restricted feeding. European Journal of Nutrition 53: 583-597.

Labussière, E., Dubois, S., Van Milgen, J. and Noblet, J., 2015. Modelling gas exchanges to partition heat production between fasting, thermic effect of feeding and physical activity in farm animals. In: Gerrits, W.J.J. and Labussière, E. (eds.) Indirect calorimetry. Techniques, computations and applications. Wageningen Academic Publishers, Wageningen, the Netherlands, pp. 73-86.

Langhans, W. and Scharrer, E., 1992. Metabolic control of eating. World Review of Nutrition and Dietetics 70: 1-67.

McDonald, T.P., Jones, D.D. Barrett, J.R., Albright, J.L., Miles, G.E., Nienaber, J.A. and Hahn, G.L., 1988. Measuring the heat increment of activity in growing-finishing swine. Transactions of the ASAE 31: 1180-1186.

McLean, J.A. and Watts, P.R., 1976. Analytical refinements in animal calorimetry. Journal of Applied Physiology 40: 827-831.

National Research Council (NRC), 1981. Nutritional energetics of domestic animals & glossary of energy terms. National Academic Press, Washington, DC, USA.

Reed, G.W. and Hill, J.O., 1996. Measuring the thermic effect of food. American Journal of Clinical Nutrition 63: 164-169.

Robinson, D.L., Goopy, J.P., Hegarty, R.S., Oddy, V.H., Thompson, A.N., Toovey, A.F., Macleay, C.A., Briegal, J.R., Woodgate, R.T., Donaldson, A.J. and Vercoe, P.E., 2014. Genetic and environmental variation in methane emissions of sheep at pasture. Journal of Animal Science 92: 4349-4363.

Schiemann, R., Nehring, K., Hoffmann, L., Jentsch, W. and Chudy, A., 1971. Energetische Futterbewertung und Energienormen. VEB Deutscher Landwirtschaftsverlag, Berlin, Germany.

Simonson, D.C. and DeFronzo, R.A., 1990. Indirect calorimetry – methodological and interpretative problems. American Journal of Physiology: Endocrinology and Metabolism 258: E399-E412.

Stubbs, R.J. and O'Reilly, L.M., 2000. Carbohydrate and fat metabolism, appetite, and feeding behavior in humans. In: Berthoud, H.R. and Seeley, R.J. (eds.) Neural and metabolic control of macronutrient intake. CRC Press LLC, Boca Raton, FL, USA, pp. 165-188.

Van den Borne, J.J.G.C., Heetkamp, M.J.W., Alferink, S.J.J. and Gerrits, W.J.J., 2015. Moving from a complete energy balance towards substrate oxidation: use of stable isotopes. In: Gerrits, W.J.J. and Labussière, E. (eds.) Indirect calorimetry. Techniques, computations and applications. Wageningen Academic Publishers, Wageningen, the Netherlands, pp. 87-114.

Van Es, A.J.H., 1978. Feed evaluation for ruminants. 1. The systems in use from May 1977 onwards in the Netherlands. Livestock Production Science 5: 331-345.

Van Milgen, J., 2002. Modeling biochemical aspects of energy metabolism in mammals. Journal of Nutrition 132: 3195-3202.

Van Milgen, J., Noblet, J., and Dubois, S., 2000. Dynamic aspects of gas exchange measurements. In: Chwalibog, A. and Jakobsen, K. (eds.) Energy metabolism in animals. Wageningen Press, Wageningen, the Netherlands, pp. 437-440.

Van Milgen, J., Noblet, J., Dubois, S. and Bernier, J.F., 1997. Dynamic aspects of oxygen consumption and carbon dioxide production in swine. British Journal of Nutrition 78: 397-410.

Wolin, M.J., 1960. A theoretical rumen fermentation balance. Journal of Dairy Science 43:1452-1459.

Chapter 7: Simultaneous measurement of energy expenditure and protein turnover in pigs

S. Moehn and R.O. Ball*
Swine Research and Technology Centre, Department of Agriculture, Food and Nutritional Science, University of Alberta, Edmonton, Alberta, T6G 2P5, Canada; soenke.moehn@ualberta.ca

Abstract

Energy and protein metabolism are closely linked and substantially influence each other. Therefore, the combined measurement of protein and energy metabolism is a logical approach. There are several methods available for measurement of energy expenditure and of protein turnover that usually can be applied simultaneously without affecting either measurement. For energy expenditure, indirect calorimetry is the most accurate and most commonly applied method, but other methods like the doubly labelled water technique or the bicarbonate recovery method have also been applied in conjunction with measurement of protein turnover. For determination of protein turnover, isotope dilution methods are generally used. The precursor method relies on the measurement of the rate of appearance of an amino acid to be metabolized, while the end product method measures the amount of tracer appearing in urea and/or ammonia. The flooding dose technique depends on the ratio of the isotope enrichment in tissues to that in plasma. Each of these methods has distinct advantages and disadvantages that include: the accuracy and precision of results, complexity and necessary skills for application, cost, and applicability in laboratory or field. Thus, the choice of method for energy and protein metabolic measurements depends firstly on the research questions, and secondly on the equipment, expertise and funds available. The preferred method for measurement of energy expenditure is indirect calorimetry, which should be combined with the end product method for determination of protein turnover. The best tracer for measurement of protein turnover is carboxyl-C labelled leucine, unless for the indicator amino acid oxidation technique, for which carboxyl-C labelled phenylalanine is preferred. If indirect calorimetry cannot be performed, the doubly labelled water technique is recommended. If determination of CO_2 production is not possible, the precursor method with ^{15}N labelled amino acids, usually glycine, is a viable alternative. However, different methods to determine energy expenditure or different tracers should be considered for research questions that make the recommended approach impractical or inappropriate. The combined measurement of protein turnover and energy expenditure is especially useful for studies in humans and large mammals where the recruitment of subjects may be difficult. The main applications have been determination of energy cost of protein synthesis under a variety of conditions. The research objectives were mainly the study of the effects of health and disease on protein and energy metabolism, assessment of intervention strategies and maximisation of production efficiency in animals. Because of the close linkage of protein and energy metabolism, their simultaneous study can provide additional insights either method alone cannot offer.

Keywords: energy expenditure, protein turnover, methods

DOI 10.3920/978-90-8686-261-0_7,

7.1 Introduction: the promise of simultaneous measurements

Protein and energy intake are key to maintenance and growth of animals and humans. Ingested protein is metabolised to synthesise the proteins essential for the functioning of an organism. Energy derived from ingested nutrients allows metabolic processes to proceed. Energy and protein metabolism are inextricably linked and substantially influence each other. Both protein synthesis and degradation are energy-dependent processes. Cellular uptake of amino acids (AA), RNA turnover and peptide bond synthesis, as well as the breakdown of protein and urea synthesis require energy (Young *et al.*, 1991), thus contributing to heat production. Conversely, energy intake can affect rates of protein synthesis (Reeds *et al.*, 1980; Sano *et al.*, 2010). However, it appears that the relationship between protein and energy metabolism is not fixed; it may be affected by the diurnal pattern of food intake (Garlick *et al.*, 1980a), body composition (Welle and Nair, 1990), health status (Berg *et al.*, 2013; Kao *et al.*, 2008) or lifestyle (Balagopal *et al.*, 2003; Pikosky *et al.*, 2008). Thus, prediction of the contribution of protein turnover to energy expenditure (Q) using calculated values, e.g. the energy cost of forming a peptide bond, may not be suitable *in vivo*.

Traditionally, traits of energy and protein metabolism have been determined separately, often because equipment was limited to one or the other area of expertise. However, because both are so closely linked, the simultaneous measurement of parameters of protein and energy metabolism can provide insights beyond those provided by the study of either area of research alone. Despite the advantages of combined measurements these have only rarely been taken advantage of for a variety of reasons including equipment requirements, analytical requirements, technical challenges and lack of expertise in interpretation of both energy and protein metabolism. The objective of this chapter is to provide the information necessary to develop and apply combined measurements of energy and protein metabolism. This review focuses on the simultaneous measurement of protein turnover and energy metabolism; for the measurement of Q alone see Chapter 1 (Gerrits *et al.*, 2015).

There are a range of methodologies available to determine protein turnover and Q that sometimes lend themselves to combined application without affecting the quality of results of either. For measurement of Q, indirect calorimetry is most often used, and is regarded as the optimal method to determine resting Q (Junghans and Chwalibog, 2001). Besides direct calorimetry (Forslund *et al.*, 1999) and serial slaughter (Möhn *et al.*, 2000), there are isotopic methods, namely doubly labelled water (Lifson and McClintock, 1966) and bicarbonate dilution technique (Elia, 1991) available to determine Q. Protein turnover, on the other hand, is predominantly measured by isotope dilution techniques that are based on either ^{15}N or stable (^{13}C) or radioactive (^{14}C) carbon labels. Hydrogen-labelled phenylalanine has also been used as a tracer in both constant infusion (Pikosky *et al.*, 2008) and flooding dose (Garlick *et al.*, 1980b) experiments. Muscle protein breakdown can be determined in some species using 3-methyl-histidine (Catzeflis *et al.*, 1985; Fitch and King, 1987), but this method will not be covered further.

Each of these methods has distinct advantages and drawbacks that make them more or less suitable for specific applications or the study of specific questions. In addition, each of the methods has specific equipment and technical requirements that sometimes coincide between energy and protein metabolism, so that both methods can be applied simultaneously. These methods have been reviewed and discussed in detail elsewhere and will only be briefly reviewed from the point of view of potential for combined measurement of protein and energy metabolism.

7.2 Methods to determine energy expenditure

Indirect calorimetry is the most commonly used method to determine Q and is regarded as the benchmark method for determination of Q. The advantages of indirect calorimetry are superior accuracy and precision of measurements and the ability to detect dynamic changes in respiratory rates (Junghans and Chwalibog, 2001) by making high-frequency measurements. Thus, indirect calorimetry lends itself to research questions that study rapidly changing respiratory rates, for example, diurnal changes, or responses to meals or exercise. Disadvantages of indirect calorimetry are the need for expensive equipment and the demand for considerable technical expertise in the conduct of experiments. Apart from the analytical instrumentation (e.g. O_2, CO_2 analysers, air flow measurement), respiration chambers or other methods of gas sampling are necessary to quantify gas exchange. This limits the use of indirect calorimetry to predominantly laboratory settings, and makes application in the field either impractical or difficult. However, indirect calorimetry has been used successfully in field studies in goats (Lachica and Aguilera, 2008) and oxen (Rometsch *et al.*, 1997). Measurement of CO_2 production is necessary to predict heat production as well as for the measurement of protein metabolism using C-labelled AAs (precursor method, see below). Therefore, the equipment needed for indirect calorimetry can be easily adapted to measure protein metabolism using the precursor method.

In contrast, it is less problematic to apply the doubly labelled water and bicarbonate dilution techniques in the field because no measurement of gas exchange is necessary and thus they lend themselves to application in grazing animals (Young, 1970). However, these methods yield only a single measurement per application so that dynamic measurements of Q are not possible. In addition, the accuracy and precision of these methods is generally less than for indirect calorimetry, especially for the bicarbonate dilution technique (Junghans and Chwalibog, 2001). When a dose of $NaH^{13}CO_3$ is applied to test subjects, the enrichment of the label in expired CO_2 is being reduced by (mostly) unlabelled CO_2 of metabolic origin. The CO_2 production is then calculated as presented in Equation 1:

$$VCO_2 = F[(\frac{Ei}{Eco_2}) - 1] \qquad (1)$$

Where F is the infusion rate of $NaH^{13}CO_3$ and Ei and Eco_2 are the ^{13}C enrichments (atom percent excess, APE) in the bicarbonate infusate and expired CO_2, respectively (Spear *et al.*,

1995). However, a part of the CO_2 released from an infusion of $NaH^{13}CO_3$ is retained in the body. Therefore, it is necessary to determine the rate of bicarbonate retention or recovery to quantify VCO_2. Determination of the bicarbonate recovery may be the major problem as it changes with physiological states and Q and may range from 50 to over 90% (El-Khouri *et al.*, 1994b; Moehn *et al.*, 2004), although recovery rates in growing animals typically range from 70 to 80% of a dose of labelled bicarbonate.

The rate of bicarbonate recovery also affects the quantification of AA oxidation when using C-labelled AAs. If Q is determined using the bicarbonate recovery method, measurement of protein turnover rates should be performed using ^{15}N-labelled AAs as applied by Li *et al.* (2008) in sheep. The use of C-labelled AAs would preclude the allocation of recovered label to bicarbonate or to AA oxidation, unless different isotopes (^{13}C and ^{14}C) are used in bicarbonate and labelled AA.

Other methods of determination of Q have been applied more rarely. Forslund *et al.* (1999) used direct calorimetry where human subjects wore a calorimetry suit (Hambraeus *et al.*, 1994) but comments that the stress of wearing the suit may have affected measurements of Q. Kita *et al.* (1993) combined the serial slaughter technique to determine energy retention and expenditure in 10-day old chicks with tracer dilution technique to determine protein turnover. Although the serial slaughter technique has been used in species other than poultry to determine energy and protein retention, we are not aware of incidences of combining it with measurements of protein turnover. A drawback of the serial slaughter technique is that it can only give an average estimate of Q over a prolonged period of time because of the tissue accretion needed to achieve accurate measurements. In case of the chicks, the serial slaughter was performed over a very short period of intense growth so that the Q was likely similar to that at the time of determination of protein turnover rates. This may not be applicable for larger animals where a minimum of 4 weeks measurement or doubling of initial slaughter weight is suggested to achieve reliable values for Q.

Calculation of Q from indirect calorimetry experiments is achieved using equations, in which coefficients derived from stoichiometry are applied to VO_2 and VCO_2, and possibly VCH_4 and urinary nitrogen excretion. A related approach is to calculate heat production based on the VO_2 and the respiratory quotient (Benedek *et al.*, 1995). For details, see Chapter 1 (Gerrits *et al.*, 2015).

Calculation of using isotopic methods uses a very different approach. The doubly labelled water technique is based on the premise that the oxygen in $^{2}H_2{}^{18}O$ equilibrates with body water and expired CO_2, while the hydrogen equilibrates with body water only. Because water and CO_2 are lost during metabolism, the elimination rates of labelled O_2 and H_2 differ after a applying a dose of $^{2}H_2{}^{18}O$. Therefore, the CO_2 production, which is related to energy expenditure, can be calculated from the different elimination rates of O_2 and H_2 from the body water space (Lifson and McClintock, 1966; Schoeller *et al.*, 1986).

Finally, Q can be calculated when using the serial slaughter technique as metabolisable energy intake minus energy retention over the entire experimental period. Energy retention, in turn is the animal (or carcass or body component of interest) energy content at the end minus that at the start of the experiment.

7.3 Methods to determine protein turnover

The most common methods to determine protein turnover are the end-product method with ^{15}N-labelled AAs or the precursor method with ^{13}C- or ^{14}C-labelled AAs as tracer. The use of ^{15}N-glycine was pioneered by Sprinson and Rittenberg (1949) using a single dose of tracer. Later, Picou and Taylor-Roberts (1969) modified the method for continuous infusion of the tracer by intravenous and intragastric routes. The principle of the method is in the dilution of the applied N-label in the excreted end products of the AA metabolism, urea and ammonia. The simplest calculation of results is based on a 2-pool model, i.e. body free AA pool and body protein pool, under assumption that the sum (flux) of AA exiting the pools (synthesis and excretion) is equal to the sum of AA entering the pools (intake, from body protein breakdown and de-novo synthesis) (Duggleby and Waterlow, 2005). Thus, once the crucial measurement of flux is known, synthesis can be calculated as flux minus the amount of end-product excreted (see below). Advantages of the end-product method are the relatively low cost of ^{15}N-glycine and the ease of urine sample collection, so that the precursor method lends itself to applications outside a laboratory setting. Furthermore, the collection of urine samples is independent of measurements of Q so that this method can be combined with any method to determine Q.

In contrast, the precursor method using C-labelled AAs relies on the determination of enrichment or radioactivity of the precursor (i.e. tracer AA) in plasma and of labelled expired CO_2 (Duggleby and Waterlow, 2005) to determine flux and oxidative disposal of the tracee, respectively. The need to quantify VCO_2 means that this method readily lends itself for combination with indirect calorimetry, whereas the need for plasma sampling is independent from the method used for measurement of Q. The use of L[1-^{14}C]leucine to measure protein synthesis *in vivo* was first described by Siekevitz and Palade (1958). The most 'generally useful' tracer (Reeds and Davis, 1999) for the precursor method is L-[1-^{13}C]leucine because L-[1-^{13}C] leucine is deaminated intracellularly to α-ketoisocaproic acid, which is released into the blood. It is assumed that the plasma enrichment of α-ketoisocaproic acid represents the intracellular enrichment of leucine. Thus, the precursor method using L-[1-^{13}C]leucine is regarded as the reference method (Wagenmakers, 1999) when the primary objective of an experiment is the determination of protein turnover rates. However, the precursor method can be used with a range of alternative tracers, depending on the primary objective of experiments. For example, in the indicator AA oxidation (IAAO) technique where the primary objective is determination of AA requirements (Pencharz and Ball, 2003), L[1-^{13}C]phenylalanine was used because of its tightly controlled pool size and rapid turnover rate (Levesque *et al.*, 2011), which assured rapid response to changing test AA intakes. Both ^{13}C-leucine (Forslund *et al.*, 1999) and

^{13}C-phenylalanine (Levesque *et al.*, 2011) have been used successfully in experiments with combined measurements of energy and protein metabolism.

The choice of tracer and its delivery mode requires careful consideration. Because the determination of protein turnover was the primary objective in the initial experiments, L-[1-^{13}C]leucine was chosen, which is regarded as the benchmark tracer for protein turnover studies (Wagenmakers, 1999). However, leucine is less suitable for IAAO studies because of its large and variable pool size. This makes leucine not sensitive when a different test AA, e.g. lysine, limits protein synthesis (Hsu *et al.*, 2006). Leucine also plays a significant role in the regulation of protein synthesis (Garlick and Grant, 1988) as well as hormonal secretion (Rocha *et al.*, 1972). Therefore, leucine may have a direct influence on the partitioning between protein synthesis and oxidation, the measurement parameter for IAAO. In contrast, phenylalanine has a comparatively small free pool size with a quick turn-over rate (Neale and Waterlow, 1974), and the size of the intracellular free phenylalanine pool is tightly regulated (Flaim *et al.*, 1982). Thus, phenylalanine oxidation was more responsive to dietary changes than lysine or leucine (Neale and Waterlow, 1974), which makes phenylalanine the preferred choice of tracer for IAAO studies.

The flooding dose technique relies on the incorporation of a large amount of tracee plus labelled tracer into body protein and provides rapid and accurate measurement of tissue protein synthesis over a short period of time (Davis and Reeds, 2001). Determination of flux rates using the flooding dose technique uses extrapolation of the decay curve of plasma enrichment, while protein synthesis relies on the ratio of label enrichment of protein-bound to plasma free tracer. Because of the short time needed for measurement, this method lends itself for determination of protein synthesis during rapidly changing states, e.g. in response to a meal or as affected by the actions of hormones (Davis and Reeds, 2001). It also is applicable to determine protein synthesis rates in very small animals, e.g. mice or young chicks (Aoyagi *et al.*, 1988, 1990).

For calculation of protein turnover, the simplest approach uses a 2-pool model (Figure 1). Parameters of protein turnover can then be calculated as presented in Equation 2 (Waterlow *et al.*, 1978):

$$\mathrm{Ra} = \mathrm{S} + \mathrm{O} = \mathrm{B} + \mathrm{I} \tag{2}$$

where the rate of appearance (Ra, sometimes also referred to as irreversible loss rate or flux) equals the sum of AA entering the pool, which at steady state is equal to the AA exiting the pool. Oxidation (O) and non-oxidative disposal (which is primarily protein synthesis, S) represent exits from the pool, and AA intake (I, including tracer amount) and AAs derived from body protein breakdown (B) and *de novo* synthesis, where applicable, represent the routes of AAs entering into the pool). Therefore, for both the precursor and end product methods, the first step in the calculation of results is the calculation of the rate of appearance of the tracee or flux. For constant infusion experiments, the rate of appearance is calculated at the isotopic steady state as in Equation 3:

$$Ra = I \times [(\frac{Ei}{Ep}) - 1] \quad (3)$$

where I is the rate of infusion of the tracer and Ei and Ep represent the enrichments (APE) of infusate and of plasma AA at steady state (Matthews *et al.*, 1980), The calculation is analogous when the ^{15}N tracer is used in constant infusion experiments. For the application of a single dose, the cumulative excretion of tracer during the entire urine collection period is used (Equation 4, Benedek *et al.*, 1995; Waterlow *et al.*, 1978):

$$Ra - E \times d/c \quad (4)$$

where E is the amount (g) of nitrogen excreted in urine as ammonia or urea, d is the dose of ^{15}N given (g), and e is the total amount (g) of isotope excreted in urine as ammonia or urea during the collection period. It should be noted that using urea or ammonia as end product usually gives different absolute rates of appearance, probably because isotopic tracers do not equilibrate equally in all organs or tissues (Duggleby and Waterlow, 2005), so that frequently the mean for both end products is reported. This compartmentalization is further illustrated by the difference in rates of appearance when different AAs are used as tracer (Duggleby and Waterlow, 2005).

To calculate the other key components using the 2-pool model, oxidation and tracee AA (or N) intake need to be known. For the precursor method, it is necessary to quantify the expired labelled CO_2 so that synthesis can be calculated as flux minus oxidation. This is only possible when using a single C-labelled tracer because VCO_2 cannot be attributed to each of multiple tracers with the same label. Similarly, tracee derived from body protein breakdown is calculated as flux minus tracee intake. For the end product method, calculations are analogous (Equation 5, Benedek *et al.*, 1995):

$$\text{N flux} = \text{N used for protein synthesis} + \text{urinary N excretion} = \text{N from protein breakdown} + \text{N intake in diet} \quad (5)$$

To conclude, there are a range of isotopic tracer methods available to measure protein turnover rates. Most of these methods can be combined with measurement of Q without causing technical or methodological conundrums. To determine protein turnover rates in a laboratory setting, the precursor method using 1-C labelled leucine appears to be the preferred approach, ideally combined with indirect calorimetry. When applying the indicator AA oxidation method, 1-C labelled phenylalanine should be used as a tracer, again combined with indirect calorimetry. If indirect calorimetry or CO_2 collection are not possible, due to lack of calorimetry equipment or working in the field, viable alternatives are the doubly labelled water technique to determine Q and the end-product method using ^{15}N-glycine to determine protein turnover rates. The choice of method to determine protein turnover and Q is, therefore, firstly governed by the research objectives, and secondly by equipment and expertise available to a researcher.

7.4 Application of combined measurement of protein and energy metabolism

In recent years, the simultaneous measurement of protein turnover and energy metabolism of the whole body was predominantly applied in humans. The major research objectives were to determine the energetics of protein turnover and its contribution to whole-body Q (Welle and Nair, 1990) and nutrient utilisation (Forslund *et al.*, 1999). Other questions were changes in the relationship of protein turnover to energy metabolism as affected by: age (Boirie *et al.*, 2001), health status (Berclaz *et al.*, 1996), exercise (Pikosky *et al.*, 2008) and nutrition (Stein *et al.*, 1991). Frequently, effectors were studied in combination to derive improved treatments of disease or better nutrition of people.

In human subjects, indirect calorimetry was the method of choice to determine Q, except for Forslund *et al.* (1999) who used both indirect and direct calorimetry. Generally, the measurements of Q were performed over short periods of time; only rarely were 24-h measurements used (Forslund *et al.*, 1999; Garlick *et al.*, 1980a), probably to limit the invasiveness of procedures.

Measurement of Q was combined with either the end-product method using ^{15}N-glycine or the precursor method using ^{13}C- or ^{14}C-leucine. With the end-product method, there appeared no preference to use single or repeated doses. In healthy adult humans, Stein *et al.* (1991) used the single dose technique, while Gougeon *et al.* (2008) applied multiple doses to determine protein synthesis during an isotopic plateau. Per kg body weight, there was little difference in resting Q or protein synthesis based on urea ^{15}N enrichment. Similarly, there was little difference in protein synthesis rates of healthy subjects that received either a single dose of ^{15}N-glycine (Vaisman *et al.*, 1992) or a 24-h primed constant infusion of ^{15}N-glycine (Fearon *et al.*, 1988).

If carbon-labelled AAs are used as the tracer to determine protein turnover rates, quantification of $^{13}CO_2$ or $^{14}CO_2$ production is necessary to calculate protein synthesis as turnover minus oxidation. Determination of VCO_2 during indirect calorimetry and the determination of $^{13}CO_2$ enrichment by isotope-ratio mass-spectroscopy (Di Buono *et al.*, 2001) or infrared absorption (Ohara *et al.*, 1998) require gas sampling. Gas sampling can be accomplished by using evacuated tubes, collection in Douglas bags (Berg *et al.*, 2013; Borel *et al.*, 1998) or using a carbon-free trapping agent, e.g. NaOH (Sano *et al.*, 2004). If ^{14}C-labelled AAs are used as a tracer, collection of CO_2 in hyamine (Winter *et al.*, 2005) or ethanolamine (Moehn *et al.*, 2004) is an option because additional non-labelled carbon does not affect scintillation counting. In addition, non-dispersive infrared analysers can be used for on-line and high-frequency measurement of $^{13}CO_2$ concentrations, or an on-line ionization chamber (MacRae *et al.*, 1988) may be used in conjunction with ^{14}C-labelled tracers.

In human studies, L-[1-^{13}C]leucine (Forslund *et al.*, 1999; Welle and Nair, 1990) or L-[1-^{14}C] leucine (Garlick *et al.*, 1980a; Giordano and Castellino, 1997) have been commonly used as tracers in primed-constant infusion protocols. The study periods were usually short, from

2.5 h (Borel *et al.*, 1998) to 5 h (Balagopal *et al.*, 2003) and sampling, other than base-line samples, was restricted to the final period of the infusion to achieve sampling during an isotopic plateau. Similarly, Q was also measured only for short periods of time. In contrast, Forslund *et al.* (1999) and Garlick *et al.* (1980a) applied a 24 h study protocol. The short infusion and sampling period in human subjects is based on the expectation that it would take 1.5 h of infusion to obtain an isotopic plateau (El Khouri *et al.*, 1995). Furthermore, El Khouri *et al.* (1994a) reported that 24 h values of leucine oxidation can be predicted with less than 4% error based on 1-h sampling periods. Repeated short periods of measurement (within a day or within months) have shown variability of 1.5 to 7.2% for resting Q and of 3.8% for leucine flux (Welle and Nair, 1990). To achieve this degree of repeatability, longer periods of study are needed in animals – typically 24 h or longer for indirect calorimetry – because animals are less likely to follow instructions (e.g. eat, rest, exercise, etc.) than humans.

An aspect to consider in application of the combined measurement of protein and energy measurement is the invasiveness of procedures. Generally, the methods employed in indirect calorimetry are low invasiveness, be it vented hoods with Douglas bags for breath collection (Berg *et al.*, 2013; Borel *et al.*, 1998) or a respiration tent (Garlick *et al.*, 1980a). Rather than the method of gas exchange measurement, it is more a question of the time of exposure to the collection methods that determines the degree of invasiveness so that the level of invasiveness will be rather determined by the measurement of protein turnover. While the collection of CO_2 after applying a carbon label will not add discomfort, isotope infusion and the need to collect blood samples to determine plasma enrichments will. However, Bross *et al.* (1989) showed that frequent oral dosing and urine sampling to determine precursor ^{13}C-enrichment yielded similar results to IV infusion and blood sampling, thus minimising invasiveness of the procedure. Urine sampling, as used for the end-product method, is usually non-invasive; urinary catheters are only used when subjects are incapable of providing samples voluntarily. Thus, the comfort of subjects does not contribute to the preference of one method over the other.

The combined measurement of whole body protein and energy metabolism has been applied less frequently in animals, possibly because experimental animals are more readily available for direct measurements than human volunteers. This applies especially to small animals, like chicks or rodents. In fact, studies in poultry (Aoyagi *et al.*, 1988, 1990; Kita *et al.*, 1983; Muramatsu and Okumura, 1985) typically used different animals to determine protein turnover and Q. This is acceptable because of the high degree of genetic homogeneity in populations of poultry or rodents. In contrast, studies in large mammals like sheep (Li *et al.*, 2008; MacRae *et al.*, 1988; Sano *et al.*, 2004; Sano and Shibasaki, 2011) or cattle (Lobley *et al.*, 1985, 1987) and pigs (Levesque *et al.*, 2011; Franco *et al.*, 2013; Samuel *et al.*, 2012) applied techniques for measurement of protein turnover and Q in the same individual animals.

As in human subjects, indirect calorimetry has been the predominant method to determine Q in large mammals. Lobley *et al.* (1985, 1987) combined indirect calorimetry for 4 days with a primed constant infusion of L-[1-^{14}C]leucine for 8-10 h to determine the relationship of protein turnover with protein retention and energy metabolism at different feeding levels. Also at the Rowett Research Institute, MacRae *et al.* (1988) confined sheep for 45 days in respiration chambers with daily determination of Q. The sheep received a primed constant infusion of L-[1-^{14}C]leucine for 8 h on days 4, 11 and 18 of clenbuterol application to determine the effect of the β-agonist on protein and energy metabolism. In contrast, Sano *et al.* (2004) and Sano and Shibasaki (2011) used the respiration equipment to quantify $^{13}CO_2$ production after primed-constant infusions of L-[1-^{13}C]leucine for 8 h in sheep. Although they used indirect calorimetry equipment in their experiment, Q was not reported because the authors' primary objective was to determine the effect of protein sources on protein turnover. To that end, Sano *et al.* (2004) and Sano and Shibasaki (2011) calculated protein turnover rates based on the enrichments in both plasma free leucine and plasma free α-ketoisocaproic acid, whereas Lobley *et al.* (1987) and MacRae *et al.* (1988) based the protein kinetic calculations on plasma free leucine only. The bicarbonate method to determine Q does not necessitate confinement of the animals so that it is suitable to determine Q under natural living conditions (Junghans and Chwalibog, 2001). Although Li *et al.* (2008) kept sheep in confinement, the combination of the bicarbonate method with a single dose application of ^{15}N-glycine circumvented the need for gas analysers. This added convenience was offset by the greater variability of Q measurements by bicarbonate dilution compared to indirect calorimetry (Junghans and Chwalibog, 2001). Nevertheless, the experimental approach by Li *et al.* (2008) offers a viable alternative to more common method combinations to study protein and energy metabolism.

In small animals, it is possible or advisable to use different methods to study protein and energy metabolism. Although the serial slaughter technique only gives an average measurement of Q over a period of time, it allows sampling of all tissues and organs and to recover sufficient sample for intended analyses. Thus, the serial slaughter technique is suitable to be combined with the flooding dose technique (Garlick *et al.*, 1980b) to determine protein synthesis, especially if differences among organs are to be determined. This does not preclude determination of whole-body protein turnover rates as shown by the research group of Muramatsu and Okumura (Aoyagi *et al.*, 1988, 1990; Kita *et al.*, 1993; Muramatsu and Okumura, 1985).

7.5 Case study: development of the simultaneous determination of protein and energy metabolism in sows

When planning a series of experiments to determine AA requirements of sows using IAAO with primed-constant infusion of L-[1-^{13}C]phenylalanine, it was decided to expand the experiments by determination of Q and protein kinetics to gain additional insight into the dynamics and underlying mechanisms of sow protein and energy metabolism. The transfer of the application of the IAAO from its established use in young pigs (Ball and Bayley, 1984) and humans (Zello *et al.*, 1990) to sows encountered difficulties beyond the mere scaling up of equipment and isotope dosages because at that time there were no published observations of protein turnover in sows. In addition, more stringent regulations for the use of radioactive isotopes and disposal of radioactively contaminated animals led to the decision to use stable isotopic tracer instead. Because the use of IAAO with stable ^{13}C tracer requires quantification of expired $^{(13)}CO_2$, it was decided to combine the tracer technique with the existing indirect calorimetry system for sows (McMillan, 2003).

Calculation of results for indirect calorimetry after Brouwer (1965) requires at least the quantification of VO_2 and VCO_2. In addition, VCH_4 and urinary N-excretion should be determined. Using the 2-pool model (Duggleby and Waterlow, 2005), parameters of protein turnover can be calculated using a stochastic model (Waterlow *et al.*, 1978). This means that it is necessary to determine VO_2, VCO_2, $^{13}CO_2$ production, and AA intake and flux, via plasma enrichment, to be able to calculate results for Q and protein turnover.

For the method development, the key objectives of planned experiments needed to be considered, which in our case were the determination of the protein turnover and Q in pregnant sows without nutritional intervention, followed by the determination of AA requirements of sows. These objectives determined the lay-out of the respiration-oxidation system, impacted the choice of isotopic tracer and affected the operational parameters.

For the design of the respiration system, it was decided to minimize the air-volume of the respiration chamber so that the air in the chamber equilibrated rapidly with the gas exchange of sows. This is not an issue if gas is collected in Douglas bags or if vented hoods with minimal dead volume are used. A rapid response of chamber $^{13}CO_2$ after application of the tracer aids the identification of isotopic plateaus, for which the 2-pool model is appropriate. Secondly, the need to accurately determine VCO_2 and ^{13}C enrichment led to a specific design of the respiration system. Air was drawn from the chamber at approximately 250 L/min and through gas meters at negative pressure to avoid loss of $^{13}CO_2$ label. The air flow rate was set to achieve a CO_2 concentration of approximately 0.5% in the chamber, with maximum concentrations not exceeding 1% CO_2. This was chosen to have sufficient CO_2 above ambient air to be measured accurately and to be low enough not to affect energy metabolism of the sows. Potential dilution of air drawn from the chamber by ambient air does not alter the amount of label drawn from the chamber, and would affect air flow rates only. Immediately after the gas meters, an aliquot of air (approximately 10 l/min) was drawn from the main air line and fed at positive pressure to gas analysers and CO_2 collection. Any loss of air at that point will not alter the gas composition of air or the ^{13}C enrichment. In fact, excess air was allowed to vent to achieve the target air-flow rates through the gas analysers (200 ml/min) and CO_2 collection.

An aliquot of chamber air was bubbled through small gas-washing bottles to collect CO_2 in 11 ml of 1 N NaOH to avoid contamination of the sample by unlabelled carbon. It was set up to collect approximately 1 mmol of CO_2 in a 30 min period to achieve a sufficient CO_2 concentration for determination of ^{13}C enrichment by mass-spectroscopy (Di Buono *et al.*, 2001). The necessary air flow was thus determined to be 150 ml/min at 0.5% CO_2, and 11 ml of NaOH were needed to completely cover the fritted gas disperser in the gas washing bottle. This collection setup assured that the NaOH was only saturated to 10% by the CO_2. The function of the CO_2 collection was verified by measuring the CO_2 concentrations before and after collection. Finding a concentration of 0% CO_2 after collection by adding a second CO_2 analyser assured that differential affinity of isotopes to bind to NaOH was not an issue.

The second aspect of CO_2 collection was to ensure sufficient ^{13}C enrichment for mass-spectroscopy. In the initial experiments in humans to determine the baseline protein turnover data, L-[1-^{13}C]leucine was used to measure protein turnover (Matthews *et al.*, 1980). However, based upon a review of current literature this tracer had not previously been used in sows and therefore the appropriate tracer dose had to be determined. A dose of 0.5 mg/kg body weight/h, chosen based on a review of data from human studies, failed, in a preliminary study, to produce $^{13}CO_2$ enrichment sufficiently above the natural background enrichment for measurement of APE (Jones *et al.*, 1985). A tracer dose of 1.0 mg/kg/h was subsequently found to be adequate, resulting in $^{13}CO_2$ of no less than 0.0003 APE above background and was used thereafter with a priming dose of 1.0 mg/kg delivered at the start of the infusion. A similar problem occurred in the AA requirement studies. Initially, a dose of 1.0 mg/kg/h of L[1-^{13}C]phenylalanine was used. However, because $^{13}CO_2$ enrichments were only marginally over the minimum APE, the dose was increased to 2.0 mg/kg/h in subsequent experiments, which resulted in an APE of 0.0005 to 0.0015 depending on the oxidation rate of phenylalanine.

Because measurement of protein turnover using 1-^{13}C-labelled AAs necessitates frequent blood sampling, we adapted a catheterization technique from human medicine to swine research. The surgical implantation of a heparin-coated catheter in the internal cephalic vein ending in a subcutaneous vascular access port (Swindle *et al.*, 2005) maximised the patency of catheters and minimized the risk of infections so that animals could be used for longer periods of time than when using conventional catheterization techniques. The vascular access port was placed 10 cm cranial to an imaginary line directly up from the cranial side of the right leg, approximately at the height of and caudal to the base of the ear. This position minimises the ability of the pig to interfere with the non-coring Huber-needle used to access the port. Access to the catheter was provided by an extension set that was externalised from the chamber to facilitate blood sampling during collection of expired air. Of 44 sows that were kept with catheters for 359 (SD 218) days, 15 retained patency until the end of studies at approximately two years after catheterisation. Early losses of catheters (less than 6 months after implantation) were caused by too shallow implantation of the injection ports (n=6) or sepsis (n=2). Twelve sows were lost due to sickness unrelated to the catheters or reproductive failure. Therefore, given appropriate access and flushing techniques, these catheters can remain patent for up to two years.

To determine the baseline protein turnover and Q, the L-[1-^{13}C]leucine infusions were maintained for 24 h. A second objective of these studies was to determine the relationship of measurements during short periods with frequent feeding to full 24-h values. The IAAO has been criticized for providing measurements in the fed state only (Bos *et al.*, 2002). Therefore, the same feeding regimen was used in the determination of baseline protein turnover and Q as in the subsequent requirement studies, i.e. sows were given half their daily feed allowance in 16, half-hourly meals. The remainder of the daily allowance was offered in the afternoon.

Thus, rates of protein turnover and Q during frequent feeding could be compared to 24-h values. Heat production during repeated small meals was 1.17 (standard error 0.02, P=0.0001) times greater than during the entire 24-h period but no significant difference was found in leucine oxidation during frequent feeding vs full 24 h values. This indicated that our primary measurement, AA oxidation, was predicted accurately during a 4-h period with half-hourly meals while a factor (1/1.17) should be applied to Q to yield valid estimates of 24-h values. In the subsequent requirement studies (Levesque *et al.*, 2011; Samuel *et al.*, 2012; Franco *et al.*, 2013,, the tracer, L-[1-^{13}C]phenylalanine, the tracer was delivered orally with the small ½-hourly meals so that plasma sampling without contamination of tracer was possible. The duration of less than 24 h was necessary to complete a requirement study in as short a time period as possible to avoid major changes in AA requirements.

7.6 Issues in calculation of results

An obvious application of the combined measurement of protein turnover and Q is to characterize the relationship between energy and protein metabolism under a range of conditions. The energy cost of protein synthesis has been calculated based on stoichiometry. It has been suggested that 5 ATP were needed to synthesize 1 peptide bond (Millward and Garlick, 1976), equivalent to 3.56 kJ/g of body protein synthesized. In contrast, Whittemore and Fawcett (1976) suggested that 8-12 Mol ATP were needed, resulting in an average energy need of 7.3 kJ/g body protein synthesis. This indicates that it is not entirely clear how much ATP is needed for protein synthesis or that a variable composition of whole-body protein may affect the energy need for synthesis. Experimental results were even more variable at between 5.35 and 21.5 kJ/g of body protein synthesis (Aoyagi *et al.* 1988).

Indirect calorimetry provides good to excellent accuracy and precision of measurements (Junghans and Chwalibog, 2001), indicating that differences in energy needed for protein synthesis are mainly caused by rates or efficiency of protein turnover or methods to measure it. The type of tracer can lead to different quantitative estimates of protein flux and synthesis (Pacy *et al.*, 1994; Pannemans *et al.* 1997), and hence the estimates of the energy needed for protein synthesis. However, it is difficult to decide which tracer or application method gives the true value for protein synthesis rates (Duggleby and Waterlow, 2005).

Boirie *et al.* (2001) and Benedek *et al.* (1995) reported similar absolute values for resting Q and protein synthesis in Caucasian and Gambian subjects, respectively. However, Benedek *et al.* (1995), using a single dose of ^{15}N-glycine, found no differences between young and elderly subjects, while Boirie *et al.* (2001) found greater leucine flux rates in young subjects when using L[1-^{13}C]leucine as a tracer. Boirie *et al.* (2001) reported greater energy cost of protein turnover in elderly (12.38 kJ/g protein) than in young adult subjects (9.87 kJ/g protein). From the study in young (8.5 years), growing subjects by Schutz *et al.* (1999), a slightly greater energy need of 15.90 kJ/g protein synthesis can be calculated. In this study, which used ^{15}N-glycine as tracer, Schutz *et al.* (1999) found no difference in protein synthesis per kg fat-free body mass between lean and obese children. In contrast, Balagopal *et al.* (2003) reported greater rates of leucine appearance and non-oxidative disposal in obese compared to lean adolescents (16 years old). It is not clear if the qualitative differences between the studies of Schutz *et al.* (1999) and Balagopal *et al.* (2003) were caused by the different tracers or age of subjects, although the tracer appears to be the more likely source of these discrepancies.

Manipulation of nutrient and energy intake in farm animals is aimed ultimately to maximize productivity and nutrient efficiency and to minimize nutrient excretion. Studies in cattle (Lobley *et al.*, 1985, 1987), sheep (Li *et al.*, 2008) and poultry (Muramatsu *et al.*, 1987) focused on the energy cost of protein synthesis. In healthy humans, the objectives of studies were more geared toward attainment and preservation of desirable body composition. Stein *et al.* (1991) varied the energy allowance in moderately obese males, while Gaine *et al.* (2005, 2006) and Pikosky *et al.* (2008) studied the interaction between varied protein intake and exercise. Because nutrient intake ranged from limiting to excess, calculation of energy expenditure for protein synthesis, which was not the primary objective of these studies, may lead to erroneous results. For application in disease, the main aim of experiments was to improve nutritional management. Because responses of protein and energy metabolism may differ widely among diseases, a discussion of these results exceeds the scope of this review.

7.7 Conclusions

Combination of measurement of Q by indirect calorimetry, doubly labelled water or bicarbonate appearance with protein turnover by isotope dilution techniques is possible without impairing the quality of either measurement, except for the combination of bicarbonate recovery with AA oxidation. Indirect calorimetry is probably the most accurate and precise method to determine Q, followed closely by the doubly labelled water method. However, both these methods require skilled personnel and are quite expensive. In contrast, the bicarbonate recovery method is cheaper and easier to apply but is less accurate than the other methods. The doubly labelled water and bicarbonate recovery methods can easily be applied in the field, while a laboratory setting is preferable for indirect calorimetry. For measurement of protein turnover, the precursor method, especially with ^{13}C AAs as tracers, is probably the method of choice, although it is quite expensive and requires complex and expensive equipment for isotope analysis. Conversely, the end-product method is easier and cheaper to apply and more suitable to use in field studies. The choice of method for energy and protein metabolic measurements depends firstly on the research questions, and secondly on the equipment, expertise and funds available. The combined measurement of protein turnover and Q is especially applicable for studies in humans and large mammals where the recruitment of subjects is more difficult than when studying small animals. The main applications are determination of energy cost of protein synthesis under a variety of conditions. In humans, the objectives are mainly the study of the effects of health and disease on protein and energy metabolism and the assessment of intervention strategies. In animals, the scope of objectives is expanded by the aim to maximise production and nutrient efficiency. Because of the close linkage of protein and energy metabolism, their simultaneous study can provide insights neither method alone can offer.

References

Aoyagi, Y., Muramatsu, T. and Okumura, J., 1990. Contribution of whole-body protein synthesis to diet-induced thermogenesis in chicks. Nutrition Research 10: 627-634.

Aoyagi, Y., Tasaki, I., Okumura, J.I. and Muramatsu, T., 1988. Energy cost of whole-body protein synthesis measured *in vivo* in chicks. Comparative Biochemistry and Physiology 91A: 765-768.

Balagopal, P., Bayne, E., Sager, B., Russell, L., Patton, N. and George, D., 2003. Effect of lifestyle changes on whole-body protein turnover in obese adolescents. International Journal of Obesity 27: 1250-1257.

Ball, R.O. and Bayley, H.S., 1984. Tryptophan requirement of the 2.5 kg piglet determined by the oxidation of an indicator amino acid. Journal of Nutrition 114: 1741-1746.

Benedek, C., Berclaz, P.-Y., Jéquier, E. and Schutz, Y., 1995. Resting metabolic rate and protein turnover in apparently healthy elderly Gambian men. American Journal of Physiology – Endocrinology and Metabolism 268: E1083-E1088.

Berclaz, P.-Y., Benedek, C., Jéquier, E. and Schutz, Y., 1996. Changes in protein turnover and resting energy expenditure after treatment of malaria in Gambian children. Pediatric Research 39: 401-409.

Berg, A., Rooyackers, O., Bellander, B.-M. and Wernerman, J., 2013. Whole body protein kinetics during hypocaloric and normocaloric feeding in critically ill patients. Critical Care 17: R158-R166.

Boirie, Y., Beaufrére, B. and Ritz, P., 2001. Energetic cost of protein turnover in healthy elderly humans. International Journal of Obesity 25: 601-605.

Borel, M.J., Buchowski, M.S., Turner, E.A., Peeler, N.B., Goldstein, N.B. and Flakoll, P.J., 1998. Alterations in basal nutrient metabolism increase resting energy expenditure in sickle cell disease. American Journal of Physiology – Endocrinology and Metabolism 274: E357-E364.

Bos, C., Gaudichon, C. and Tomé, D., 2002. Isotopic studies of protein and amino acid requirements. Current Opinion in Clinical Nutrition and Metabolic Care 5: 55-61.

Bross, R., Ball, R.O. and Pencharz, P.B., 1989. Development of a minimally invasive protocol for the determination of phenylalanine and lysine kinetics in humans during the fed state. Journal of Nutrition 128: 1913-1919.

Brouwer, E. 1965. Report of Sub-Committee on Constants and Factors. In: Blaxter, K.L. (ed.) Proceedings of the 3rd Symposium on Energy Metabolism. EAAP Publication No. 11. Academic Press, New York, NY, USA, pp. 441-443.

Catzeflis, C., Schutz, Y., Micheli, J.-L., Welsch, C., Arnaud, M.L. and Jequier, E., 1985. Whole body protein synthesis and energy expenditure in very low birth weight infants. Pediatric Research 19: 679-687.

Davis, T.A. and Reeds, P.J., 2001. Of flux and flooding: the advantages and problems of different isotopic methods for quantifying protein turnover *in vivo*: II. Methods based on the incorporation of a tracer. Current Opinion in Clinical Nutrition and Metabolic Care 4: 51-56.

Di Buono, M., Wykes, L.J., Ball, R.O. and Pencharz, P.B., 2001. Total sulfur amino acid requirement in young men as determined by indicator amino acid oxidation with L-[1-C-13] phenylalanine. American Journal of Clinical Nutrition 74: 756-760.

Duggleby, S.L. and Waterlow, J.C., 2005. The end-product method of measuring whole-body protein turnover: a review of published results and a comparison with those obtained by leucine infusion. British Journal of Nutrition 94: 141-153.

Elia, M., 1991. Energy equivalents of CO_2 and their importance in assessing energy expenditure when using tracer techniques. American Journal of Physiology – Endocrinology and Metabolism 260: E75-E88.

El-Khoury A.E., Sánchez, M., Fukagawa, N.K., Gleason, R.E. and Young, V.R., 1994b. Similar 24-h pattern and rate of carbon dioxide production, by indirect calorimetry vs. stable isotope dilution, in healthy adults under standardized metabolic conditions. Journal of Nutrition 124: 1615-1627.

El-Khoury, A.E., Fukagawa, N.K., Sánchez, M., Gleason, R.E., Tsay, R.H. and Young, V.R., 1995. The 24-h kinetics of leucine oxidation in healthy adults receiving a generous leucine intake via three discrete meals. American Journal of Clinical Nutrition 62: 579-590.

El-Khoury, A.E., Fukagawa, N.K., Sánchez, M., Tsay, R.H., Gleason, R.E., Chapman, T.E. and Young, V.R., 1994a. The 24-h pattern and rate of leucine oxidation, with particular reference to tracer estimates of leucine requirements in healthy adults. American Journal of Clinical Nutrition 59: 12-20.

Fearon, K.C.H., Hansel, D.T., Preston, T., Plumb, J.A., Davies, J., Shapiro, D., Shenkin, A., Caiman, K.C. and Burns, H.J.G., 1988. Influence of whole body protein turnover rate on resting energy expenditure in patients with cancer. Cancer Research. 48: 2590-2595.

Fitch, W.L. and King, K.C., 1987. Protein turnover and 3-methylhistidine excretion in non-pregnant, pregnant and gestational diabetic women. Human Nutrition: Clinical Nutrition 41C: 327-339.

Flaim, K. E., Peavy, D.F., Everson, W.V. and Jefferson, L.S., 1982. The role of amino acids in the regulation of protein synthesis in perfused rat liver. I. Reduction in rates of synthesis resulting from amino acid deprivation and recovery during flow through perfusion. Journal of Biological Chemistry 57: 2932-2938.

Forslund, A.H., El-Khoury, A.E., Olsson, R.M., Sjödin, A.M., Hambraeus, L. and Young, V.R., 1999. Effect of protein intake and physical activity on 24-h pattern and rate of macronutrient utilization. American Journal of Physiology – Endocrinology and Metabolism 276: E964-E976.

Franco, D.J., Josephson, J.K., Moehn, S., Pencharz, P.B. and Ball, R.O., 2013. Isoleucine requirement of pregnant sows. Journal of Animal Science 91: 3859-3866.

Gaine, P.C., Pikosky, M.A., Martin, W.F., Bolster, D.R., Maresh, C.M. and Rodriguez, N.R., 2006. Level of dietary protein impacts whole body protein turnover in trained males at rest. Metabolism Clinical and Experimental 55: 501-507.

Gaine, P.C., Viesselman, C.T., Pikosky, M.A., Martin, W.F., Armstrong, L.E., Pescatello, L.S. and Rodriguez, N.R., 2005. Aerobic exercise training decreases leucine oxidation at rest in healthy adults. Journal of Nutrition 135:1088-1992.

Garlick, P.J. and Grant, I., 1988. Amino acid infusion increases the sensitivity of muscle protein synthesis *in vivo* to insulin. Effect of branched-chain amino acids. Biochemical Journal 254: 579-584.

Garlick, P.J., Clugston, G.A., Swick, R.W. and Waterlow, J.C., 1980a. Diurnal pattern of protein and energy metabolism in man. American Journal of Clinical Nutrition 33: 1983-1986.

Garlick, P.J., McNurlan, M.A. and Preedy, V.R., 1980b. A rapid and convenient technique for measuring the rate of protein synthesis in tissues by injection of [^{3}H]phenylalanine. Biochemical Journal 192: 719-723.

Gerrits, W.J.J., Van den Borne, J.J.G.C. and Labussière, E., 2015. Deriving heat production from gaseous exchange: validity of the approach. In: Gerrits, W.J.J. and Labussière, E. (eds.) Indirect calorimetry. Techniques, computations and applications. Wageningen Academic Publishers, Wageningen, the Netherlands, pp. 19-34.

Giordano, M. and Castellino, P., 1997. Correlation between amino acid induced changes in energy expenditure and protein metabolism in humans. Nutrition 13: 309-312.

Gougeon, R., Morais, J., Chevalier, S., Pereira, S., Lamarche, M. and Marliss, E.B., 2008. Determinants of whole-body protein metabolism in subjects with and without type 2 diabetes. Diabetes Care 31: 128-133.

Hambraeus, L., Sjödin, A., Webb, P., Forslund, A., Hambraeus, K. and Hambraeus, T., 1994. A suit calorimeter for energy balance studies on humans during heavy exercise. European Journal of Applied Physiology 68: 68-73.

Hsu, J. W-C., Kriengsinyos, W., Wykes, L.J., Rafii, M., Goonewardene, L.A., Ball, R.O. and Pencharz, P.B., 2006. Leucine is not a good choice as an indicator amino acid for determining amino acid requirements in man. Journal of Nutrition 136: 958-964.

Jones, P.J.H., Pencharz, P.B., Bell, L. and Clandinin, M.T., 1985. Model for determination of 13C substrate oxidation rates in humans in the fed state. American Journal of Clinical Nutrition 41: 1277-1282.

Junghans, P. and Chwalibog, A., 2001. Complementory applicalton of stable isotope techniques and indirect calorimetry for determining energy expenditure. In: Chwalibog, A. and K. Jacobsen (eds.). Energy metabolism of animals. EAAP Publication No. 103. Wageningen Academic Press, Wageningen, the Netherlands, pp. 425-428.

Kao, C.C., Guntupalli, K.K., Bandi, V. and Jahoor, F., 2008. Whole-body CO_2 production as an index of the metabolic response to sepsis. Shock 32: 23-28.

Kita, K., Muramatsu, T. and Okumura, J., 1993. Effect of dietary protein and energy intakes on whole-body protein turnover and its contribution to heat production in chicks. British Journal of Nutrition 69: 681-688.

Lachica, M. and Aguilera, J.F., 2008. Methods to estimate the energy expenditure of goats: from the lab to the field. Small Ruminant Research 79: 179-182.

Levesque, C.L., Moehn, S., Pencharz, P.B. and Ball, R.O., 2011. The threonine requirement of sows increases in late gestation. Journal of Animal Science 89: 93-102.

Li, L., Oddy, V.H. and Nolan, V.J., 2008. Whole-body protein metabolism and energy expenditure in sheep selected for divergent wool production when fed above or below maintenance. Australian Journal of Experimental Agriculture 48: 657-665.

Lifson, N. and McClintock, R., 1966. Theory of use of the turnover rates of body water for measuring energy and material balance. Journal of Theoretical Biology 12: 46-74.

Lobley, G.E., Connell, A. and Buchan, V., 1987. Effect of food intake on protein and energy metabolism in finishing beef steers. British Journal of Nutrition 57: 457-465.

Lobley, G.E., Connell, A., Mollison, G.S., Brewer, A., Harris, C.I. and Buchan, V., 1985. The effects of a combined implant of trenbolone acetate and oestradiol-17β on protein and energy metabolism in growing beef steers. British Journal of Nutrition 54: 681-694.

MacRae, J.C., Skene, P.A., Connell, A., Buchan, V. and Lobley, G.E., 1988. The action of the β-agonist clenbuterol on protein and energy metabolism in fattening wether lambs. British Journal of Nutrition 59: 457-465.

Matthews, D.E., Motil, K.J., Rohrbaugh, D.K., Burke, J.F., Young, V.R. and Bier, D.M., 1980. Measurement of leucine metabolism in man from a primed, continuous infusion of L-[1-^{13}C]-leucine. American Journal of Physiology - Endocrinology and Metabolism 238: E473-E479.

McMillan, D.J. 2003. Development of low protein diets for sows: effect on performance and energy metabolism. MSc Thesis, University of Alberta, Edmonton, AB, Canada.

Millward, D.J. and Garlick, P.J., 1976. The energy cost of growth. Proceedings of the Nutrition Society 35: 339-349.

Moehn, S., Bertolo, R.F.P., Pencharz, P.B. and Ball, R.O., 2004. Pattern of carbon dioxide production and retention is similar in sows when fed hourly, but not when fed a single meal. BMC Physiology 4: 11.

Möhn, S., Gillis, A.M., Moughan, P.J. and De Lange, C.F.M., 2000. Influence of lysine and energy intakes on body protein deposition and lysine utilization in growing pigs. Journal of Animal Science 78: 1510-1519.

Muramatsu, T. and Okumura, J., 1985. Whole-body protein turnover in chicks at early stages of growth. Journal of Nutrition 115: 483-490.

Muramatsu, T., Aoyagi, Y., Okumura, J. and Tasaki, I. 1987. Contribution of whole-body protein synthesis to basal metabolism in layer and broiler chickens. British Journal of Nutrition 57: 269-277.

Neale, R.J. and Waterlow, J.C., 1974. The metabolism of 14C-labelled essential amino acids given by intragastric or intravenous infusion to rats on normal and protein-free diets. British Journal of Nutrition 32: 11-25.

Ohara, S., M. Kato, M. Asaka and T. Toyota. 1998. The UBiT-100 $^{13}CO_2$ infrared analyzer: comparison between infrared spectrometric analysis and mass spectrometric analysis. Heliobacter 3: 49-53.

Pacy, P.J., Price, G.M., Halliday, D., Quevedo, M.R. and Millward, D.J., 1994. Nitrogen homeostasis in man: the diurnal responses of protein synthesis and breakdown and amino acid oxidation to diets with increasing protein intakes. Clinical Science 86: 103-118.

Pannemans, D.L.E., Wagenmakers, A.J.M., Westerterp, K.R., Schaafsma, G. and Halliday, D., 1997. The effect of an increase of protein intake on whole-body protein turnover in elderly women is tracer dependent. British Journal of Nutrition 127: 1788-1794.

Pencharz, P.B. and Ball, R.O., 2003. Different approaches to define individual amino acid requirements. Annual Review of Nutrition 23: 101-116.

Picou, D. and Taylor-Roberts, T., 1969. The measurement of total protein synthesis and catabolism and nitrogen turnover in infants in different nutritional states and receiving different amounts of dietary protein. Clinical Science 36: 283-296.

Pikosky, M.A., Smith, T.J., Grediagin, A., Castaneda-Sceppa, C., Byerley, L., Glickman, E.L. and Young, A.J., 2008. Increased protein maintains nitrogen balance during exercise-induced energy deficit. Medicine and Science in Sports and Exercise 40: 505-512.

Reeds, P.J. and Davis, T.A., 1999. Of flux and flooding: the advantages and problems of different isotopic methods for quantifying protein turnover *in vivo*: I. Methods based on the dilution of a tracer. Current Opinion in Clinical Nutrition and Metabolic Care 2: 23-28.

Reeds, P.J., Cadenhead, A., Fuller, M.F., Lobley, G.E. and McDonald, J.D., 1980. Protein turnover in growing pigs. Effects of age and food intake. British Journal of Nutrition 43: 445-455.

Rocha, D.M., Faloona, G.R. and Unger, R.H. 1972. Glucagon-stimulating activity of 20 amino acids in dogs. Journal of Clinical Investigation 51: 2346-2352.

Rometsch, M., Roser, U., Becker, K. and Susenbeth, A.,1997. Heart rate and respiratory adjustments during work of increasing intensity in Hinterwaelder and Zebu oxen. Animal Science 64: 233-238.

Samuel, R.S., Moehn, S., Pencharz, P.B. and Ball, R.O., 2012. The dietary lysine requirement of sows increases in late gestation. Journal of Animal Science 90: 4896-4904.

Sano, H. and Shibasaki, S., 2011. Effects of urea and soybean meal supplementation on rates of plasma leucine turnover and protein synthesis in sheep fed concentrate-based diets, Journal of Applied Animal Research, 39: 239-244.

Sano, H., Kajita, M. and Fujita, T., 2004. Effect of dietary protein intake on plasma leucine flux, protein synthesis, and degradation in sheep. Comparative Biochemistry and Physiology Part B 139: 163-168.

Sano. H., Murakami, S., Sasaki, S. and Al-Mamun, M. 2010 Effects of dietary energy intake and cold exposure on kinetics of plasma phenylalanine, tyrosine and protein synthesis in sheep, Archives in Animal Nutrition 64: 47-55.

Schoeller, D.A., Ravussin, E., Schutz, Y., Acheson, K.J., Baertschi, P. and Jequier, E., 1986. Energy expenditure by doubly labeled water: validation in humans and proposed calculation. American Journal of Physiology – Regulatory, Integrative and Comparative Physiology 250: R823-R830.

Schutz, Y., Rueda-Maza, C.M., Zaffanello, M. and Maffeis, C., 1999. Whole-body protein turnover and resting energy expenditure in obese, prepubertal children. American. Journal. of Clinical Nutrition 69: 857-862.

Siekevitz, P. and Palade, G.E., 1958. A cytochemical study on the pancreas of the guinea pig III. *In vivo* incorporation of leucine-1-C^{14} into the proteins of cell fractions. Journal of Biophysical Biochemistry and Cytology 4: 557-566.

Spear, M.L., Darmaun, D., Sager, B.K., Parsons, W.R. and Haymond, M.W. 1995. Use of [^{13}C]bicarbonate infusion for measurement of CO_2 production. American Journal of Physiology – Enclocrinology and Metabolism 268: E1123-E1127.

Sprinson, D.B. and Rittenberg, D., 1949. The rate of interaction of the amino acids of the diet with the tissue proteins. Journal of Biological Chemistry 180: 715-726.

Stein, T.P., Rumpler, W.V., Leskiw, M.J., Schluter, M.D., Staples, R. and Bodwell, C.E., 1991. Effect of reduced dietary intake on energy expenditure, protein turnover, and glucose cycling in man. Metabolism 40: 478-483.

Swindle, M.M., Nolan, T., Jacobson, A., Wolf, P., Dalton, M. and Smith, A.C., 2005. Vascular access port (VAP) usage in large animal species. Contemporary Topics in Laboratory Animal Science 44:7-17.

Vaisman, N., Clarke, R., Rossi, M., Goldberg, E., Zello, G.A. and Pencharz, P.B., 1992. Protein turnover and resting energy expenditure in patients with undernutrition and chronic lung disease. American Journal of Clinical Nutrition 55: 63-69.

Wagenmakers, A.J.M., 1999. Tracers to investigate protein and amino acid metabolism in human subjects. Proceedings of the Nutrition Society 58: 987-1000.

Waterlow, J.C., Garlick, P.J. and Millward, D.J., 1978. Protein turnover in mammalian tissues and in the whole body. North Holland Publ. Co., Amsterdam, the Netherlands.

Welle, S. and Nair, K.S., 1990. Relationship of resting metabolic rate to body composition and protein turnover. American Journal of Physiology – Endocrinology and Metabolism 258: E990-E998.

Whittemore, C.T. and Fawcett, R.H., 1976. Theoretical aspects of a flexible model to simulate protein and lipid growth in pigs. Animal Production 22: 87-96.

Winter, T.A., O'Keefe, S.J., Callanan, M. and Marks, T., 2005. The effect of severe undernutrition and subsequent refeeding on whole-body metabolism and protein synthesis in human subjects. Journal of Parenteral and Enteral Nutrition 29: 221-228.

Young, B.A., 1970. Application of the carbon dioxide entry technique to measurement of energy expenditure by grazing cattle. In: Schürch, A. and Wenk, C. (eds.) Energy metabolism of farm animals. EAAP Publication No. 13. Juris Druck and Verlag, Zürich, Switzerland, pp. 237-241.

Young, V.R., Yu, Y.M. and Fukagawa, N.K., 1991. Protein and energy interactions throughout life: Metabolic basis and nutritional implications. Acta Paediatrica Scandinavica 375 Suppl.: 5-24.

Zello, G.A., Pencharz, P.B. and Ball, R.O., 1990. Phenylalanine flux, oxidation and conversion to tyrosine in humans studied with L-[1-^{13}C]-phenylalanine. American Journal of Physiology – Endocrinology and Metabolism 259: E835-E843.

Chapter 8: Quantifying physical activity heat in farm animals

W.J.J. Gerrits[1], M.J.W. Heetkamp[2], E. Labussière[3,4] and J.B. Van Klinken[5]*
[1]Animal Nutrition Group, Wageningen University, P.O. Box 338, 6700 AH Wageningen, the Netherlands; [2]Adaptation Physiology Group, Wageningen University, P.O. Box 338, 6700 AH Wageningen, the Netherlands; [3]INRA-UMR 1348 Pegase, 35590, Saint-Gilles, France; [4]Agrocampus Ouest – UMR1348 PEGASE, 35000 Rennes, France; [5]Department of Human Genetics, Leiden University Medical Center, P.O. Box 9600, 2300 RC Leiden, the Netherlands; walter.gerrits@wur.nl

Abstract

The time dependent character of data generated by modern calorimetry equipment provides the unique opportunity to monitor short term changes in energy expenditure related to physical activity, feeding pattern and other experimental interventions. When timed recordings of physical activity are available, total heat production can be decomposed into heat production due to physical activity and resting metabolism. Based on additional measurements and assumptions, further separation of the thermic effect of feeding is possible. When separating total heat production into its components, the separation of activity energy is a key step, even if activity energy is not the main interest of the study. This chapter provides an overview of a current methodology used to measure physical activity within the setting of indirect calorimetry. An approach to decompose total heat production into activity related and resting metabolic rate, the latter including the short term effect of feeding, is presented. In regression analysis, the energetic costs of each unit of physical activity, as well as the resting metabolic rate are assumed to be constant. Improved methodology is available using penalized least squares regression to model time dependent variation in the resting metabolic rate. The impact of experimental variables and settings, e.g. sampling frequency and chamber washout time are discussed. In addition, housing conditions, including ambient temperature, and feeding methods can impact on the relationship between physical activity and total heat production, affecting the reliability with which total heat production can be decomposed into resting metabolic rate and heat production related to physical activity. Finally, single housing conditions, often difficult to avoid in indirect calorimetry experiments, can modify behaviour, induce stress and therefore substantially influence activity related heat production and the results of calorimetry experiments. The effect of single vs group housing in growing pigs extracted from previously unpublished data and from the literature, suggest that single housing may increase activity heat production possibly because of induced stress. From behavioural and calorimetry data obtained in calves it is also demonstrated that the energetic costs of stereotyped behaviours can be very high.

Keywords: physical activity, activity energy, energy expenditure

DOI 10.3920/978-90-8686-261-0_8,

8.1 Introduction

Indirect calorimetry can be conducted in different experimental settings. Depending on the study objectives, subjects may be housed individually or in groups and may be allowed to move freely or movements may be restricted. This impacts the way physical activity (PA) can be recorded, but can also influences the optimal procedures used to estimate the individual components that make up total heat production (Q). Traditionally, Q is subdivided into basal metabolic rate (BMR), heat production for physical activity (AQ) and thermic effect of food (TEF). In addition, the energy expenditure due to thermoregulation (TR) represents a considerable part of Q if the ambient temperature lies outside the thermoneutral range. Energy expenditure related to inflammation and immune responses may, depending on its nature, be part of BMR, or can be considered separately.

$$Q = BMR + AQ + TEF + TR \tag{1}$$

Only Q is directly measured using calorimetry, and additional measurements, experimental interventions and dedicated computational techniques are required to disentangle its components. The first step of component analysis consists of separating the physical activity related heat production from the other components. The non-activity related component is often referred to as resting metabolic rate (RMR) (Blaxter, 1989; Bursztein, 1989), which includes a short, but also a long term component of TEF (Chapter 4; Labussière *et al.*, 2015). The estimation of AQ is possible because the short term changes in Q are mainly due to changes in PA, which means that AQ can be estimated from the time correlation between the recorded intensity of PA and Q. Assuming that the intensity of PA linearly correlates with the AQ, methods for Q decomposition focus on estimating the RMR and caloric cost of activity (CCA), given the recorded time sequences of Q and PA:

$$Q(t) = RMR(t) + CCA(t) \times PA(t) + e(t) \tag{2}$$

with $e(t)$ the residual heat production. Importantly, since the TEF, RMR and PA vary with time, the time dependency is stated explicitly in Equation 2. The CCA, which is the amount of energy required to fuel a unit of activity, depends on the type of activity and on the device that is used to measure PA. Consequently, CCA can change in time when different types of PA succeed one another.

Awareness of the relationships between the type of activity sensors, the procedure to quantify activity heat, and experimental settings and their influence on (behaviour) activity is important for drawing conclusions from calorimetry studies. This chapter focuses on ways to record PA in various calorimetry settings. Subsequently, computational techniques to separate Q into its components will be addressed. Finally, three examples will be presented, demonstrating how housing conditions can, by inducing stress, affect estimates of AQ.

8.2 Activity measurements

Several devices exist for quantifying the level of spontaneous PA, such as photocell sensors, piezo-electric force transducers or simple weighing scales, microwave radar systems, video-tracking systems and multi-axis accelerometers. For the purpose of separating AQ from RMR, any device recording PA in arbitrary units can be used, provided that the data are quantitative. Beam interruption measurements have been used to discriminate posture changes, typically in large animal species (Van den Borne *et al.*, 2006), but interruption of multiple beams is now commonly applied in housing systems of rodents to continuously record x,y coordinates of freely moving, individually housed subjects. For group-housed animals, radar devices have been used extensively, as these can be mounted above a pen or cage, completely covering the housing area. Changes in the frequency of the reflected radar waves on the surface of moving subjects are converted into electrical pulses. Important calibration steps for such devices include a sensitivity check, to prevent them from responding to irrelevant events in the chamber. Influences of variation in the distance between moving subjects and the radar device on its recordings should be checked and prevented. If needed, multiple devices can be used (Heetkamp *et al.*, 1995).

Force sensors or weighing devices are commonly used for the recording of PA of animals housed in metabolism cages, but it can also be applied for group-housed subjects, mounting the complete pen (pigs, Quiniou *et al.*, 2001) or cage (rats, Heetkamp *et al.*, 2003) on force sensors or on a weighing scale. The advantage of such device over devices that generate pulses only is that the response, like AQ, is proportional to the mass of movement. Whereas this advantage seems obvious, this was not corroborated by our work in group housed, adult rats (Heetkamp *et al.*, 2003). In this study, cages containing 3 rats were placed on a weighing device inside a calorimetry chamber, recording the weight of the cage every 0.7 s. Physical activity was calculated from the successive weight recordings either as cumulative short term changes in weights over time or as the cumulative number of weight changes exceeding 0.1 g. The former approach would account for the mass of movement, and the latter accounts for activity counts only. We found no improved correlation between Q and PA measured by the cumulative weight changes over the cumulative number of weight changes (Figure 1), and nonlinearity of both relationships was evident.

Beam interruption for the detection of posture changes has been combined with force sensors in pigs (Labussière *et al.*, 2013) and calves (Labussière *et al.*, 2008) and with radar devices in calves by Van den Borne *et al.* (2006). This allows PA patterns to be analysed within a posture, potentially accounting for differences in the cost of each unit of PA between postures. It has also been demonstrated that, at least in larger animal species, posture changes can also be detected from continuous force sensor data (Figure 2, Heetkamp *et al.*, 2007). With the increased availability of video-tracking software and 3-D movement sensors, activity signals can be separated into classes representing differences in the energetic costs per unit of PA (CCA in Equation 2). Although this has been applied in humans (e.g. Westerterp, 1999) this has, to the best of our knowledge, not yet been done in animals.

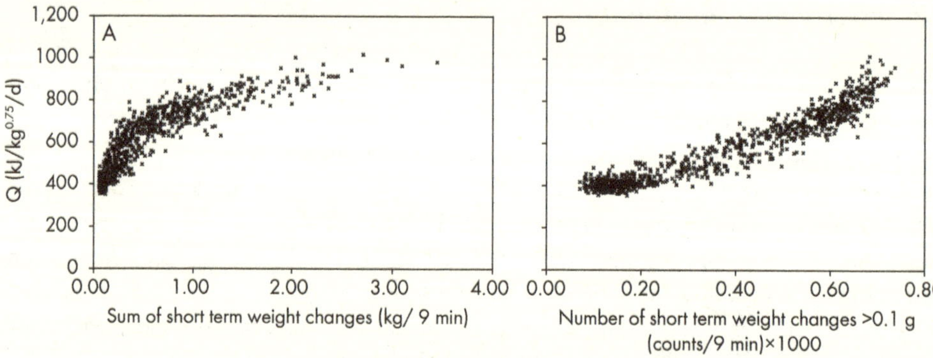

Figure 1. Relationship between physical activity recorded as (A) the sum of short-term weight changes, or (B) the number of short term weight changes, and total heat production in a single group of rats (data from Heetkamp et al., *2003).*

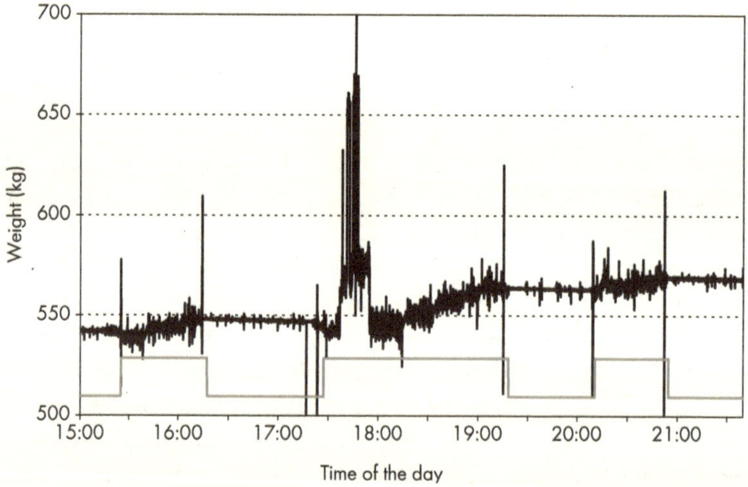

Figure 2. Extracting posture changes (lower line with low level and high level representing lying and standing, respectively) from short term variation in weight registration (top line, scaling on left Y-axis) in a dairy cow (data from Heetkamp et al., *2007).*

For component analysis it is important that the activity signal – after appropriate preprocessing – strongly correlates with the heat production, preferably with an R^2 of more than 0.70 (Figures 1 and 4; Van Klinken *et al.*, 2013). This depends on the sensitivity and accuracy of the activity registration, but also on the cycle and washout time of the chamber. In the setup described by Labussière *et al.* (2013), using force sensors and continuous differential measurement of gas concentrations, the R^2 of this relationship varies between 0.8 and 0.9. In a setting of group housed pigs (room volume 27,500 l, washout time τ=90 min, multiplexing, 6 min interval), using radar for activity registration, typical R^2 values between 0.60 and 0.70 can be obtained (W.J.J. Gerrits, unpublished data). Whether an activity sensor generates data that is suitable for Q decomposition can be checked by comparing the residual signal $e(t)$ between periods of activity and rest – ideally the variance in $e(t)$ should remain constant. In contrast, if the variance in $e(t)$ during activity is much larger than during rest then the resolution of the activity data is not high enough to accurately estimate AQ.

8.3 Separating activity heat from total heat production

8.3.1 Preprocessing data

Decomposition of Q can usually not be performed on raw data, since then the correlation between the PA and Q time sequence is sub-optimal. The most important step in data preprocessing is to take into account the fact that the time pattern in the respiratory exchange is dampened due to gas mixing in the metabolic chamber and the delay related to sample tubing and analytical equipment, while activity measurements are instantaneous. This dampening effect is proportional to the washout time τ of the system, which in the case of ideal mixing is equal to the chamber volume divided by the ventilation air flow. If τ is small, typically under 10 min, then the same delay can be induced on the activity data by means of a first order filter (Van Klinken *et al.*, 2013):

$$PA^*(i) = a \times PA^*(i-1) + b \times PA(i) \tag{3}$$

with i the time index in the sequence, PA the instantaneous measurements, PA^* the corrected PA sequence, and a and b the filter coefficients:

$$a = \exp\left(-\frac{T}{\tau}\right) \qquad b = 1 - \exp\left(-\frac{T}{\tau}\right) \tag{4}$$

where T is the sample time of activity measurements. In the case τ is large, such as for human metabolic chambers or those for large mammals, one cannot apply the washout effect on the PA time sequence as this will filter out the high frequency variations related to subsequent activity bouts, making Q decomposition infeasible. In this situation the instantaneous Q needs to be calculated – a mathematical procedure called deconvolution (Arch *et al.*, 2006) – and component analysis can be performed on the instantaneous Q and PA. The term deconvolution refers to computational methods to reverse the effect of convolution on recorded data. As discussed earlier, in the case of indirect calorimetry, it intends to estimate the instantaneous Q of the subject housed within a calorimetry chamber, while the recorded gas exchange is dampened and delayed due to among others the mixing process of the exhaled air with the air inside the chamber and sample tubing and analytical equipment. Typically, restoring the original signal, i.e. the timed Q of the subject inside the chamber, leads to noisy data, predominantly depending on the frequency of the gas measurement and τ. In addition, changes in air volume of the chamber between successive measurements contribute to an increase in noise of the data. Such changes result from short term variation in air pressure, temperature and relative humidity. When pressure, temperature and air moisture content are continuously recorded, these volume changes can be taken into account. An example of the effect of deconvoluting data in a large calorimeter, including variation in chamber volume, is presented in Figure 3. Alternatively, modelling approaches have been published (Chapter 4, Labussière *et al.*, 2015) in which the timed production of CO_2 and consumption of O_2 by the subject in the calorimeter is predicted from measured differential concentrations of CO_2 and O_2 and airflow, taking into account chamber volume, along with other, continuously recorded climate related parameters. This dynamic model implicitly deconvolutes the recorded data.

In certain cases the relationship between the recorded intensity of PA and Q will be nonlinear. Nonlinearity can be detected by inspecting the scatterplot of the PA against Q (Figure 1 and 4) or against the residuals $e(t)$ from component analysis. Since component analysis is based on the assumption that the relationship between activity and AQ is linear, nonlinearities need to be corrected by transforming the activity data in order to assure that the energy components are estimated accurately. The type of function used for transforming the activity data can be chosen based on the shape of the point cloud. In the case the function contains free parameters, the values of these parameters can be selected by minimizing the variance in the residual signal $e(t)$ (Figure 4).

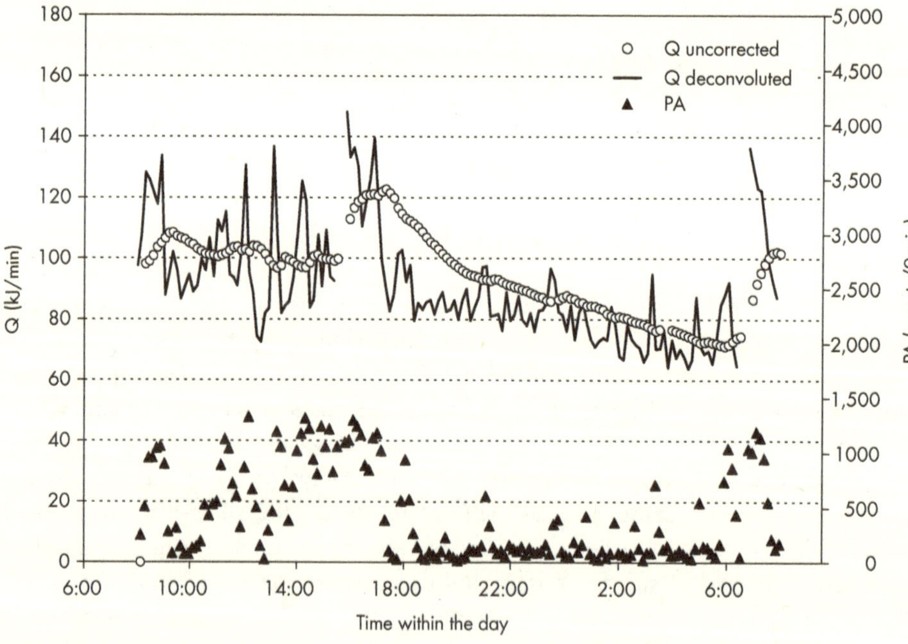

Figure 3. Effect of deconvolution on heat production (Q) data in an example dataset of a group of 12 growing pigs in an open circuit respiration chamber (volume 73,000 l, washout time τ=100 min, multiplexing, 9 min interval). Uncorrected heat production calculated directly from ventilation rate and measured O_2 and CO_2 in inflowing and outflowing air. Physical activity signal (PA) is plotted on the right Y axis.

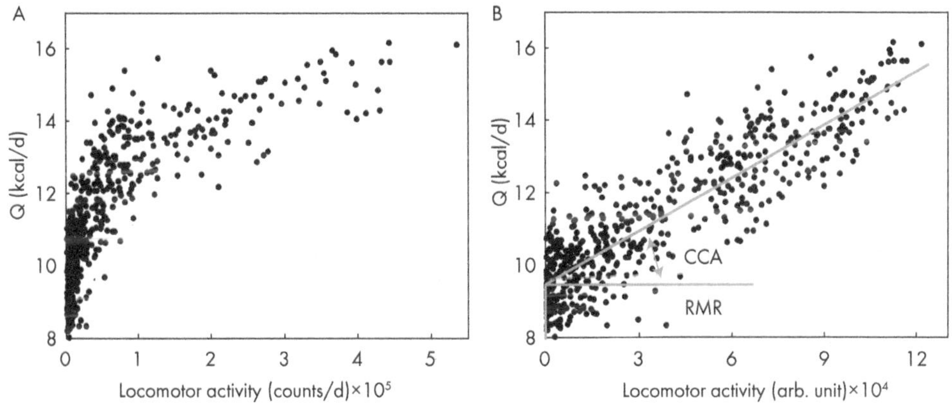

Figure 4. Typical relationship between total heat production and locomotor activity in single housed mice. (A) Q has a nonlinear relation with the raw locomotor activity, which was measured as the number of infrared beam breaks. (B) After transforming the activity data with a power function (PA – θ)^p the relationship is linearized. CCA = caloric cost of activity; RMR = resting metabolic rate.

8.3.2 Regression analysis

Having properly aligned Q and PA time sequences and having linearised the relationship between Q and PA, the central step of component analysis corresponds to the estimation of the RMR, CCA and AQ (Equation 2). The traditional approach for Q decomposition is to perform linear regression for each subject of Q against PA (Ravussin *et al.*, 1986; Verstegen *et al.*, 1987; Wenk and Van Es, 1976). In this case the intersection of the regression line with the y-axis corresponds to the average RMR and the slope of the line to the CCA (Figure 4). One of the main disadvantages of the regression approach is that it assumes that the RMR is constant over time, which does not consider diurnal variations in heat production and changes due to the TEF. Also, from a statistical perspective, not taking into account the time variation in the RMR has the downside of increasing the amount of uncertainty in the CCA and RMR estimates and of introducing a bias into these estimates (Van Klinken *et al.*, 2012).

An improvement to this approach is to fit separate lines, with the same slope, for data obtained during light and dark periods of the day (Gentry *et al.*, 1997) or to model the time dependency of the RMR with more complex functions (Brown *et al.*, 1991). Recently we proposed a regression based component analysis in which the time-dependent RMR is modelled with penalised spline (P-spline) functions (Van Klinken *et al.*, 2012). This enables unbiased estimates of the time variation in the RMR, but also increases the accuracy with which the CCA and AQ are estimated. In addition, this method has demonstrated to be relatively robust to low sample frequencies and noisy activity measurements, which is important when performing component analysis on data generated by multiplexed systems.

The difference between ordinary and penalised splines regression comes down to how RMR(t) is modelled in Equation 2. For ordinary regression RMR(t) is equal to $1 \cdot \beta$, with 1 a sequence of all ones and β the regression coefficient that is estimated in the model, which corresponds to the average RMR. In contrast, in penalized splines regression RMR(t) is modelled as $Z \cdot \beta$, with Z a matrix containing the spline basis functions and β the corresponding spline coefficients that need to be estimated. Note that the separate elements in β do not have biological meaning; conversely, the product $Z \cdot \beta$ corresponds to the estimated time sequence of the RMR. The time variation that can be captured in RMR(t) depends on the number of knots that is chosen in the spline function, which determines the length of β. More knots will allow one to model quicker changes in the RMR, but this comes at the price of estimating a larger number of spline coefficients and therefore a higher degree of uncertainty in the estimate of RMR. As a rule of thumb, to prevent overfitting, the number of knots should not be higher than 0.2 times the number of data points.

8.3.3 Kalman filtering

An alternative method for estimating the time-dependent RMR from Q and PA is Kalman filtering, as was originally proposed by Even *et al.* (1991). Kalman filtering is based on the assumption that the unknown state variables, here RMR and CCA, represent normally distributed stochastic processes. The relation between the state variables and the measured Q and PA time sequences are expressed in a state space model, from which the time-dependent estimates of the RMR and CCA can be obtained. The accuracy of the RMR and CCA estimates depends on the choice of several parameters of the filter, which must be set in advance by the user. Importantly, for Kalman filtering to give optimal results in this setting, Q and PA data have to be sampled with a high frequency, such as those generated by single channel indirect calorimetry systems in which data is registered continuously (Even *et al.*, 2012).

8.3.4 Mathematical modelling of activity heat and thermic effect of feeding

Alternatively, the time variations in Q can, using a modelling approach, be decomposed into AQ, TEF and a daily constant component. This component comprises the resting, fasting Q and the long term component of TEF (Van Milgen *et al.*, 1997). In this approach (Chapter 4, Labussière *et al.* (2015) for further details), each component is considered to generate instantaneous O_2 consumption and CO_2 production inside the chamber, and mixing of these gases within the chamber and time delay with gas analysers are calculated taking into account the physical characteristics of the chamber (temperature, relative humidity, ventilation rate). VO_2 and VCO_2 linked to AQ are considered to be proportional with daily-fixed coefficients to one estimate of PA. The latter is the electrical signals from force transducers that are placed under the cage where the animals are housed. The TEF is approached as an increase (followed by a decrease) in VO_2 and VCO_2 after feed intake according to a gamma function, whereas the gas exchange linked to the remaining, constant component described above is supposed to be constant within the day. Parameter estimation requires high-frequency measurements of PA and gas concentrations and data generated by multiplexed systems are not adequate. It is performed for each day of measurement, and the average values of parameters over at least 5 days are used to calculate AQ in particular from volumes of gas and cumulative signal from force transducers using the Brouwer equation (1965).

8.3.5 Influence of experimental variables and settings

As discussed previously, the type of activity sensor can have a huge impact on the accuracy with which the RMR, CCA and AQ can be estimated. In addition, the frequency with which Q and PA are sampled and the size of the chamber affect the performance of the aforementioned computational methods, and optimization of these experimental settings is vital for obtaining accurate results. Finally, the relationship between Q and PA is heavily influenced by the choice of housing and feeding methods.

8.3.5.1 Sampling frequency

The frequency with which the respiratory exchange can be sampled depends on the type of indirect calorimetry system. For single channel systems, a dedicated set of O_2 and CO_2 analysers measures the gas exchange in a single chamber. For such systems, Q can be determined at any desired frequency, with as only constraint technical limitations related to data storage and data handling. On the other hand, multiplexing systems are often used as well, using a single set of analysers measuring the air from multiple chambers in succession. This setting puts a limit on the time resolution that can be attained since the tubing between the multiplexer and the sensors, sometimes containing a demoisturiser, needs to be cleared from old air every time the system switches to the next chamber. Clever solutions to maximize the sampling frequency in multiplexing systems have been described by Melanson *et al.* (2010). Conversely, for most activity sensors no limits exist with regard to the sampling frequency, so PA can and should typically be measured with a high time resolution.

Sampling frequency is an important parameter that affects both the accuracy and the amount of details that can be obtained with decomposition algorithms. If the sampling frequency is low, fewer data points are available, which increases the variability in the estimated AQ and RMR. Especially the accuracy with which the high frequency components in the RMR, i.e. the rapid changes, can be estimated is affected by changes in the sampling frequency (Van Klinken *et al.*, 2012). As a result, measures of the dynamic response of metabolism on (experimental) interventions, such as the TEF, are suffering more from low sample rates than, for instance, the BMR (Van Klinken *et al.*, 2013). In addition, frequently sampled data is necessary for performing deconvolution.

8.3.5.2 Chamber washout time

As discussed earlier, the time pattern in the VO_2 and VCO_2, measured at the chamber's outlet is dampened because of the mixing process of the exhaled air with the incurrent air. In the ideal case, gas mixing can be modelled as a linear low-pass filter on the instantaneous respiratory exchange, and the same transformation can be applied on the PA time sequence in order to align it with Q. Unfortunately, the dampening effect attenuates the high frequency variations in Q that are due to activity, meaning that decomposition algorithms – which rely on the time correlation between the PA and Q – become less robust. The amount of dampening is directly related to τ, i.e. to the proportion of the chamber size to the rate of the air flow (Van Klinken, 2013). Consequently, application of the low-pass filter on the PA sequence remains relatively accurately only when τ is relatively small, such as for rodent indirect calorimetry systems, whereas for systems with large chamber sizes component analysis can only be performed on the instantaneous PA and Q data, where the latter has to be calculated using deconvolution.

8.3.5.3 Choice of housing conditions and feeding method

Changes in Q due to PA are much quicker than the time variation in the other components, enabling to identify the PA related component in Q using the techniques described above. A clear relationship between Q and PA provides the basis for this separation. The time variation in PA is, however, strongly influenced by housing conditions and feeding method chosen. For example, in group housed subjects, variation in PA in time is dampened as activity patterns between individuals differ. This may be enlarged when food is available *ad libitum*. In general, meal feeding increases the variation in PA with time, as feeding bouts always coincide with an increase in PA, preceding the increase in TEF. Therefore, in a setting with group-housed subjects, combined with *ad libitum* availability of food, the correlation between Q and PA may be below the desirable R^2 of 0.70. In an experiment with group-housed growing pigs (5 pigs/group), measuring PA by radar, we typically found an R^2 of 0.60 when pigs were given *ad libitum* access to feed, whereas this increased ($P<0.01$) to 0.66 when pigs were fed twice daily. It further increased to 0.71 during a 24 h period of fasting (W.J.J. Gerrits, unpublished observations). In the case of *ad libitum* access to feed, it may be considered to include a separate day of measurement in which PA is induced at times when TEF is low, to enable a reliable estimate of the energetic costs of physical activity (CCA).

8.4 Influence of housing conditions on activity and behaviour

Decomposing Q into its components has a considerable added value for indirect calorimetry data and contributes greatly to the interpretation and biological meaning of the results obtained. The biological question addressed usually determines the experimental setting chosen, and researchers should be aware of the potential effects of their conditions on PA and behaviour. In this paragraph, attention is paid to relationship between ambient temperature and AQ, and on the impact of individual housing on AQ and behaviour.

8.4.1 Ambient temperature

When using indirect calorimetry, proper mixing of the air within the chamber and adequate systems for enthalpy regulation (Chapter 11, Gaughan *et al.*, 2015) offer the opportunity to study the impact of changes in environment on Q and AQ. If variations in relative humidity are not very common, variations in temperature are more frequent because they address targeted issues linked to animal adaptation to its environment. Variation in ambient temperature can modify behaviour of the animals by modifying locomotion activity, duration in standing or lying postures and number of posture changes. Additionally, physiological adaptations to cold (intense shivering) and heat stress (panting) generate muscular work which contributes to AQ. The way physical activity is measured strongly affects the estimate of ambient temperature effect on AQ. In particular, sensitivity of the detectors that are used to measure physical activity is key to estimate the contribution a very small movements to AQ, such as those related to panting and shivering. Using force sensors and mathematical modelling of daily dynamics of heat production, a rather large range of ambient temperatures have been studied in growing pigs with BW that varied from 25 to 75 kg (Figure 5). Irrespective of body weight (BW), the minimal AQ varied from 7 to 10% of metabolic energy (ME) intake in the pigs included in these studies. At ambient temperatures above the thermoneutral zone, ME intake decreased but a larger portion of ME is dissipated as AQ (up to 100% increase relative to the minimal loss), mainly related to panting that increased AQ. It appeared that the increase in AQ related to panting completely compensated for the decreased time spent in standing position (Renaudeau *et al.*, 2013). Even though shivering is used as a way to produce Q and maintain body temperature, its contribution to AQ is rather low and the increase averaged only 50% of the minimal loss when ambient temperature decreased down to 12 °C (Figure 5).

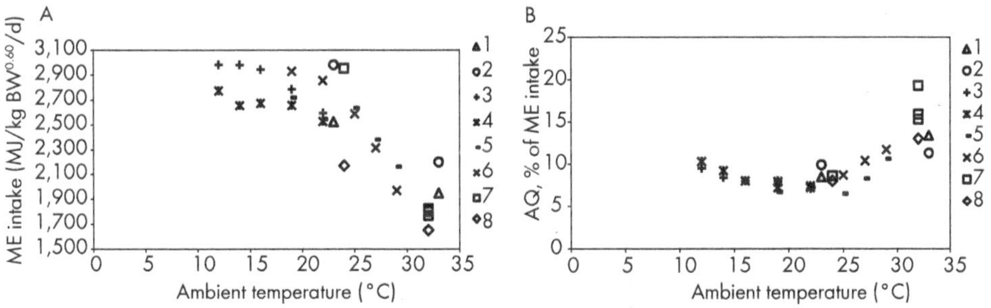

Figure 5. Relationship between ambient temperature (°C) and (A) metabolisable energy intake (ME) and (B) heat production related to physical activity (AQ) in growing pigs. (1) Collin et al., *2001a, 25 kg body weight (BW) piglets, individual housing, temperature from 23 to 33 °C; (2) Collin* et al., *2001b, 25 kg BW piglets, group housing, temperature from 23 to 33 °C; (3) Quiniou* et al., *2001, 50 kg BW growing pigs, group housing, temperature from 12 to 22 °C; (4) Quiniou* et al., *2001, 75 kg BW growing pigs, group housing, temperature from 12 to 22 °C; (5) Quiniou* et al., *2001, 50 kg BW growing pigs, group housing, temperature from 19 to 29 °C; (6) Quiniou* et al., *2001, 75 kg BW growing pigs, group housing, temperature from 19 to 29 °C; (7) Renaudeau* et al., *2013, 60 kg BW growing pigs, individual housing, temperature from 24 to 32 °C; (8) Campos* et al., *2014, 65 kg BW growing pigs, group housing, temperature from 24 to 32 °C.*

8.4.2 Individual vs group housing

The biological question addressed determines the experimental setting chosen, often requiring subjects to be housed individually. In an indirect calorimetry setting, this usually coincides with social isolation, i.e. without visual, olfactory and auditory contact with conspecifics. Separation from a group, and the single-housing of animals of gregarious species, for even short periods, can be a significant stress factor. For these reasons, EU guidelines prescribe that, 'should single-housing in a small enclosure be necessary, because of a special experimental paradigm, the duration and extent of confinement should be justified by the experimenter, balancing the likely effect on the well-being of the animal against the scientific value and requirements of the experiment' (EU Cons 123 (2006) 3). In attempting to minimize chamber washout time, as mentioned above, these housing requirements are conflicting with the desired precision of the measurements, and careful procedures should be in place. For cats, this is extensively discussed in Chapter 13 by Gooding *et al.* (2015). The stress induced by single-housing may alter heat production. Although most experimenters take great care in keeping the environmental conditions constant within a trial, scientific answers obtained may depend on the conditions under which the experiment was conducted. These effects are rarely the subject of study. In Table 1, the effect of individual vs group housing in pigs was extracted from various studies conducted in our lab. In all studies presented, the comparison of housing systems was made under identical nutritional conditions (i.e. identical diet composition and feed intake) and ambient temperatures. In the study of Geverink et al (2004), conducted with adult gilts, prolonged individual housing (5 mo) increased AQ by 60%, i.e. from 16 to 23% of Q. This increase coincided with an altered circadian salivary cortisol rhythm and increased stomach wall damage in the individually housed gilts. In the study by Rijnen (2003) in growing

Table 1. Comparison of activity related heat production (AQ) in an individual vs group housing setting in pigs.

Individual housing		Group housing		*P*-value	Reference
No pigs/no observations	AQ, kJ/(BW$^{0.75}$/d)	No pigs/no observations	AQ, kJ/(BW$^{0.75}$/d)		
36/12	105	36/12	65	<0.001	1
10/10	70	140/10	78	0.18	2
18/18	116	50/10	83	<0.001	3

[1] Geverink *et al.*, 2004; gilts of 160-170 kg body weight (BW); housing either in groups of 3 per chamber, or individual housing of 3 gilts on metabolism cages in one chamber.

[2] Rijnen, M., 2003; growing pigs of 40-50 kg BW, either housed individually (one pig/chamber) or in groups of 14 in 2 pens in one chamber.

[3] Blok *et al.*, 2015: growing pigs of 20-30 kg BW, either housed individually (one pig/chamber) or in groups of 5 in one chamber.

pigs, no effects of housing conditions on AQ were observed. The work of Bosch *et al.*, reported by Blok *et al.* (2015), also in growing pigs, revealed that single housing increases AQ by 40%, or from 12 to 17% of Q, when compared with a group housed setting of 5 pigs/pen. It indicates that, although single-housed animals have less room to move around, AQ is not reduced. If anything, it is more likely to be increased, and this increase may well be related to the stress imposed by the housing conditions.

When performing indirect calorimetry studies in calves, individual housing is commonly applied (see e.g. Labussière *et al.*, 2008; Van den Borne *et al.*, 2006). Milk-fed calves often develop abnormal oral behaviours, including tongue playing and excessive oral manipulation of the pen, supposedly due to chronic stress from a limited opportunity to eat and ruminate on roughage, from housing conditions, or both. Bokkers *et al.* (2012) monitored plasma cortisol and behaviour of veal calves following their restraint in a metabolic cage, as commonly applied in calorimetry studies. Plasma cortisol concentrations returned to baseline within 48 h after the restraint, but abnormal oral behaviours did not, although large inter-individual variation was observed. In an earlier calorimetry experiment in milk-fed calves (120 kg of BW) by Van den Borne *et al.* (2004, 2006), calorimetry was combined with behavioural observations, using 6 min scan sampling for 54 h per calf. Physical activity was recorded by a radar device and posture was recorded by infrared beam interruption. A massive difference in the time spent idling was observed between standing (5%) and lying (47%), and stereotyped behaviours, mainly manipulation of the bucket or pen fixtures and tongue playing, were substantially increased during standing. Performing stereotyped behaviours during standing increased Q almost threefold when compared with idling (Figure 6). From these observations, it appears that the energetic costs for stereotyped behaviours are considerable. If calorimetry measures are to be conducted in single-housing, great care should be taken to minimize the occurrence of such behaviours.

In summary, this chapter provides an overview of current methodology used to measure physical activity within the setting of indirect calorimetry. Several approaches to decompose total heat production into activity related and resting metabolic rate are discussed. In conventional approaches, the energetic costs of each unit of physical activity, as well as the resting metabolic rate are assumed to be constant. Improved methodology is available using penalised least squares regression, to model time dependent variation in the resting metabolic rate. The impact of experimental variables and settings, e.g. sampling frequency and chamber washout time have been discussed. In addition, housing conditions, including ambient temperature, and feeding methods can impact on the relationship between physical activity and total heat production, affecting the reliability with which total heat production can be decomposed into activity related heat production and resting metabolic rate. Finally, individual housing conditions, often difficult to avoid in indirect calorimetry experiments, can, by influencing behaviour and stress, substantially influence activity related heat production and thereby influence the results of calorimetry experiments.

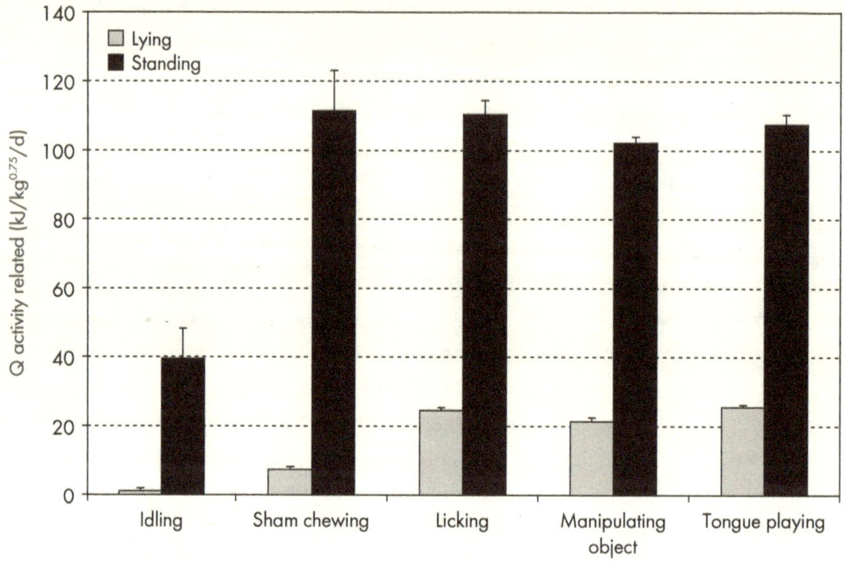

Figure 6. Activity related heat production (AQ) in milk-fed calves, during idling or when performing various stereotyped behaviours, either when lying or when standing (data from Van den Borne et al., *2004, 2006).*

References

Arch, J.R., Hislop, D., Wang, S.J. and Speakman, J.R., 2006. Some mathematical and technical issues in the measurement and interpretation of open-circuit indirect calorimetry in small animals. International Journal of Obesity 30: 1322-1331.

Blaxter, K., 1989. Energy metabolism in animals and man. Cambridge University Press, Cambridge, UK.

Blok, M.C., Brandsma, G., Bosch, G., Gerrits, W.J.J., Jansman, A.J.M., Fledderus, J. and Everts, H., 2015. A new Dutch net energy formula for feed and feedstuffs for growing and fattening pigs. CVB Documentation Report nr 56. WUR Livestock Research, Dept. Animal Nutrition, Wageningen, the Netherlands.

Bokkers, E.A.M., Webb, L.E., Engel, B., Berends, H., Gerrits, W.J.J. and Van Reenen, C.G., 2012. How does solid feed provision and display of abnormal oral behaviours affect salivary cortisol in veal calves? In: Waiblinger, S., Winckler, C., Gutmann, A. (eds.) ISAE2012. Proceedings of the 46th Congress of the International Society for Applied Ethology. Quality of life in designed environments? Wageningen Academic Publishers, Wageningen, the Netherlands, p. 217.

Brouwer, E., 1965. Report of sub-comittee on constants and factors. In: Blaxter, K.L. (ed.) Energy metabolism. Academic Press, London, UK, pp. 441-443.

Brown, D., Livesey, G. and Dauncey, M.J., 1991. Influence of mild cold on the components of 24 hour thermogenesis in rats. Journal of Physiology 441: 137-154.

Bursztein, S., 1989. Energy metabolism, indirect calorimetry, and nutrition. Williams & Wilkins, Baltimore, MD, USA.

Campos, P.H.R.F., Labussière, E., Hernandez-Garcia, J., Dubois, S., Renaudeau, D. and Noblet, J., 2014. Effects of ambient temperature on energy and nitrogen utilization in lipopolysaccharide-challenged growing pigs. Journal of Animal Science 92: 4909-4920.

Collin, A., Van Milgen, J., Dubois, S. and Noblet, J., 2001a. Effect of high ambient temperature on energy utilization in piglets. Journal of Animal Science 79: 1849-1857.

Collin, A., Van Milgen, J., Dubois, S. and Noblet, J., 2001b. Effect of high temperature on feeding behavior and heat production in group-housed young pigs. British Journal of Nutrition 86: 63-70.

EU Cons 123, 2006. Appendix A of the European Convention for the Protection of Vertebrate Animals Used for Experimental and Other Scientific Purposes (ETS No. 123). Guidelines for Accommodation and Care of Animals. Article 5 of the Convention, Strasbourg, 15 June 2006, 109 pp.

Even, P.C. and Nadkarni, N.A., 2012. Indirect calorimetry in laboratory mice and rats: principles, practical considerations, interpretation and perspectives. American Journal of Physiology – Regulatory, Integrative and Comparative Physiology 303: R459-R476.

Even, P.C., Perrier, E., Aucouturier, J.L. and Nicolaidis, S., 1991. Utilisation of the method of Kalman filtering for performing the on-line computation of background metabolism in the free-moving, free-feeding rat. Physiology and Behaviour 49: 177-187.

Gaughan, J.B., Heetkamp, M.J.W. and Hendriks, P., 2015. Indirect calorimetry: assessing animal response to heat and cold stress. In: Gerrits, W.J.J. and Labussière, E. (eds.) Indirect calorimetry. Techniques, computations and applications. Wageningen Academic Publishers, Wageningen, the Netherlands, pp. 213-230.

Gentry, J.L., Swinkels, J.W., Lindemann, M.D. and Schrama, J.W., 1997. Effect of hemoglobin and immunization status on energy metabolism of weanling pigs. Journal of Animal Science 75: 1032-1040.

Geverink, N.A., Heetkamp, M.J.W., Schouten, W.G.P., Wiegant, V.M. and Schrama, J.W., 2004. Backtest type and housing condition of pigs influence energy metabolism. Journal of Animal Science 82: 1227-1233.

Gooding, M.A., McKnight, L.L., Ramsey, J.J. and Shoveller, A.K., 2015. Conducting indirect calorimetry studies in companion animals. In: Gerrits, W.J.J. and Labussière, E. (eds.) Indirect calorimetry. Techniques, computations and applications. Wageningen Academic Publishers, Wageningen, the Netherlands, pp. 247-274.

Heetkamp, M.J.W., Gerrits, W.J.J., Van Knegsel, A.T.M. and Van den Brand, H., 2007. Comparing techniques to estimate energy expenditure on physical activity in individually housed dairy cows. In: Ortigues-Marty, I. (ed.) Energy and protein metabolism and nutrition. EAAP publication No.124. Wageningen Academic Publishers, Wageningen, the Netherlands, pp. 211-212.

Heetkamp, M.J.W., Schrama, J.W., De Jong, L., Swinkels, J.W., Schouten, W.G. and Bosch, M.W., 1995. Energy metabolism in young pigs as affected by mixing. Journal of Animal Science 73: 3562-3569.

Heetkamp, M.J.W., Van den Brand, H., Degen, S.B. and Gerrits, W.J.J., 2003. Comparing two techniques to estimate energy expenditure on physical activity in group housed animals. In: Souffrant, W.B. and Metges, C.C. (eds.) Progress in research on energy and protein metabolism. EAAP publication No. 109. Wageningen Academic Publishers, Wageningen, the Netherlands, pp. 507-511.

Labussière, E., Dubois, S., Van Milgen, J. and Noblet, J., 2013. Partitioning of heat production in growing pigs as a tool to improve the determination of efficiency of energy utilization. Frontiers in Physiology 4: 146.

Labussière, E., Dubois, S., Van Milgen, J. and Noblet, J., 2015. Modelling gas exchanges to partition heat production between fasting, thermic effect of feeding and physical activity in farm animals. In: Gerrits, W.J.J. and Labussière, E. (eds.) Indirect calorimetry. Techniques, computations and applications. Wageningen Academic Publishers, Wageningen, the Netherlands, pp. 73-86.

Labussière, E., Dubois, S., Van Milgen, J., Bertrand, G. and Noblet, J. 2008. Fasting heat production and energy cost of standing activity in veal calves. British Journal of Nutrition 100: 1315-1324.

Melanson, E.L., Ingebrigtsen, J.P., Bergouignan, A., Ohkawara, K., Kohrt, W.M. and Lighton, J.R.B., 2010. A new approach for flow-through respirometry measurements in humans. American Journal of Physiology – Regulatory, Integrative and Comparative Physiology 298: R1571-R1579.

Quiniou, N., Noblet, J., Van Milgen, J. and Dubois, S. 2001. Modelling heat production and energy balance in group-housed growing pigs exposed to low or high ambient temperatures. British Journal of Nutrition 85: 97-106.

Ravussin, E., Lillioja, S., Anderson, T.E., Christin, L. and Bogardus, C., 1986. Determinants of 24-hour energy expenditure in man – methods and results using a respiratory chamber. Journal of Clinical Investigation78: 1568-1578.

Renaudeau, D., Frances, G., Dubois, S., Gilbert, H. and Noblet, J., 2013. Effect of thermal heat stress on energy utilization in two lines of pigs divergently selected for residual feed intake. Journal of Animal Science 91: 1162-1175.

Rijnen, M., 2003. Dietary fiber in pigs. PhD thesis. Wageningen Institute of Animal Sciences, Wageningen University, Wageningen, the Netherlands.

Van den Borne, J.J.G.C., Van der Heijden, S.J.F.M., Oorsprong, H., Bokkers, E.A.M., Bolhuis, J.E. and Gerrits, W.J.J., 2004. High energetic costs of stereotyped behaviour in preruminant calves. Journal of Dairy Science 87 Suppl. 1: 251.

Van den Borne, J.J.G.C., Verstegen, M.W.A., Alferink, S.J.J., Giebels, R.M.M. and Gerrits, W.J.J., 2006. Effects of feeding frequency and feeding level on nutrient utilization in heavy preruminant calves. Journal of Dairy Science 89: 3578-3586.

Van Klinken, J.B., Van den Berg, S.A.A. and Willems van Dijk, K., 2013. Practical aspects of estimating energy components in rodents. Frontiers in Physiology 4: 94.

Van Klinken, J.B., Van den Berg, S.A.A., Havekes, L.M. and Willems van Dijk, K., 2012. Estimation of activity related energy expenditure and resting metabolic rate in freely moving mice from indirect calorimetry data. PLoS ONE 7: e36162.

Van Milgen, J., Noblet, J., Dubois, S. and Bernier, J.F. 1997. Dynamic aspects of oxygen consumption and carbon dioxide production in swine. British Journal of Nutrition 78: 397-410.

Verstegen, M.W.A., Van der Hel, W., Brandsma, H.A., Henken, A.M. and Bransen, A.M., 1987. The Wageningen respiration unit for animal production research: a description of the equipment and its possibilities. In: Verstegen, M.W.A. and Henken, A.M. (eds.) Energy metabolism of farm animals: effects of housing, stress, and disease. Martinus Nijhoff, Dordrecht, the Netherlands, pp. 21-48.

Wenk, C. and Van Es, A.J.H., 1976. Eine Methode zur Bestimmung des Energieaufwandes für die Körperliche Aktivität von wachsenden Küken. Schweizerische Landwirtschaftlicher Monatshefte 54: 232.

Westerterp, K.R., 1999. Body composition, water turnover and energy turnover assessment with labelled water. Proceedings of the Nutrition Society 58: 945-951.

Chapter 9: Calorimeters for enteric methane measurements

E. Kebreab
Department of Animal Science, University of California, Davis, CA 95616, USA; ekebreab@ucdavis.edu

Abstract

Methane (CH_4) is a natural by-product of microbial fermentation of carbohydrates and amino acids in the rumen and the hindgut of farm animals. Methane is a powerful greenhouse gas, therefore, there has been increasing interest in reducing methane emissions from livestock. Calorimetry chambers have been built, in part, to measure emissions and assess effects of mitigating effects of dietary manipulations. The objective of this chapter is to review various types of indirect calorimetry chambers developed to measure CH_4 emissions from livestock. Indirect calorimetry chambers allow measurement of gaseous exchange to high degree of accuracy. However, confinement in small spaces results in altered animal behaviour, including reduced feed intake. Modifications that use polycarbonate has improved animal welfare concerns while maintaining high accuracy of measured gases. Principles of indirect calorimetry were used to build ventilated hoods (head box) to measure gas exchange. Units that are suitable for group of animals have also been constructed to measure methane emissions based on the same principles. All of the above techniques cannot be used for animals on pasture, therefore, tracer based techniques such as the sulphur hexafluoride or GreenFeed systems have been developed. Compared to indirect calorimetry methods, tracer methods do not measure whole animal emissions and there is high animal-to-animal variability associated with measurements. So, frequent sampling and more animals are required to improve accuracy of measurements. Indirect calorimetry methods are considered to be the gold standard in measuring methane emissions. However, for pasture based systems, tracer methods are more suited and will continue to be used.

Keywords: greenhouse gases, enteric fermentation, livestock

9.1 Introduction

The livestock contribution to global anthropogenic greenhouse gas (GHG) emission ranges from 7 to 18% depending on accounting approaches and scope (Hristov *et al.*, 2013). Enteric fermentation and manure storage are the main sources of methane (CH_4) that contribute directly to the total GHG emissions from livestock production systems. According to estimates for the United States, in 2011, CH_4 emissions from enteric fermentation and manure management accounted for 23.4 and 8.9% of anthropogenic CH_4 emissions, respectively (EPA, 2013). Life-cycle assessments of livestock systems have shown that on-farm emissions, the largest of which

DOI 10.3920/978-90-8686-261-0_9, © Wageningen Academic Publishers 2015

is CH_4 from enteric fermentation, represent the main contribution to carbon footprint of dairy or beef supply chains (e.g. Thoma *et al.*, 2013). Methane is produced by methanogenic archaea (methanogens) in the fermentative fore-stomach (rumen) and large intestine of cattle. Methanogens primarily utilise hydrogen as an energy source and carbon dioxide (CO_2) to produce CH_4 and water (Attwood and McSweeney, 2008). Enteric CH_4 production in cattle represents a dietary energy loss which has been measured to range between 1 and 11% of the gross energy intake (Moraes *et al.*, 2014). The majority of CH_4 emission is through eructation, with only 2-3% of the total CH_4 emissions coming from flatulence (Murray *et al.*, 1976).

Globally considerable effort has been spent to reduce CH_4 emissions from ruminants because it potentially offers a dual purpose of increasing the amount of energy available to the animal and reducing the environmental impact of livestock production. Hristov *et al.* (2013) comprehensively reviewed mitigation options to reduce CH_4 emissions from livestock and recommended that dietary manipulation and chemical compounds that serve as alternative hydrogen sink hold promise. In order to assess effectiveness of mitigation options, CH_4 emissions should be measured accurately. There have been various attempts to develop methods of measuring CH_4 emissions suitable to experimental and production conditions. The methods range from whole-animal respiration calorimetry chambers with various types of gas analysers to tracer techniques and whole-barn mass-balance techniques (Johnson and Johnson, 1995; McLean and Tobin, 1987; Young *et al.*, 1975). Some of these methods have been described (e.g. Harper *et al.*, 2011; Johnson and Johnson, 1995; Kebreab *et al.*, 2006) and can be broadly categorized as chamber, tracer, and micrometeorological techniques. Indirect calorimetry chambers are considered to be the gold standard (Johnson and Johnson, 1995) in measuring CH_4 as well as O_2 and CO_2 from animals. The objective of this chapter is to review various types of indirect calorimetry chambers developed to measure heat and gaseous exchange from livestock with particular focus on CH_4 gas emissions.

9.2 General description of respiration chambers for measuring methane

Respiration chambers developed to measure CH_4 emissions range from sealed stainless steel whole animal chambers to more portable covering part of the animals to flexible tunnel-like enclosures that may house a number of animals at the same time. All of these methods involve the creation of air flow that passes over one or more animals from an incoming to an outgoing opening. Outside air is circulated around the whole animal (chambers), head (hoods) or mouth and nose (masks) and expired air mixed with incoming fresh air are intermittently sub-sampled and analysed or aliquots collected for later analyses. Gaseous exchange is determined by measuring the total airflow through the system and the difference in gas concentrations in the inflowing and outflowing air streams. Gaseous composition of the ingoing and outgoing air from the respiration chamber can be measured using methods such as dual-channel infrared (CH_4, CO_2) and paramagnetic (O_2) analysers (Cammell *et al.*, 1986; Chapter 3, Lighton *et al.*, 2015). Other chambers of similar design and function measure the rate of air flow and the concentration of various gases at the incoming and outgoing vents with direct measure analysers, including flame ionization and photo-acoustic technology (e.g. Hamilton *et al.*, 2010; Stackhouse *et al.*, 2011; Sun *et al.*, 2008).

9.2.1 Whole animal indirect calorimetry chambers

Typically, chambers for large animals are constructed of steel and have an air conditioning system to maintain a temperature range of 18±2 °C and relative humidity of 60±10%. Various papers describe in detail the construction of indirect calorimetry chambers including air flow, ventilation rate, sensors for various gases, etc. (Cammell *et al.*, 1981; Hellwing *et al.*, 2012). For example Cammell *et al.* (1981) offer a guide for construction of open-circuit calorimeters for measuring gaseous exchange and calculating heat production in sheep and young cattle. Similarly, Miller and Koes (1988) published a method for construction and operation of an open-circuit indirect calorimetry system for small ruminants. Chambers are normally calibrated before any experiment by releasing known amounts of CH_4 in each chamber. For example, Grainger *et al.* (2007) calibrated their chambers in 3 stages. In stage 1, the chambers were operated with no source of gas (empty chamber and no gas release). This allowed evaluation of any bias error in the analytical system. In stage 2, the mass flow controllers were used to release CH_4 into the exhaust duct. This results in near instantaneous evaluation of the entire analytical system that included the turbines, sample lines, connections, solenoid valves, pumps, gas analyser, data recorder, and the calculation of the flux rates. It was expected that near-perfect recovery should result from this test, because leakage of gas along the exhaust duct was unlikely. Finally, in stage 3, CH_4 was released inside the chamber and recovery test conducted. The authors carried out a total of 8 calibrations using stages 1, 2, and 3 to characterise recovery rates. There was a maximum of 1% difference between released and recovered CH_4.

Although used extensively, whole-animal open-circuit indirect respiration chamber systems are expensive to construct and maintain. Feed intake is normally reduced in this type of environment because cows are effectively locked in the chamber for several days and exhibit anxiety. Several institutions now require that animals be kept in pairs during measurement or the device should allow the animals to see and hear others around them. Some of these systems have been adapted to use polycarbonate, which is a transparent material and cheaper to construct. Hellwing *et al.* (2012) described a low-cost system using transparent polycarbonate chambers with simpler construction and air-conditioning designs compared to the traditional systems (Figure 1). Hellwing *et al.* (2012) placed the chambers in the barn where other animals are located so that experimental animals feel they are in their normal environment. The disadvantage of this system is that background concentrations of gases are highly enriched and variable so need to be measured constantly. Dust could also cause problems with sample lines and ducts. Similarly, Hassanat *et al.* (2013) reduced anxiety of dairy cows by using chambers that were equipped with windows and speakers so the cow in the chamber can see and hear other cows. Future construction of calorimetry chambers will take advantage of improved

Figure 1. An example of a modern type of whole animal indirect calorimetry chamber made from polycarbonate on a steel frame (Foulum, Denmark). Photo courtesy of E. Kebreab.

technology in order to create a controlled environment whereby only imposed treatments vary without affecting animal welfare and behaviour.

Whole animal indirect calorimetry chambers are used to obtain accurate and frequent measurements of enteric CH_4 production. However, the method is time consuming, expensive (even with polycarbonate) and not applicable in some production systems such as cattle on pasture.

9.2.2 Ventilated hood chambers

Easily transportable chamber-based methods focus solely on eructated gaseous emissions (those coming from the mouth and nostrils of the animal) and heat exchange (Kelly *et al.*, 1994; Odongo *et al.*, 2008). These methods allow the animal to eat, drink, and lay down while secured to a head chamber. They are a cheaper and more flexible alternative to whole enclosure chambers, but they are still useful to determine the effects of various treatment options on eructated gaseous emissions. Odongo *et al.* (2008) described in detail the construction of ventilated hood chambers. The authors used polypropylene sheeting material mounted on steel frames. On one end of the chamber they had an opening through which the animal's head was placed. A tightly woven tarp material was attached to the opening in the hood and secured around the animal's neck using a nylon drawstring, such that the animal could stand or lie down in the stanchions with its head in the hood. On the other end of the hood is a door which is used to offer feed for the animal or clean the water trough. Modern ventilated hood chambers are made of polycarbonate material to allow an outside view for the animals (e.g. Place *et al.*, 2011, Figure 2). On the side of the hood there is an inlet to allow outside air to enter the hood. On the top of the hood, there is an outlet port which connects to instrumentation devices by a sampling line. Methane and other gases are then measured from the sampled air using various types of analysers such as gas chromatography. The portability of the chambers are an advantage as they can be moved in barns or stored with relative ease.

The main disadvantages of ventilated hood chambers are their high labour costs, their inability to measure hindgut CH_4 (although this is a tiny fraction) and their unsuitability for use in pasture. Similar to whole animal chambers, welfare and behavioural concerns arise because of restricted movement. A face mask that measures CH_4 from eructated gases has been developed, but it does not allow the animal to eat, drink, or behave normally (Liang *et al.*, 1989).

9.2.3 Polythene tunnel

As mentioned earlier, chambers made of steel or polycarbonate and ventilated hood chambers measure one animal (or a small group of animals) at a time in confinement. Therefore, they are not suitable for studies that are pasture-based or use multiple animals to simulate conditions on a farm. Using similar principles to indirect calorimetry chambers, Lockyer and Jarvis (1995) developed a system constructed from a large polythene tunnel with two small wind tunnels to blow air into and draw air from a larger tunnel. Each tunnel consists of a steel duct, approximately 1.5×0.4 m, housing a variable speed co-axial fan and a vane anemometer head. Air flow through each tunnel can be controlled at rates of up to 0.9 m^3/s and it was calculated from measurements of air speed. The concentration of CH_4 in the air entering and leaving

Figure 2. A cow standing while her eructated and respired emissions are being measured in the ventilated hood chamber (University of California, Davis, CA, USA).

the tunnel is measured using a gas chromatograph fitted with a flame ionization detector (for detailed description see Lockyer and Jarvis, 1995). Using the same system, Murray *et al.* (1999) reported that a peak CH_4 concentration rise to about 10 ml/l in the tunnel can be detected with an accuracy of about 0.4% over the measurement range. Evaluation of the system showed that it recovers 92.6% (standard error = 1.55), which is less CH_4 than traditional chambers (Murray *et al.*, 1999). The advantages of the system are that it allows free movement of animals inside the tunnel and is inexpensive to build. However, control of temperature inside the tunnel when ambient temperatures are high is not possible. Up until recently the polythene tunnel method has been confined to using sheep due to space limitations. However, recent modifications to the system has allowed researchers to use the method for cattle. For example, Cooprider *et al.* (2011) used a cattle pen enclosure or 'biobubble' (Figure 3) that consisted of a domelike structure measuring 22×11 m with dirt floors, pipe fencing, feed bunks, and a water trough to measure CH_4 emissions from a group of beef cattle. In the structure, inlet air (through a cooling pad) and outflow air (through fans with ventilation openings) samples were taken to measure CH_4 and other gaseous concentrations. Air ventilation flow rate was also measured. Emission rate (l/animal) was calculated using the following equation:

$$\text{Emission rate} = \frac{\dfrac{\text{flow} \times \text{net gas} \times 10^{-6}}{\text{volume}} \times \text{molecular weight}}{\text{n}} \qquad (1)$$

Figure 3. Environmental chamber or 'bio-bubble' used to measure methane emissions and other gas exchanges from a group of animals (University of California, Davis, CA, USA).

where *flow* was the ventilation flow rate in l/h, *net gas* was the concentration differential (difference between inlet and outflow in μl/l), *volume* was the volume of 1 molar ideal gas in l/mole (= 24.04 at 20 °C), *molecular weight* was in g/mole, and *n* was the number of animals in each 'biobubble'. Airflow was determined by the rotations per minute of the fans and the static differential pressure between inside and outside the structure recorded using data loggers on 10-min intervals. Temperature and relative humidity inside the structure were recorded on a data logger to correct the airflow rate. Concentrations of target gases in the structure were continuously measured for CH_4 using a TEI Model 55C Direct Methane, Non-Methane Hydrocarbon Analyzer (Thermo Environmental Instruments, Waltham, MA, USA) and for N_2O using an INNOVA Model 1412 Photoacoustic Field Gas-Monitor (INNOVA, AirTech Instruments, Ballerup, Denmark). The overall system, including 'biobubble', sensors, sampling devices, and analysers was evaluated by introducing 3 l/min pure CH_4 gas into the structure, mixing it with the ventilation air, and measuring the CH_4 concentration in the outflow of the 'biobubble.' Cooprider *et al.* (2011) reported that the recovery of the introduced CH_4 was 100±10% with the system. Although the system is not suited for pasture-based measurements, it can simulate on-farm conditions in most confined systems, particularly, feedlots. Welfare and behaviour concerns related to restriction is not an issue because the animals are kept in groups with unlimited access to feed and water. In addition, despite the relative low accuracy in methane recovery, it offers an opportunity to measure CH_4 from animals and manure in similar conditions where animals are kept in commercial operations. The typical duration of measurements is about 4 weeks with 4 chambers running simultaneously.

9.3 Other methods for measuring methane emissions

There are several other methods used to determine CH_4 emissions. Although they are not built based on the principle of calorimetry chambers, they have been extensively used in pasture-based production systems and often compared to indirect calorimetry techniques. Therefore, before reviewing comparison of techniques, a brief description of the most widely techniques for measuring CH_4 emissions is given below.

9.3.1 Tracer techniques

The most commonly used inert gas tracer is sulphur hexafluoride (SF_6; Chapter 10, Pinares-Patiño *et al.*, 2015). The technique as described by Johnson *et al.* (1994a) involves placing a permeation tube containing SF_6 in the rumen, collecting samples from the animal's nose and mouth and determining CH_4 and SF_6 concentrations by gas chromatography. Methane production is then calculated as the ratio of CH_4 and SF_6 concentrations multiplied by release rate of SF_6 from the permeation tube.

A tracer technique using ethane (C_2H_4) to measure rumen gas kinetics in grazing dairy cows is described by Moate *et al.* (1997). They reported that C_2H_4 does not have an effect on the rumen fermentation pattern and was not metabolized. They continuously injected C_2H_4 into the rumen and simultaneously collected rumen gas, which was then analysed for C_2H_4 and principal rumen gases, such as CH_4, hydrogen, CO_2, hydrogen sulphide and oxygen to study gas kinetics in rumen headspace. Total CH_4 production can be calculated by dividing the proportion of CH_4 by the proportion of C_2H_4 in the collected gas and multiplying the fraction by total C_2H_4 infused into the rumen (Mbanzamihigo *et al.*, 2002).

Tracer technique is particularly useful for free ranging cattle because it allows direct CH_4 emission estimates as the animal is grazing at pasture, which in the case of beef cattle could be 5 to 12 months per year (McCaughey *et al.*, 1997). Another advantage is that minimal training is required to adapt animals to the use of the apparatus. Limitations include whole animal emissions are not measured, there is high animal-to-animal variability associated with measurements (Pinares-Patino *et al.*, 2011), long withdrawal time of animals after the gas has been released, milk produced may need to be discarded and training is required in handling tracer gases. For detailed information on tracer technique please refer to Chapter 10 (Pinares-Patiño *et al.*, 2015).

9.3.2 Mobile carbon dioxide and methane analysers

Madsen *et al.* (2010) proposed to use CO_2 as internal marker. The method is based on the assumption that CO_2 production can be estimated from heat production, which in turn can be calculated from the difference between intake of metabolisable energy and energy in products, i.e. weight gain and milk production. The above knowledge combined with simultaneous measurements of the concentration of CH_4 and CO_2 is the basis for the method of quantifying CH_4 production from individual animals. The method may be useful for barn-reared animals where diet is known. It still needs a high frequency measurements to match the variation with feeding events. The method needs to be evaluated using the chamber technique and has the potential to provide an easy, mobile and cheap measurement method, particularly for inventory purposes and large commercial operations. Limitations include difficulty of adapting the system to use on pasture based animals and if not automated only snapshots of emissions can be measured.

An internet-interfaced system that measure CH_4, CO_2 and other metabolic emissions from individual animals and track changes in emissions over time has been developed by Zimmerman *et al.* (2011). The system (GreenFeed) is based on a headstall unit to restrict and control atmospheric mixing and avoid expired gases being lost from the feeding station. It also has a radio-frequency identification of each animal, and a feed or water dispenser so that the animal voluntarily keeps its head in the correct position to obtain quantitative, representative metabolic gas measurements. A fan unit for unidirectional air circulation and sensors to measure expired and eructated gases along with air flow-rate through the headstall manifold are installed. Using a tracer gas (propane), sensors measure CH_4, CO_2, water vapour, molecular hydrogen, hydrogen sulphide, and air flow rate which is captured in a data acquisition and control system with a remote data link to transmit data to a specified location. The system has been installed in several countries around the world (e.g. Australia, UK, USA, Brazil, etc.), which is particularly suited to measure emissions for pasture based systems (e.g. Waghorn *et al.*, 2013). However, several researchers are using the system under confined system as well (e.g. Weber *et al.*, 2013). The system uses 'bait' to entice animals to visit the GreenFeed. Although the pellets used as 'bait' are given in small quantities, they should either be of similar quality as the base feed or need to be accounted for during calculation (as feed input).

9.4 Comparison of the indirect calorimetry technique with other methods

The whole animal chamber indirect calorimetry method is considered to be the gold standard in measuring CH_4 emissions because recovery of gases is very high with minimal loss from the animal. Therefore, other methods including ventilated hood chambers, polyethene tunnels and CH_4 measurement techniques that use tracers are evaluated against the indirect calorimetry technique.

9.4.1 Indirect calorimetry vs tracers

Grainger *et al.* (2007) compared CH_4 emissions estimated using the indirect calorimetry and SF_6 tracer techniques. The authors used 16 lactating cows placed in a chamber that were fitted with the SF_6 tracer apparatus to simultaneously collect CH_4 emissions from the mouth and rectum, simulating whole animal emissions. The authors reported that measurements with the SF_6 apparatus were more variable across days and across cows compared with chamber method measurements. However, these differences were small and, overall, CH_4 emission measurements from both techniques were similar. The differences may be greater in field conditions because of interferences resulting from wind, dust, etc. Boadi *et al.* (2002) and Pinares-Patiño *et al.* (2011) also compared the SF_6 technique with typical respiration chamber measurements. They found that the SF_6 technique provides measurements that have higher across- and within-animal variation. Hence, this method may require many animals and replications to produce acceptable results. Johnson *et al.* (1994b) compared CH_4 production by cattle in chambers with values obtained using the SF_6 technique. Although they found CH_4 estimates from the SF_6 technique to be numerically lower (by 7%), the difference was not significant. Most of the techniques use specialized equipment for determination of the concentration of CH_4 and other gases. Therefore, the differences in CH_4 estimates were likely due to sample concentrations rather than due to the inaccuracy of measuring equipment.

9.4.2 Indirect calorimetry vs polythene tunnel

Murray *et al.* (1999) compared CH_4 emissions from sheep in a polythene tunnel system and sheep in a respiration chamber. They found that CH_4 production measured in the respiration chamber was 12.9% greater than CH_4 production measured in the polythene tunnel system (in l/kg dry matter intake). Given that recovery of added CH_4 was above 95% for both systems and that no systematic differences in measurement errors were detected, the authors suggested that differences in CH_4 emissions were due to effects of housing during sample collection.

9.4.3 Indirect calorimetry vs other tracer techniques

There is a considerable interest in developing technologies that are less expensive and practical compared to whole animal indirect calorimetry chambers. Hellwing *et al.* (2013) compared the system developed by Madsen *et al.* (2010), described in a previous section, with a whole animal indirect calorimetry chamber. The authors reported that although the Madsen *et al.* (2010) method explained 55% of the variation in measured CH_4 production, it underestimated the CH_4 production measured by whole animal indirect calorimetry chamber. Hellwing *et al.* (2013) concluded that although the method of Madsen *et al.* (2010) is easier to implement, it needs to be improved further. Potential improvement in the CH_4:CO_2 ratio method for estimation of CH_4 production can come from better prediction of CO_2 production from cows' based on their live weight and milk production.

When compared on a daily basis, Hammond *et al.* (2013) reported that CH_4 emission estimates from indirect calorimetry chambers were not statistically different to those estimates obtained by the GreenFeed technique described in Section 9.3.2. The mean CH_4 emission (g/d) estimates in 2 experiments were 215 vs 198, and 209 vs 208 for indirect calorimetry chamber and GreenFeed, respectively. The authors further reported that CH_4 emission patterns determined by chambers and GreenFeed were comparable. However, diet effects were significant when CH_4 was measured by indirect calorimetry chambers, but not with GreenFeed. They attributed this discrepancy to a lower number of visits to GreenFeed or the timing of the visits. Methane emissions patterns throughout a day follows feed intake pattern with a lag of an hour or two. For example, Figure 4 shows CH_4 emission pattern when heifers were fed three treatments and

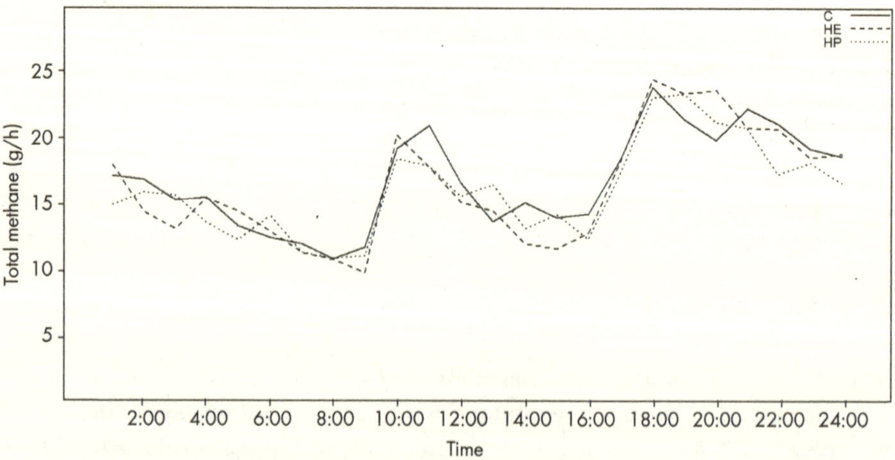

Figure 4. Daily methane emission patterns of heifers fed three treatments: control (CON), high energy (HE) and high protein (HP) diets using ventilated hood chambers (Kebreab et al., *2012).*

measured using ventilated hood chambers (Kebreab *et al.*, 2012). Samples were collected every 10 seconds but averaged over a minute before plotting. It is obvious that time and number of visits (sampling) will have an impact on extrapolation of emissions to a daily basis. Therefore, Hammond *et al.* (2013) concluded that systems such as the GreenFeed system are capable of estimating CH_4 emissions from livestock, but deployment and replication must be considered carefully to ensure adequate numbers of measurements are obtained.

Standardisation of techniques is required to build a database of emissions from livestock production, which will allow better estimation of CH_4 emissions accounting for measurement errors. Data from studies of enteric CH_4 emissions from cattle on various diets and measured with different techniques need to be collated into this database, which could then be used as a platform to develop prediction models and to identify gaps in knowledge. In addition, measurements taken at different spatial scales need to be standardized, especially when they are used to downscale or upscale emission values.

9.5 Conclusions

The literature suggests that measurements of emissions from confined animals using various techniques agree with an acceptable level of variation. The traditional chamber technique, although reliable, might influence emission rates by changing animals' behaviour such as reduced feed intake and anxiety. However, modern modifications of the whole animal chamber technique have solved the issue of animal welfare. In addition, the cost of building whole animal chambers have come down, therefore, construction of more chambers with transparent polycarbonate and steel frames is expected to continue. Such devices would still be unable to be used in grazing systems, therefore, other mobile systems based on tracer techniques, such as the GreenFeed and SF_6 would continue to be used for pasture systems for the foreseeable future.

References

Attwood, G. and McSweeney, C., 2008. Methanogen genomics to discover targets for methane mitigation technologies and options for alternative H_2 utilisation in the rumen. Australian Journal of Experimental Agriculture 48: 28-37.

Boadi, D.A., Wittenberg, K.M. and Kennedy, A.D., 2002. Validation of the sulphur hexafluoride (SF_6) tracer gas technique for measurement of methane and carbon dioxide production by cattle. Canadian Journal of Animal Science 82: 125-131.

Cammell, S.B., Beever, D.E., Skelton, K.V. and Spooner, M.C., 1981. The construction of open-circuit calorimeters for measuring gaseous exchange and heat production in sheep and young cattle. Laboratory Practice 30: 115-119.

Cammell, S.B., Thomson, D.J., Beever, D.E., Haines, M.J., Dhanoa, M.S. and Spooner, M.C., 1986. The efficiency of energy utilisation in growing cattle consuming fresh perennial ryegrass (*Lolium perenne* cv. Melle) or white clover (*Trifolium repens* cv. Blanca). British Journal of Nutrition 55: 669-680.

Cooprider, K.L., Mitloehner, F.M., Famula, T.R., Kebreab, E., Zhao, Y. and Van Eenennaam, A.L., 2011. Feedlot efficiency implications on greenhouse gas emissions and sustainability. Journal of Animal Science 89: 2643-2656.

Environmental Protection Agency (EPA), 2013. Inventory of U.S. greenhouse gas emissions and sinks: 1990-2011. EPA 430-R-13-001. Environmental Protection Agency, Washington, DC, USA.

Grainger, C., Clarke, T., McGinn, S.M., Auldist, M.J., Beauchemin, K.A., Hannah, M.C., Waghorn, G.C., Clark, H. and Eckard, R.J., 2007. Methane emissions from dairy cows measured using the sulfur hexafluoride (SF_6) tracer and chamber techniques. Journal of Dairy Science 90: 2755-2766.

Hamilton, S.W., DePeters, E.J., McGarvey, J.A., Lathrop, J. and Mitloehner, F.M., 2010. Greenhouse gas, animal performance, and bacterial population structure responses to dietary monensin fed to dairy cows. Journal of Environmental Quality 39: 106-114.

Hammond, K., Humphries, D.J., Crompton, L.A., Kirton, P., Green, C., Reynolds, C.K., 2013. Methane (GreenFeed) compared emissions from growing dairy heifers estimated using an automated head chamber to respiration chambers or SF_6 techniques. In: Proceedings of the 5th Greenhouse Gasses in Agriculture Conference, Dublin, Ireland, p. 391.

Harper, L.A., Denmead, O.T. and Flesch, T.K., 2011. Micrometeorological techniques for measurement of enteric greenhouse gas emissions. Animal Feed Science and Technology 166/167: 227-239.

Hassanat, F., Gervais, R., Julien, C., Massé, D.I., Lettat, A., Chouinard, P.Y., Petit, H.V. and Benchaar, C., 2013. Replacing alfalfa silage with corn silage in dairy cow diets: Effects on enteric methane production, ruminal fermentation, digestion, N balance, and milk production. Journal of Dairy Science 96: 4553-4567.

Hellwing, A.L.F., Lund, P., Madsen, J. and Weisbjerg, M.R., 2013. Comparison of enteric methane production predicted from the CH_4/CO_2 ratio and measured in respiration chambers. Advances in Animal Biosciences 4: 557.

Hellwing, A.L.F., Lund, P., Weisbjerg, M.R., Brask, M. and Hvelplund, T., 2012. Technical note: test of a low-cost and animal-friendly system for measuring methane emissions from dairy cows. Journal of Dairy Science 95: 6077-6085.

Hristov, A.N., Oh, J., Lee, C., Meinen, R., Montes, F., Ott, T., Firkins, J., Rotz, A., Dell, C., Adesogan, A., Yang, W., Tricarico, J., Kebreab, E., Waghorn, G., Dijkstra, J. and Oosting, S., 2013. Mitigation of greenhouse gas emissions in livestock production – a review of technical options for non-CO_2 emissions. In: Gerber, P.J., Henderson, B. and Makkar, H.P.S. (eds.). FAO Animal Production and Health Paper No. 177. FAO, Rome, Italy.

Johnson, K.A. and Johnson, D.E., 1995. Methane emissions from cattle. Journal of Animal Science 73: 2483-2492.

Johnson, K.A., Huyler, M.T., Westberg, H.H., Lamb, B.K. and Zimmerman, P., 1994a. Measurement of methane emissions from ruminant livestock using a SF6 tracer technique. Environmental Science Technology 28: 359-362.

Johnson, K.A., Huyler, M.T., Westberg, H.H., Lamb, B.K. and Zimmerman, P., 1994b Measurement of methane emissions from ruminant livestock using a sulfur hexafluoride tracer technique. In: Aquilera, J.F. (ed.) Energy metabolism of farm animals. EAAP Publication No. 76. CSIC Publishing Service, Granada, Spain, pp. 335-338.

Kebreab, E., Clark, K., Wagner-Riddle, C. and France J., 2006. Methane and nitrous oxide emissions from Canadian animal agriculture: a review. Canadian Journal of Animal Science 86: 135-158.

Kebreab, E., Cruz, G.D., Hai, P.H., Anantassok, N., Polyorach, S., Beelen, P. and Rosa H.D., 2012. Methane emission profile in dairy heifers fed rice straw. In: Proceedings of the 1st International Conference on Animal Nutrition and Environment, 14-15 Sep. 2012, Khon Kaen, Thailand, pp. 9-14.

Kelly, J.M., Kerrigan, B., Milligan L.P. and McBride, B.W., 1994. Development of a mobile, open circuit indirect calorimetry system. Canadian Journal of Animal Science 74: 65-72.

Liang, J.B., Terada, F. and Hamaguchi, I., 1989. Efficacy of using the face mask technique for the estimation of daily heat production of cattle. In: Van Der Honing, Y. and Close, W.H. (eds.) Energy metabolism of farm animals. Pudoc, Waginingen, the Netherlands, pp. 348-351.

Lighton, J.R.B., 2015. Metabolic measurement techniques: baselining, mathematical correction of water vapour dilution and response correction. In: Gerrits, W.J.J. and Labussière, E. (eds.) Indirect calorimetry. Techniques, computations and applications. Wageningen Academic Publishers, Wageningen, the Netherlands, pp. 57-72.

Lockyer, D.R. and Jarvis, S.C., 1995. The measurement of methane losses from grazing animals. Environmental Pollution 90: 383-390.

Madsen, J., Bjerg, B.S., Hvelplund, T., Weisbjerg, M.R. and Lund, P., 2010. Methane and carbon dioxide ratio in excreted air for quantification of the methane production from ruminants. Livestock Science 129: 223-227.

Mbanzamihigo, L., Fievez, V., Da Costa Gomez, C., Piattoni, F., Carlier, L. and Demeyer, D., 2002. Methane emissions from the rumen of sheep fed a mixed grass-clover pasture at two fertilization rates in early and late season. Canadian Journal of Animal Science 82: 69-77.

McCaughey, W.P., Wittenberg, K. and Corrigan, D., 1997. Methane production by steers on pasture. Canadian Journal of Animal Science 77: 519-524.

McLean, J.A. and Tobin, G., 1987. Animal and human calorimetry. Cambridge University Press, New York, NY, USA.

Miller, W.H. and Koes, R.M., 1988. Construction and operation of an open-circuit indirect calorimetry system for small ruminants. Journal of Animal Science 66: 1042-1047.

Moate, P.J., Clarke, T., Davies, L.H. and Laby, R.H., 1997. Rumen gases and load in grazing dairy cows. Journal of Agricultural Science 129: 459-469.

Moraes, L.E., Strathe, A.B., Fadel, J.G., Casper, D.P. and Kebreab, E., 2014. Prediction of enteric methane emissions from cattle. Global Change Biology 20: 2140-2148.

Murray, P.J., Moss, A., Lockyer, D.R. and Jarvis, S.C., 1999. A comparison of systems for measuring methane emissions from sheep. Journal of Agricultural Science 133: 439-444.

Murray, R.M., Bryant, A.M. and Leng, R.A., 1976. Rates of production of methane in the rumen and large intestine of sheep. British Journal of Nutrition 36: 1-14.

Odongo N.E., AlZahal, O., Las, J.E., Kramer, A., Kerrigan, B., Kebreab, E., France, J. and McBride B.W., 2008. Data capture: development of a mobile open-circuit ventilated hood system for measuring real-time gaseous emissions in cattle. In: France, J. and Kebreab, E. (eds.), Mathematical modelling in animal nutrition. CAB International, Wallingford, UK, pp. 225-240.

Pinares-Patiño, C.S., Lively, F. and Lassey, K.R., 2015. The sulphur hexafluoride tracer technique for estimating enteric methane emissions from ruminants. In: Gerrits, W.J.J. and Labussière, E. (eds.) Indirect calorimetry. Techniques, computations and applications. Wageningen Academic Publishers, Wageningen, the Netherlands, pp. 185-212.

Pinares-Patino, C.S., Lassey, K.R., Martin, R.J., Molano, G., Fernandez, M., MacLean, S., Sandoval, E., Luo D. and Clark H., 2011. Assessment of the sulphur hexafluoride (SF_6) tracer technique using respiration chambers for estimation of methane emissions from sheep. Animal Feed Science and Technology 166/167: 201-209.

Place, S.E., Pan, Y., Zhao Y. and Mitloehner, F., 2011. Construction and operation of a ventilated hood system for measuring greenhouse gas and volatile organic compound emissions from cattle. Animals 1: 433-446.

Stackhouse, K.R., Pan, Y., Zhao, Y. and Mitloehner, F.M., 2011. Greenhouse gas and alcohol emissions from feedlot steers and calves. Journal of Environmental Quality 40: 899-906.

Sun, H., Trabue, S.L., Scoggin, K., Jackson, W.A., Pan, Y., Ahao, Y., Malkina, I.L., Koziel, J.A. and Mitloehner, F.M., 2008. Alcohol, volatile fatty acid, phenol, and methane emissions from dairy cows and fresh manure. Journal of Environmental Quality 37: 615-622.

Thoma, G., Popp J., Nutter, D., Shonnard, D., Ulrich, R., Matlock, M., Kim, D.S., Neiderman, Z., Kemper, N., East, C. and Adom, F., 2013. Greenhouse gas emissions from milk production and consumption in the United States: a cradle-to-grave life cycle assessment circa 2008. International Dairy Journal 31 (Suppl. 1): S3-S14.

Waghorn, G., Garnet, E.J., Pinares, C.S. and Zimmerman, S., 2013. Implementation of GreenFeed for estimating methane in a dairy herd grazing pasture. In: Proceedings of the 5th Greenhouse Gasses in Agriculture Conference. Dublin, Ireland, p. 436.

Weber, K.L., Rendon, J.A., Antwi, C., Van Eenennaam, A.L. and Kebreab E., 2013. Methane production is not associated with sire groups or residual feed intake in feedlot Angus cattle. In: Proceedings of the 5th Greenhouse Gasses in Agriculture Conference. Dublin, Ireland, p. 348.

Young, B.A., Kerrigan, B. and Christopherson, R.J., 1975. A versatile respiratory pattern analyzer for studied of energy metabolism of livestock. Canadian Journal of Animal Science 55: 17-22.

Zimmerman, P., Zimmerman, S., Utsumi, S. and Beede, D., 2011. Development of a user-friendly online system to quantitatively measure metabolic gas fluxes from ruminants. Journal Animal Science 89 (E-Suppl. 1)/Journal Dairy Science 94 (E-Suppl. 1): 759.

Chapter 10: The sulphur hexafluoride tracer technique for estimating enteric methane emissions from ruminants

C.S. Pinares-Patiño[1], F. Lively[2] and K.R. Lassey[3]*
[1]Integrated Agricultural Systems, Agriculture Flagship, CSIRO, GPO Box 1600, Canberra ACT 2601, Australia; [2]AFBI Hillsborough, Large Park, Hillsborough, County Down BT26 6DR, United Kingdom; [3]Lassey Education & Research, 4 Witako St., Lower Hutt 5011, New Zealand; cesar.pinarespatino@csiro.au

Abstract

Methane (CH_4) emission from ruminant livestock, the single most important anthropogenic source of emissions of this gas, has both environmental and nutritional implications. Methane emissions can be accurately measured using enclosure facilities, but if data are to be extrapolated to production situations, caution needs to be exercised as there are limitations. The sulphur hexafluoride (SF_6) tracer technique for estimating CH_4 emissions exploits the knowledge that most of CH_4 is excreted at the mouth and nostrils of the animal. Estimations are based on known release rate of SF_6 from a permeation tube introduced into the reticulo-rumen of an animal and the ratio of CH_4/SF_6 concentrations in aliquots of excreted gases sampled unobtrusively. The technique remains the method of choice for emission estimations in conditions being similar to those under production. Nevertheless, the accuracy and precision of the technique has been debated. Most of the evaluations of the tracer technique involving sheep and cattle indicate that it agrees well with the respiration chamber methods in their overall mean emission estimates, but a greater variation for SF_6 tracer emission estimates remains a common feature. Growing evidence suggests that the large variability of tracer-based CH_4 emission estimates is related to the experimental conditions and improper handling of the technique (e.g. inaccurate assessment of background SF_6 and CH_4 concentrations), and more recently the instability of the sample flow rate has been associated with inaccuracy and imprecision of the technique. Fortunately, the identified and suspected sources of variation are controllable. Here, we discuss some potential issues related to the technique and recommend measures of quality assurance and control when applying the technique.

Keywords: CH_4, SF_6, tracer, ruminants, greenhouse gas

DOI 10.3920/978-90-8686-261-0_10, © Wageningen Academic Publishers 2015

10.1 Introduction

Methane (CH_4) emissions from ruminant livestock have both nutritional and environmental implications. Methane is a by-product of microbial fermentation that takes place in the digestive tract, especially in the rumen, and hence is a loss of digestible energy. Globally, CH_4 emissions from ruminant livestock constitute the single largest anthropogenic source of emissions of this greenhouse gas (Steinfeld *et al.*, 2006).

Methane emissions from individual animals can be accurately measured using animal enclosure facilities (e.g. respiration chambers). However, these methods are expensive and the extent to which results can be extrapolated to less controlled systems (e.g. grazing) has been questioned, stimulating the development of alternative measurement/estimation techniques. The sulphur hexafluoride (SF_6) tracer technique for estimating enteric CH_4 emissions was developed in the mid-1990s (Zimmerman, 1993) for use with penned and free ranging ruminants. Since then, it has been widely used around the world with some modifications of the original recommendations, especially regarding sample collection equipment and sample handling. The SF_6 tracer technique has been debated in regards to its accuracy and precision. Most of the criticisms have arisen from studies where the SF_6 technique was compared against respiration chambers with penned or enclosed animals, where the principles of tracer usage may have not been fully adhered to. Lassey (2007) has critically reviewed the above concerns and more recently Deighton *et al.* (2014) have proposed modifications to the technique in order to improve its reliability. Here, we develop the discussion further, particularly regarding the reliability of the SF_6 tracer technique for estimating CH_4 emissions from individual animals under 'production' conditions. Finally, recommendations on quality control are provided to avoid potential setbacks and mistakes when using this technique.

10.2 Historical drivers for development of techniques for enteric methane measurements

Techniques for enteric CH_4 emission measurements on individual animals can be classified into direct and indirect methods. Direct methods include total or partial enclosure of animals, whereas indirect methods include the use of tracers (isotopic and non-isotopic) and estimations based on rumen fermentation characteristics and short-term measurements. Direct measurements of CH_4 emissions have developed alongside indirect calorimetry and nutritional energetic methods (Johnson *et al.*, 2003).

Energy excretion in ruminant CH_4 accounts for 2-12% of the gross energy intake (Johnson and Johnson, 1995). This was the main reason for the considerable research aimed at increasing feed efficiency that started in the 1800s (e.g. Armsby, 1904; Kellner, 1910), receiving renewed interest in the 1960s and 1970s (Blaxter and Clapperton, 1965; Czerkwaski, 1969; Demeyer and van Nevel, 1975; Wolin, 1960), which used animal calorimetry facilities (Figure 1). From the 1990s and continuing today, research on ruminant CH_4 has been principally driven by the need to reduce implications of CH_4 on climate change (Clark *et al.*, 2005; Martin *et al.*, 2010). Development of alternative measurement methods has been prompted by the question on the extent to which enclosure-based results can be extrapolated to free ranging animals, with the SF_6 tracer technique as the most widely adopted method.

A new surge in building respiration chamber facilities for measurement of CH_4 emission has been witnessed (Figure 1) over the last 15 years. Such facilities, ranging in sophistication from the basic instrumented enclosure to highly automated systems have been commissioned around the world (e.g. Pinares-Patiño and Waghorn, 2014). These new facilities share common features such as flexibility and mobility, less expensive than the conventional ones but accurate, and compliant with the strict regulations of animal welfare. More recently, the GreenFeed system (C-Lock Inc, Rapid City, SD, USA) has been developed. This method relies on voluntary animal access to a feeding station, where computer-controlled delivery of palatable feed entices the animal to prolong occupation in order to capture gaseous emissions, with daily emissions estimated from short-term instantaneous measurements of emissions throughout the day and across days.

Large between-animal variation in CH_4 emissions was first reported by Lassey *et al.* (1997) and genetic basis for this variation has been demonstrated (Pinares-Patiño *et al.*, 2013b). Accordingly, animal geneticists around the world have shown great interest in researching animal breeding to reduce ruminant livestock CH_4 emissions. Developing this mitigation avenue requires large numbers of animals to be phenotyped under production conditions, and for which high throughput, non-invasive and cheap methods of measurement are needed (Oddy *et al.*, 2014). Short-term measurement systems seem to meet this need, and among the proposed methods are the portable accumulation chambers (PAC) for sheep (Goopy *et al.*,

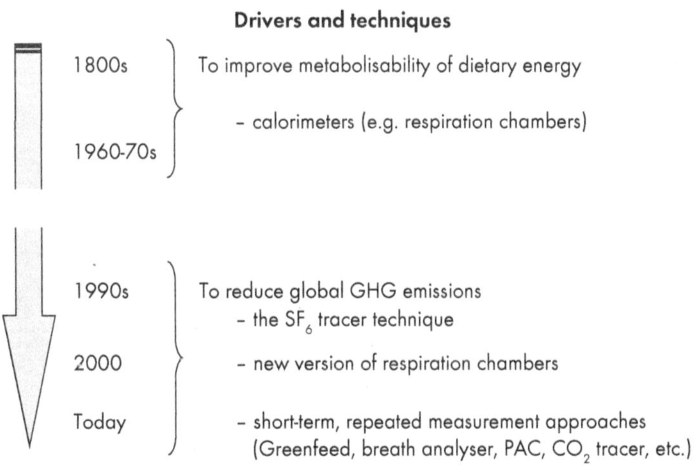

Figure 1. Development of techniques for measurement of methane emissions, drivers and methods (from Pinares-Patiño and Clark (2008) and courtesy of C. Martin (INRA Clermont-Ferrand, France).

2011) and the breath sample analyses developed at the University of Nottingham for intensive farmed dairy cattle (Garnsworthy *et al.*, 2012). In an alternative approach, Madsen *et al.* (2010) has proposed an indirect method of estimation of CH_4 emissions based on short term measurements of CH_4/CO_2 ratio of concentrations in breath and estimated CO_2 emissions as an internal tracer. Certainly, methods based on short-term measurements need thorough scrutiny for accuracy of emission rankings of individual animals.

The application of the SF_6 tracer technique for CH_4 emission estimation has extended to almost all domestic ruminant species and production systems, including those less widely farmed such as camelids, yaks and reindeer. The SF_6 tracer technique, in some cases, has also been slightly modified to suit the animal's anatomy and production systems. For example, the extended sample collection period (5 or 10 days) instead of daily collections (Pinares-Patiño *et al.*, 2012) has proved convenient for extensive grazing systems (Dini *et al.*, 2012). Collection canisters and sample flow restrictors are varied, but the principles of the technique remain the same. Another variant of the original protocol is the collection of samples directly from the rumen (Bayat *et al.*, 2015). This later modification to the protocol requires gas leakage throughout the rumen cannulae be minimised. Exhalation routes of excretion of CH_4 not accounted for may be a weakness of this variant.

10.3 Enteric methane production and routes of excretion

Murray *et al.* (1976) conducted studies with rumen and caecum fistulated sheep fed hourly on lucerne hay. The sheep were ruminally infused with isotopically-labelled CH_4 (^{3}H-CH_4 and ^{14}C-CH_4) and gases collected using both masks and intra-ruminally. These authors reported that the rumen accounted for 87% of total enteric CH_4 production, and that 95% of rumen-produced CH_4 was excreted by eructation. Of the 13% of CH_4 produced in the hindgut, 89% was excreted through the lungs and 11% through the anus. No transfer of CH_4 between rumen and hindgut pools was observed. Torrent and Johnson (1994), using the rumen emptying technique in sheep, found that the hindgut accounted for 12% of the total CH_4 production (12%), whereas using similar technique Hofmeyr *et al.* (1984) reported a much lower contribution (7%) from the hindgut. Based on volatile fatty acids (VFA) profiles in rumen and caecum, Immig (1996) suggested that between 6 and 14% of the daily CH_4 production would be produced in the hindgut. More recently, Reynolds *et al.* (2013) have reported that milk is a negligible source of excretion of CH_4. Studies involving cattle (Hoernicke *et al.*, 1965; Reynolds *et al.*, 2013) have reported that in this species ruminal absorption of CH_4 is substantial compared to that observed in sheep (Murray *et al.*, 1976).

From the studies in sheep and cattle described above, it can be concluded that >94% of the total digestive tract CH_4 production in ruminants is excreted via the animal's mouth and nostrils. Thus, if an external tracer source introduced into the animal forestomach is also excreted at the same excretion points as CH_4 (i.e. via mouth and nostrils) at predictable rates, the continuous sampling of aliquots of the excreted gases (containing the tracee and tracer gases) would yield a good estimate of the total excretion of CH_4. Indeed, not only is SF_6 a biologically inert gas (Johnson *et al.*, 1994a), but there is evidence (Vlaming, 2008) that faeces and urine are negligible as routes of excreted SF_6. Production of CH_4 in the hindgut and likely differences in ruminal absorption and solubility of CH_4 and SF_6 suggests that eructated and absorbed (exhaled) excretion pathways may differ in their CH_4/SF_6 ratios of concentrations, hence affecting the accuracy of the SF_6 tracer technique (Berends *et al.*, 2014). However, studies with cattle (e.g. Colvin *et al.*, 1957; Hoernicke *et al.*, 1965) have revealed that most of the eructated gases are inhaled into the lungs and subsequently exhaled. It is unknown whether this phenomenon is also present in other ruminants. The inhalation and subsequent exhalation of most of the eructated gases could be regarded as processes aiding the proper mixing of tracee and tracer gases.

10.4 The need for a measurement method matching methane emission features

Enteric CH_4 production is a complex process that is not fully understood. It is a result of interactions between the rumen ecosystem and the amount and quality of the diet, which are in turn modulated by the animal behaviour and factors relating to the animal itself (Figure 2). Environmental conditions (e.g. weather) also can modify the interactions. These complex interactions cannot be mimicked in respiration chambers, hence the need for CH_4 quantification methods appropriate to the range of production situations. Nevertheless, addressing specific nutritional implications of CH_4 emission and evaluation of some CH_4 mitigation approaches require highly controlled methods, being the respiration chamber the most appropriate method for such purposes.

Figure 2. Enteric methane production, a result of complex interactions between diet and rumen environment, modulated by animal factors and influenced by specific environmental conditions. (Courtesy of N. Swainson, AgResearch Limited, New Zealand).

Large variation between individuals in CH_4 yield (i.e. emissions per unit of dry matter intake (DMI)) is a common feature of CH_4 measurements using the SF_6 tracer technique. This was first reported by Lassey *et al.* (1997) for grazing sheep in a study that estimated feed intake from total faecal collection and *in vitro* feed indigestibility and tracer CH_4 emissions estimated from daily sample collection over five days (Figure 3A). In more controlled situations, where CH_4 emissions were measured using respiration chambers over two consecutive days and during two periods (14 d interval) and feed intake measured accurately, variation between-animals in CH_4 yield was also large (Figure 3B). A practical implication of the observed between-animal variation in CH_4 emissions, and the fact that such variation has a genetic component, is that it is important to consider the appropriate number of experimental animals when comparing treatment effects.

Day-to-day variation in CH_4 emissions is commonly reported from tracer and respiration chamber measurements (Figure 3). This variation seems larger under free ranging situations compared to pen-fed conditions (Hammond *et al.*, 2009; Pinares-Patiño and Clark, 2008), which may be due to the accentuated effect of factors that influence feed intake and CH_4 production (e.g. forage availability, weather, animal to animal interaction, etc.). Methane emissions also vary within day (Figure 4), reflecting feed intake and fermentation rates. Emissions are larger following feeding than pre-feeding. Moreover, due to the influence of feed intake rate the postprandial increase in CH_4 emissions is highly variable among animals (Derno *et al.*, 2013). The implications of the between-day and within-day variation in CH_4

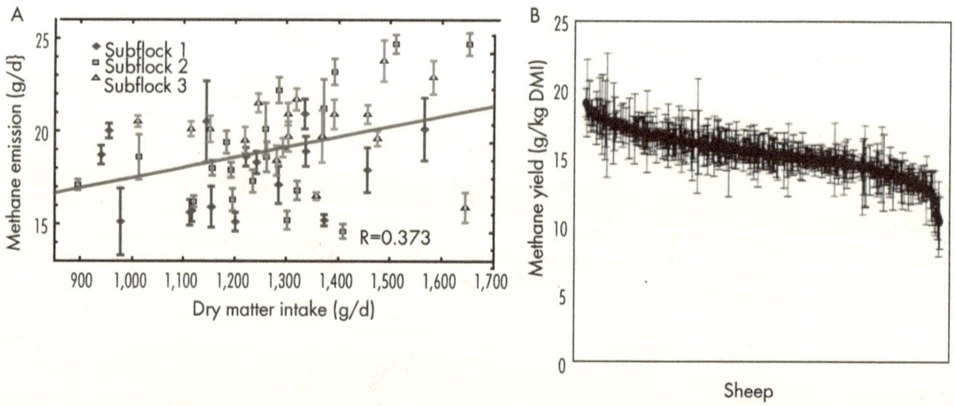

Figure 3. Between-animal variation in CH_4 emissions (vertical bars denote day-to-day variation, ± 1 standard deviation). (A) Absolute CH_4 emissions from individual grazing sheep (y-axis; measured by the tracer technique), with corresponding feed DMI (x-axis; estimated from faecal output and feed indigestibility). Between-sheep variation was responsible for 86% of total variation in daily emissions (Lassey et al.*, 1997). (B) Methane yield (y-axis) from individual sheep (n= ~1,200) in respiration chambers, with individual animals ranked from highest to lowest emitters (x-axis). Sheep were fed on lucerne pellets at 2.0 maintenance requirements (data supporting Pinares-Patiño* et al.*, 2013b).*

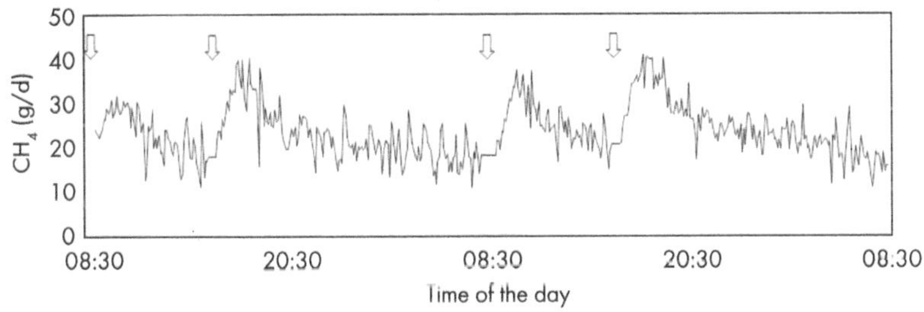

Figure 4. Within-day variation in emissions of CH_4 from a sheep fed twice a day on fresh cut pasture forage. Emissions were measured at 5 min intervals over two consecutive days. Arrows indicate feed delivery times (from Pinares-Patiño et al., *2011b).*

emissions are that emission measurement methods need to account for emissions occurring across the 24 h feeding cycle and across days, with parallel measurement/estimation of feed intake (the most important factor influencing CH_4 production).

10.5 The SF_6 tracer technique

In ruminants, most of the enteric CH_4 is sourced in the rumen and excreted via eructation and exhalation (see previous section). The SF_6 tracer technique exploits this knowledge. The technique involves introducing a 'permeation tube' into the reticulo-rumen of an animal that releases the SF_6 gas at a low but steady rate into the rumen gas head space. Assuming that both SF_6 and CH_4 have similar excretion pathways and dilution rates with the background air (Johnson *et al.*, 1994a), the CH_4 emission rate (QCH_4, g/d; both eructed and exhaled) is calculated as:

$$QCH_4 = QSF_6 \times \frac{[CH_4]}{[SF_6]} \times 1{,}000{,}000 \times 16/146$$

where QSF_6 is the known rate of SF_6 permeation (expressed as g/d) from the particular permeation tube, $[CH_4]$ and $[SF_6]$ denote the net concentrations (volume/volume basis) of the two gases (i.e. in excess of concentrations in background air). The gas chromatography analysis output for $[CH_4]$ is in ppm, whereas for $[SF_6]$ is in ppt, which are molar ratios in dried sample gas. The factor 1,000,000 allows for SF_6 concentration (ppt) to be converted to ppm. The multiplier 16/146 (i.e. 1 mol CH_4 = 16 g CH_4 and 1 mol SF_6 = 146 g SF_6) allows the expression of QCH_4 in g/d given that QSF6 is also in g/d.

As stated earlier, there is evidence (Colvin *et al.*, 1957; Dougherty and Cook, 1962; Hoernicke *et al.*, 1965) that most (70-99%) of the eructed gases are first inhaled into the lungs, and then exhaled along with respired gases. The hindgut is a secondary source of CH_4, whereas SF_6

is sourced only in the rumen. Nevertheless, because most of the hindgut CH_4 production is absorbed into blood and exhaled throughout the lungs and eructated gases are inhaled and then exhaled with ruminally- and hindgut-absorbed gases, it is assumed that the SF_6 tracer mixes with both eructed and exhaled CH_4, hence, SF_6 traces both ruminally and hindgut sourced CH_4.

The calibrated tracer source, the collection of gas and its subsequent analysis are the major components of the SF_6 tracer technique (Lassey *et al.*, 1997). A brass permeation tube with a permeable Teflon™ membrane at one end and containing the SF_6 gas in the brass tube is prepared (Figure 5). The SF_6 tubes, while at rumen-like temperature (39 °C), are calibrated by serial weighing for at least eight weeks to establish a steady permeation rate, and inserted in the rumen about 7 days before the gas samplings begin.

Trials in New Zealand use either sheep tubes or cattle tubes, distinguished by size: tare weights ~32 and 57 g, respectively. The respective charges are ~0.8 and ~2.2 g of SF_6, with corresponding release rates in the ranges 0.7-1.7 and 1.8-3.5 mg SF_6/d for sheep and cattle, respectively (Lassey *et al.*, 2001). When the weekly tube mass is regressed against time, the slope of the linear fitting gives the release rate with R^2 values above 99.95%. This *in vitro* release rate is assumed to continue *in rumen* once the tube has been orally introduced into the animal. However, there is evidence that release rates gradually decline over several months. Therefore, it is recommended that experiments be conducted within approximately one month of deployment (Lassey *et al.*, 2001). For experiments lasting beyond the recommended period, it is advised that permeation tubes are recovered and re-calibrated and the newer permeation rate used in the estimation of CH_4 emissions.

The SF_6 gas is nine times heavier than CH_4. Nevertheless, because of the forceful processes of eructation and exhalation, it is assumed that excreted CH_4 and SF_6 are equally diluted with background air and therefore the $[CH_4]/[SF_6]$ ratio of concentrations is unaffected by such dilution (Lassey, 2007). However, in highly enclosed environments with poor ventilation, the CH_4 and SF_6 concentrations in the background air may vary across horizontal and vertical axes

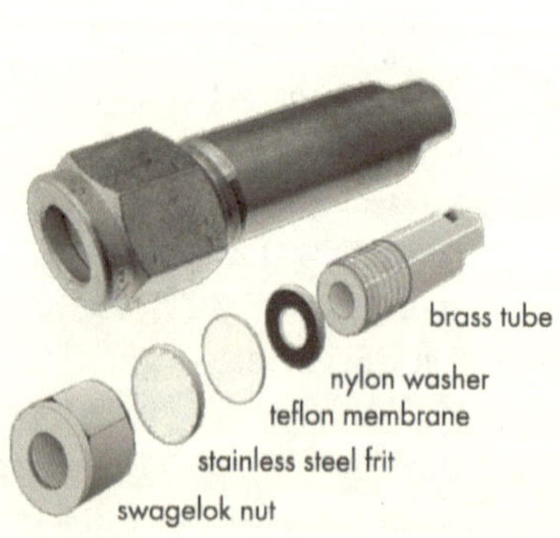

Figure 5. Brass permeation tube showing the components (from Lassey et al., *2001).*

(Williams *et al.*, 2011), hence influencing emission estimates if background concentrations are not properly accounted for.

With the SF_6 tracer technique, the gases eructated and exhaled by the animal are unobtrusively sampled (from ~2 cm above the nostrils of the animal) over ~24 h periods during 5-7 days. Samples are drawn into a pre-evacuated canister via a sample line consisting of small diameter nylon tubing with in-line flow restrictor, all mounted on a harness (Johnson *et al.*, 2007; Lassey *et al.*, 1997). The most common sample flow restrictor is a short stainless steel capillary with a section crimped to achieve the desired sample flow. The flow restriction unit enables a sample delivery rate resulting in sample volumes over the collection period (usually 24 h) of about half of the canister volume.

Minor variations to sample collection lines and collection canisters have been applied. For example, the sample flow restrictor can be a ball bearing unit placed at the sample inlet (Gere and Gratton, 2010; Pinares-Patiño *et al.*, 2012) and collection canisters can be of different shapes to comfortably accommodate to the animal morphology and whether the animal is grazing or pen-fed. Yoke-shape canisters are preferred for grazing animals, whereas shoulder canisters are preferable for animals needing to access feeding bins. Canisters should be light weight, typically made of polyvinyl chloride (PVC) or stainless steel. The volume of PVC yoke-shape canisters are commonly 1.7 or 2.5 l for sheep and cattle, respectively, with matched sample delivery rates of ~0.6 or 0.9 ml/min for 24 h sampling, respectively.

Background air samples are collected in parallel to sample collection from animals (Figure 6). These background samples should represent the air to which the experimental animals are exposed. Thus, for free ranging ruminants, static background samplers should be located upwind (Figure 6). Alternatively, background samples obtained using animals acting as mobile background samplers (Pinares-Patiño *et al.*, 2013a) seem more representative than the static background samples, especially under high stocking density conditions and changing wind direction. For experiments with limited ventilation, such as in animal houses, sites for background air samplers need careful consideration due to the large gradients of CH_4 and/or SF_6 concentrations that can prevail (Lassey, 2013; Williams *et al.*, 2011).

At the end of each sampling period in the field, the collection canisters need to be checked for the residual vacuum. Only those samples at close to the target residual vacuum (½ of the full vacuum) are processed further. Abnormally high vacuums can indicate blockage of the sample line resulting in limited sample collected, while leakages can result in near-atmospheric pressure. At the laboratory, the collection canisters are usually over-pressured (diluted) with pure nitrogen gas to ~1.5 atm. Alternatively, a sub-sampling process can be carried out especially when a rapid turnaround of collection canisters are required. Samples can also be directly analysed using a purpose-designed sample extractor (Lassey *et al.*, 2014), without the need of dilution or sub-sampling. The concentrations of CH_4 and SF_6 gases are determined by a gas chromatography using flame ionisation detector and electron capture detector, respectively. Final concentrations of CH_4 and SF_6 gases (after dilution) are typically

Figure 6. Application of the SF_6 tracer technique with cattle grazing on a pasture with high standing biomass. The collection canister outside and upwind of the grazing strip collects a background air sample (static background). With ruminants grazing at high stocking densities or when wind direction is unpredictable, it is preferable to have animals as background sample collectors (mobile background). In the latter case, sample inlet is located near the hip of the animal.

found in the range 7-70 ppm and 100-600 ppt, respectively, for sheep (Lassey *et al.*, 1997). By comparison, clean air in 2014 contains ~1.8 ppm CH_4 and ~6.7 ppt SF_6 (K.R. Lassey, personal communication)

Because enteric CH_4 arises from the fermentation of feed in the rumen and hindgut, it is advisable that CH_4 emission determinations are associated with parallel measurement/estimation of feed intake and feed quality (e.g. digestibility and chemical composition).

10.6 Does SF_6 fulfil the ideal tracer gas principles?

A tracer technique enables an emissive flux of enteric gas to be quantified without the need to measure total emissions. Instead, only an undetermined fraction of that emission needs to be sampled.

The ideal external tracer for estimation of enteric CH_4 emissions has a constant and known (or predictable) release rate and would be sourced and emitted alongside the source of the target gas (the tracee, CH_4). Importantly, the tracer should be chemically, physically and biologically stable; meaning that it is not removed, transformed or augmented during passage from source to sampling point, on the basis that the target gas also behaves conservatively. A conservative tracer will also be scalable — i.e. its concentration in the sample is directly proportional to its source strength. Therefore, both tracer and tracee are sampled with identical efficiencies.

Further, because animal measurements need to comply with safety regulations, animal welfare and product quality, the tracer should be non-flammable, non-toxic, non-allergenic, inert to the rumen microbial ecosystem, and with no residues in animal products.

A tracer of enteric CH_4 also needs to be applicable to all the ruminant production systems (e.g. intensive in barns, feedlot pens, and intensive and extensive grazing). Thus, adsorption of the tracer on animal hair/wool, harnesses and sample collection gear should be nil or negligible. In addition, concentrations of tracer and tracee in the experimental neighbourhood should be small compared to those in samples (Lassey, 2013).

A tracer for enteric CH_4 needs to fulfil the criteria of 'insensibility'; i.e. it does not increase the physical features of the rumen environment, does not modify intra-ruminal gas pressure and processes of excretion of fermentation gases and does not modify rumen fermentation. To comply with this criterion, the tracer should be released at very low rates, which implies that the tracer gas has to be detectable at low concentrations in the collected sample. The need for the tracer gas to be released at very low rates also means that concentration of this gas in the background air needs to be very low or below detection limit.

The SF_6 gas properties not only enable the charge of permeation tubes by desublimation using liquid nitrogen, but also a highly predictable permeation rate. The SF_6 is a proven tracer with a low detection limit of a few ppt (e.g. Clemons *et al.*, 1968). It is also inert in a rumen environment (Johnson *et al.*, 1994a) and safe to be used in animal (and human) studies (Morel *et al.*, 2000). Nevertheless, the US Environmental Protection Agency (USEPA, 2002) has suggested an upper limit of exposure of SF_6 at 1000 ppm, which is approximately 1000 times above the likely concentration of SF_6 that can be found in the rumen head gas space of a sheep deployed with a permeation tube with 1.0 mg/d release rate.

Macmüller and Hegarty (2006) have attempted to develop alternative tracers for enteric CH_4 emission. Of the almost 300 potential tracers screened, most were eliminated due to potential health risks, differences in their properties relative to CH_4, unsuitability to be delivery by permeation tubes, and high purchase and analytical costs. Ethane was identified as potential alternative tracer to the SF_6. Ethane solubility is similar to that of CH_4, able to be frozen in liquid nitrogen for permeation tube filling, and detectable at ppb levels by photoacoustic analyser, but in order to match detection limit, a permeation tube with much higher release rate than for SF_6 would be needed. Unfortunately, establishing a predictable release rate of ethane from the high release rate permeation tubes failed (Macmüller and Hegarty, 2006). Radioisotopes of CH_4 have been used to measure CH_4 production and excretion pathways (Murray *et al.*, 1976), but their uses are not practical or desirable. On the other hand, stable isotopes of CH_4 might be a desirable option, but reliability of permeation rate of CH_4 from permeation tubes was shown to be poor (Hegarty, 2005). In addition, using stable isotopes (e.g. $^{13}CH_4$) to trace CH_4 might be prohibitive due to the costs associated with purchase and analysis.

The permeation tube (containing the tracer) generally resides in the reticulum of the animal, whereas most of CH_4 arising from fermentation takes place at the intermediate and upper strata of the rumen contents. Most of the ruminal gases are eructated from the rumen head gas space, with both gases expected to be ejected and dispersed from the mouth and nostrils in an identical fashion. The physics of these processes do not discriminate among the constituent gases. Therefore, the eructed CH_4 and SF_6 are expected to be detected in the same proportion as their presence in the rumen headspace, with small augmentations from the lung that originate from ruminal and hindgut absorption and gases passed from the rumen into the hindgut. Thus, although absorption rates of CH_4 and SF_6 may not be same, the rumen and lung are the combined 'sources' of both tracer and tracee, with the inhalation of most of the eructed gases aiding to the homogenisation of concentrations arising from the two sources.

Whether breath sample collections of the SF_6 tracer technique is representative of the gas composition in the rumen head gas space remains to be thoroughly evaluated. Using rumen-fistulated dairy cows, Martin *et al.* (2012), using rumen-cannulated cattle, conducted simultaneous collection of rumen head space and breath samples over a 8-h period (following feeding), with rumen samples collected directly via cannulae at hourly intervals, whereas there was a single 8-h breath sample. Expectedly, concentrations of both CH_4 and SF_6 in breath diluted samples were much lower than in rumen, but the CH_4/SF_6 ratio of concentrations and therefore the calculated CH_4 emissions for breath samples were significantly higher (by 9%) than the corresponding values for rumen samples (mean of hourly collections). Gas composition in the rumen head space of fistulated cows has been shown to be greatly altered in comparison to normal concentrations (Moate *et al.*, 2013b). Assuming that rumen fistula does not alter the mixing ratios of CH_4 and SF_6, it can be suggested that the higher CH_4 emissions calculated using the breath sample compared to that of rumen sample may indicate the contribution of the lower gut CH_4 source as well as rumen-absorbed CH_4.

Hoernicke *et al.* (1965), using tracheostomised cattle, found that 25-94% of the total CH_4 emissions (flatus not included) before feeding were excreted via absorption and subsequent exhalation, whereas this pathway accounted only for 9-43% of total CH_4 emissions immediately after feeding. Whether SF_6 excretion pathways mirror those of ruminal CH_4 remains to be demonstrated. Although solubility, diffusion and absorption of CH_4 and SF_6 are different (Levitt *et al.*, 1973; Meyer *et al.*, 1980; Schimmel *et al.*, 2004), these processes can be regarded to be unimportant for the excretion of both gases because excretion processes will equilibrate with time after tracer administration. Vlaming (2008) reported negligible concentrations of SF_6 in faeces and urine of cattle bearing SF_6 permeation tubes. Thus, aided by the inhalation of most of the eructated gases and equilibration of absorption and exhalation, SF_6 is an adequate tracer for CH_4.

Compared to the rate of CH_4 production, the *in rumen* release rate of SF_6 (~0.02 ml/h in cattle) is extremely small on volume/volume basis. Hence, chances for SF_6 to get trapped within crevices (rumen papillae) and feed particles are proportionally higher than for CH_4. In fact, high frequency breath sample collection (3 times/h) conducted by Lassey *et al.* (2011) revealed occasionally extraordinarily high or low SF_6 concentrations. However, over a 24 h cycle there was a high correlation between CH_4 and SF_6 excretions, with peaks of SF_6 concentrations matching those of CH_4 and apparently coinciding with eructation events.

It should also be noted that SF_6 is released from its host tube by permeation through a permeable membrane (Figure 5), a process that is known to be sensitive to temperature (~3.0-3.7% per °C in the range 35-43 °C: Deighton *et al.*, 2013c; Gere *et al.*, 2014). Thus, rumen temperatures systematically different from tube calibration temperature (39 °C) can lead to systematic error in emission estimates. In practice, ruminal temperature will suffer sudden and temporary changes depending on temperature of drinking water and feed (specially grazed herbage) and rate of fermentation.

Finally, although background concentrations of CH_4 and SF_6 are generally low, those concentrations can be enriched with both horizontal and vertical gradients within a building (Williams *et al.*, 2011). Thus, if the background air concentrations are not properly accounted for, such as by using ill-sited background samplers, accuracy of the tracer technique may be compromised (Lassey, 2013). These considerations can also apply to high stocking densities in well ventilated experimental settings. It is not known how large the impact of enriched background air is in altering gas concentrations in the rumen head gas space. However, based on the concentrations of N_2 in rumen head gas space (Moate *et al.*, 1997), it can be assumed that air intake (background air) into the rumen contributes a negligible amount to the concentrations of rumen 'generated' gases.

10.7 Accuracy and precision of the SF_6 tracer technique

The SF_6 tracer technique was evaluated by its developers (Johnson *et al.*, 1994b), who compared 55 measurements made with the tracer (with cattle in pens) to those obtained from 25 measurements using respiration chambers. It was found that the tracer mean estimate was 93% of that in the chambers, but these values were not significantly different. Subsequently, Pinares-Patiño (2000) carried out three separate evaluation trials using sheep. The two initial studies, which were carried out with sheep in a building with limited ventilation and using permeation tubes beyond five months of deployment, produced implausible results; whereas the third study, which used fresh permeation tubes and appropriate ventilation, showed that the mean absolute CH_4 emission (g/d) and CH_4 yield (g/kg DMI) determined by the tracer method was 96 and 89% of the corresponding chamber values, respectively.

The SF_6 tracer technique has been extensively used in New Zealand both to assist emissions inventory development and to investigate mitigation of CH_4 emissions from the predominantly pastoral livestock farming. Reviews of the New Zealand CH_4 emission dataset pertaining to the SF_6 tracer (Hammond *et al.*, 2009; Pinares-Patiño and Clark, 2008) concluded that the SF_6 tracer CH_4 emission estimates matched well the values obtained in respiration chambers. However, the SF_6 tracer technique emission estimates were associated with larger between-

and within-animal variability than the values measured in respiration chambers. In addition, there was an apparent positive association between release rate of SF_6 from the permeation tube and CH_4 emission estimates (Pinares-Patiño and Clark, 2008). The latter issue ascribed to the tracer technique arose from studies (Pinares-Patiño *et al.*, 2008b; Vlaming *et al.*, 2007) involving rumen-fistulated cattle housed indoors and deployed with sets of permeation tubes having low, medium, high or very high release rates of SF_6, as well as from the statistical analysis of large dataset of CH_4 emission estimates from grazing dairy cattle deployed with permeation tubes having a range of SF_6 release rates (Pinares-Patiño *et al.*, 2008b).

The studies of Vlaming *et al.* (2007) and Pinares-Patiño *et al.* (2008b) with rumen-fistulated cattle were carried out within a building with artificial ventilation aided by natural ventilation during the day (doors open). There is recent evidence that suggests that these two studies were likely in breach of the SF_6 tracer principles. For example, Williams *et al.* (2011) reported that CH_4 and SF_6 can accumulate at non-homogeneous ratios across the horizontal and vertical axes of a cattle building and even along the height of the feeding bins, despite adequate natural ventilation. In turn, Beauchemin *et al.* (2012) using rumen fistulated cattle found overestimated and highly variable CH_4 emission estimates which was ascribed to the dissimilar CH_4:SF_6 ratios between samples collected at the nostrils and those around the cannulae (external). Furthermore, Moate *et al.* (2013b) reported highly disturbed composition of gases in the rumen head gas space of fistulated cattle. In rumen fistulated animals, absorbed CH_4 (from rumen and hindgut) would not be properly traced because the fraction of SF_6 lost via cannulae will be higher than that of the total CH_4, hence overestimation (and variability) of CH_4 emission estimates will be exacerbated. Thus, unless leakage severity is minimised, rumen-fistulated animals will yield unreliable CH_4 emission estimates.

Pinares-Patiño and co-workers (Pinares-Patiño, 2000; Pinares-Patiño *et al.*, 2008a, 2011a; C.S. Pinares-Patiño, unpublished data) conducted five studies to compare the SF_6 tracer technique (with sheep housed in metabolic crates outside chambers) against the respiration chamber method, on the basis of CH_4 yield (g/kg DMI) estimates. Except for two (Pinares-Patiño *et al.*, 2008a, 2011a), these studies found that CH_4 yield estimates by the tracer method were ~35% higher than chamber values and without exception tracer estimates had higher (twice-thrice) among- and within-animal variation in CH_4 yield than chamber values. The comparative studies associated with the worst performance of the SF_6 tracer technique took place early in the process of adoption of the technique in New Zealand, which used buildings with limited ventilation and permeation tubes with long (>5 months) deployment in the animals. Williams *et al.* (2011), based on their observation of disproportionate accumulation of gases across a cattle building and also inside and outside feeding bins, recommended that care must be applied to background air sampling. If concentrations of either gas (CH_4 and SF_6) are different from that to which animals are exposed, biased estimation of emissions could occur. Lassey (2013) discussed the likely impact of inhaling gases enriched by the exhalations of neighbours (with different CH_4/SF_6 ratios) within the housing environment, suggesting that this also could account for the positive association reported between CH_4 emission estimates and release rate of SF_6 when permeation tubes with a large range of release rate (2.624-5.689 mg/d) were used under pen-feeding situations (Pinares-Patiño *et al.*, 2008b; Vlaming *et al.*, 2007) and grazing at high stocking density (Pinares-Patiño *et al.*, 2008b).

Various authors (Grainger *et al.*, 2010; Muñoz *et al.*, 2012; Pinares-Patiño *et al.*, 2008a; Swainson *et al.*, 2009) have reported that overestimation of CH_4 emissions increase as time of deployment of permeation tubes in the rumen increases, which has been ascribed to the decrease in release rate of SF_6 with time of deployment (Lassey *et al.* 2001). Lassey *et al.* (2001) speculated that departure of rate of mass loss from the pre calibrated linearity to be related to the surface interactions between SF_6 and the Teflon membrane. This hypothesis was recently confirmed by an elegant experimental design conducted by Deighton *et al.* (2013a), who reported that 160 days exposure of Teflon membrane to SF_6 was associated with 36% reduction in SF_6 release rate. To account for effects of decrease in release rate of SF_6 on emission estimates, it would be advisable to recover the tubes soon after a trial has finished and measure the release rate again. However, this may be impractical in most cases. Assuming that the rumen environment has nil effect on release rate (Deighton *et al.*, 2013a), keeping surveillance tubes in laboratory may aid in fitting a quadratic equation to the mass loss and use it to estimate the release rates of the tubes in use (Lassey, *et al.*, 2001). Alternatively, the Michaelis-Menten kinetics has been proposed by Moate *et al.* (2013a) to accurately describe the release rate of SF_6 up to a period of 600 days.

Experiments requiring prolonged *in vivo* deployment of permeation tubes will not only result in overestimated emissions, but also in increased variability, both within- and between-animals. Figure 7 shows an extreme situation for overestimated CH_4 emissions from sheep when the SF_6 measurements were conducted with old permeation tubes (~24 months after filling) and approaching (~2 months) their 'use-by-dates' (dates of exhaustion of tube-borne non-gaseous SF_6: Lassey *et al.*, 2001). In this study, the sheep were fed on standard conserved forage and measurements conducted over 12 experimental periods; i.e. emissions from each individual sheep, eating the same amount of feed, were measured over 6 periods by the SF_6 technique and 6 periods in chambers, each period involving three measurement days, with tracer and chamber methods alternated from period to period.

In sheep, studies comparing the SF_6 tracer technique (in well ventilated environments) against the standard respiration chamber method found no significant differences between the estimates of CH_4 emission. In cattle, however, comparative evaluation of the SF_6 tracer technique have been carried out with the SF_6 tracer technique deployed in respiration chambers using beef (McGinn *et al.*, 2006) and lactating dairy cattle (Grainger *et al.*, 2007; McCourt *et al.*, 2008). Agreement and correspondence between the SF_6 tracer method and respiration chamber were reported as expected. Comparative studies with the tracer technique deployed outside chambers were conducted earlier by Boadi *et al.* (2002) with heifers and more recently by Muñoz *et al.* (2012) and Deighton *et al.* (2013b) with dairy cattle. These studies found good agreement and concordance between the methods in terms of mean emissions, but tracer-based estimates remained associated with larger variability than chamber-measured values.

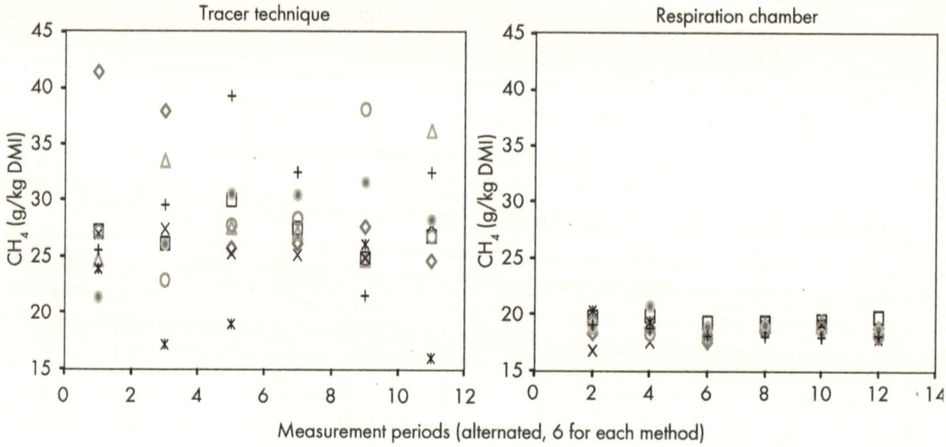

Figure 7. Comparison of CH_4 yield (g/kg dry matter intake) estimates using the SF_6 tracer technique and respiration chambers (C.S. Pinares-Patiño, unpublished data). Eight sheep (labelled by symbols) were measured in alternated periods either by the SF_6 tracer or in respiration chambers while fed on a standard diet. There were 6 periods for each method. The SF_6 tracer technique was conducted with permeation tubes approaching their 'use-by-dates' (when non-gaseous SF_6 in the tube is exhausted) hence produced aberrant emission estimates of high variability between periods. For tubes charged in New Zealand, the 'use-by-date' occurs when the SF_6 load falls to 150 and 600 mg SF_6 in sheep and cattle tubes, respectively (Lassey et al.*, 2001).*

The SF_6 tracer technique relies on a calibrated flow restrictor (a capillary tube) and an evacuated canister to collect representative breath samples (24-h feeding cycles) from an animal. The rate of sample inflow is adjusted to the volume of the collection canister aiming at obtaining ~50% of residual vacuum (i.e. ~0.50 atm) at the end of the collection period. A major assumption of the technique was that sample flow rate would be stable during the collection period. The stability of the sample flow rate was challenged only few years ago by Gere and Gratton (2010); i.e. during filling of an evacuated container using a flow regulator, the rate of sample inflow will decrease as the pressure difference between the ends of the capillary decreases due to vacuum exhaustion. Accordingly, these authors recommended an alternative inflow regulator based on a ball-bearing system. However, in applying the tracer technique, the above issue was only recently experimentally addressed (Deighton *et al.*, 2014; Gere *et al.*, 2014). Deighton *et al.* (2014) demonstrated that the system based on capillary tube was associated with a rapid decline in flow rate and hence inaccuracy of the CH_4 emission estimates, a fact that was postulated to be due to the dissimilar within-day emission patterns of CH_4 and SF_6, highly variable and relatively stable, respectively. The latter asseveration was based on observations of hourly emissions of CH_4 and SF_6 from dairy cows fed *ad libitum* in respiration chambers, with inflow and outflow chamber airstreams analysed for CH_4 and SF_6. Accordingly, Deighton *et al.* (2014) proposed the use of an orifice plate as a flow restrictor instead of a capillary.

In contrast to the findings of Deighton *et al.* (2014), Lassey *et al.* (2011), using high frequency (every 20 min) inline breath sample analysis from sheep fed twice a day at restricted levels, reported not only highly variable within-day emission patterns of both CH_4 and SF_6, but concentrations of these gases were highly correlated, suggesting that the effect of the instability of sample flow rate (declining) throughout the capillary would have little impact on the reliability of emission estimates. The discrepancy between the above two studies may be attribute to the differences in feeding levels, animal species and experimental settings involved. The study of Deighton *et al.* (2014) used a large respiration chamber (41.5 m^3) with a low exchange of air (1 m^3/min) (Grainger *et al.*, 2007) and the animals were fed *ad libitum*. Although it is expected that *ad libitum* feeding leads to smaller within-day variation in CH_4 emissions than restricted feeding, the relatively constant emissions of SF_6 may had been an artefact arising from the 'smoothing' effect of chamber air mixing and air exchange on the measured emissions of SF_6. Nevertheless, the change in flow restriction unit from capillary-based to orifice plate proposed by Deighton *et al.* (2014) should be adopted especially in sample analysis systems involving prior dilution with N_2, where gas concentrations will be low due to dilution.

In summary, comparisons of CH_4 emissions using the SF_6 technique and respiration chambers have generally reported good agreement between their mean estimates. It appears that the large between-animal and within-animal variability in CH_4 emission estimates using the SF_6 technique are associated both with the experimental conditions and the technique itself. For example, in experiments conducted in housed environments or high stocking density grazing situations, variability of estimations will be associated with the richness of CH_4 and SF_6 in the background to which each individual is exposed to, which will in turn be dependent on location of a particular animal (if in a pen), social interaction and changes in wind speed and direction (grazing). Accounting for background concentrations in the SF_6 tracer technique is still group-based. Technique-related variability are associated with the sudden changes in release rate of the SF_6 tracer due to changes in rumen temperature as well as 'aging' of tubes (SF_6 action on permeation unit), cross-contamination of samples (residual SF_6 in collection canisters), likely dissimilar sampling efficiencies of CH_4 and SF_6 in poor ventilated environments, and the recently hypothesised biased 24-h sample (due to declining sample flow rate). Variability in emissions from grazing animals will always be larger than at pen-fed situations, amid to the larger variability in feed intake, intake rate, social interaction and feed choice.

The claimed association between SF_6 release rate and estimated CH_4 emission (e.g. Pinares-Patiño *et al.*, 2008b), seems also be associated with the experimental settings leading for example to improper accounting of background concentrations under high stocking density. Recent studies conducted at Agri-Food and Biosciences Institute (Belfast, UK) (F. Lively, personal communication) found no relationship between permeation rate of SF_6 and CH_4 yield estimates (Figure 8). Similarly, preventing all known possible causes of bias, Deighton *et al.* (2014) reported tracer CH_4 emission estimates (per unit of feed intake) of similar magnitude and variation as for the chamber method (22.3±1.44 and 21.9±1.65 g/kg DMI, respectively) for pen-fed dairy cattle. Hence, quality assurance and quality control measures are paramount for improving accuracy and precision of the SF_6 tracer technique (Section 10.8).

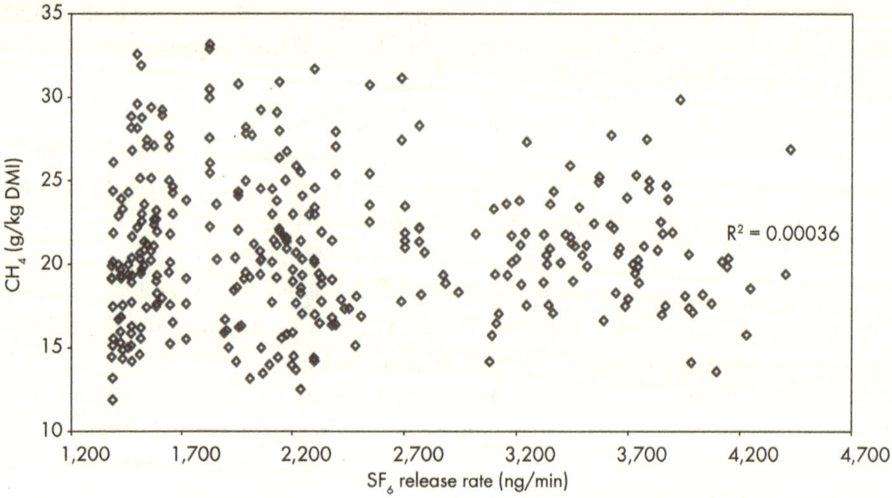

Figure 8. Methane yield (g/kg dry matter intake) by grazing dairy heifers and beef steers estimated using the SF_6 tracer technique. Data are from studies conducted in summer and autumn 2011. No apparent relationship between CH_4 yield and SF_6 release rate is observed. (F. Lively, AFBI Hillsborough, personal communication).

Experimental settings for applying the SF_6 tracer technique always differ from those encountered in respiration chambers. Consequently, it proves difficult to match chamber-determined emission values with the SF_6 tracer technique, unless the tracer technique is deployed inside the chambers. The SF_6 tracer technique is best suited to open and well ventilated environments such as free ranging situations. As recognised by Ulyatt *et al.* (1999), variability in day-to-day CH_4 emissions that arise from changing environmental conditions, feeding patterns, and social interactions cannot be reproduced in chambers as sources of variation. Similar difficulties were also encountered when the polytunnel system and the chamber techniques were compared (Murray *et al.*, 1999). Under grazing conditions, the CH_4 emission estimates using the SF_6 tracer technique agrees well with the mean herd/flock emission estimates based on micro-meteorological methods (Judd *et al.*, 1999; Laubach *et al.*, 2013, 2014; Leuning *et al.*, 1999).

10.8 Quality assurance and quality control when applying the SF_6 tracer technique

Complying with the principles of the SF_6 tracer technique involves strict control and monitoring of the activities both at the laboratory and the field not only during the period of measurements but also before and after (Pinares-Patiño *et al.*, 2014). Here, we summarise some recommendations to avoid errors and reduce variation when applying the technique:

Permeation tubes

- The batches of materials used in the manufacture of permeation tubes should be properly recorded. Batches should not be mixed to supply permeation tubes to be deployed in an experiment.
- The SF_6 should be charged in a moisture-free (and, ideally, CO_2-free) environment to avoid trapping of condensates (e.g. Lassey *et al.*, 2001). The liquid nitrogen under which the tubes are charged with SF_6 should be occluded from the permeation tube during SF_6 charging.
- The permeation tubes should be calibrated over at least a two-month period, while maintained at a constant temperature (39 °C) (Deighton *et al.*, 2013c). Deighton *et al.* (2013a,d) reported no effect of rumen fluid or water on release rate of SF_6 as suggested earlier by Rochette *et al.* (2012).
- Weighing should be conducted once to twice a week using a high accuracy balance (0.1 mg). Recorded weighing times should be accurate to within 30 min, so that large batches may need subdivision for the purposes of weighing. The weights during the first 14 days post-fill should be ignored when establishing the release rate by regression.
- Only those permeation tubes with high R^2 (>0.999) regression line fitting should be used in the experiment. A representative sample of permeation tubes should be kept in the laboratory and monitored as controls.
- In any trial permeation tubes should be chosen to have a narrow range of release rates and release rates should be balanced across experimental treatments.
- Tubes are administered orally with care to prevent damage by chewing. Tube and corresponding animal identification should be accurately recorded. After tube administration, animals should be monitored in case some un-swallowed tubes are expelled. Orally administered tubes locate in the reticulum of either sheep or cattle. Location can be confirmed if tubes are recovered at slaughter.
- 'Use-by-date' can be calculated for every tube, based on its release rate and the time it takes for the initial SF_6 load to reach a residual load comprising only gaseous SF_6 ('minimum useful load of 344 mg SF_6 per ml of internal volume': Lassey *et al.*, 2001). In practice, knowing the 'use-by-date' of a tube is not of much importance. No experiment should use a permeation tube beyond ~3-4 months following charge because release rate decreases with time.
- Following tube administration, experimental sample collection should be initiated after one week. Experiments should be conducted within about 1 month of tube dosing without adjustment of SF_6 permeation rate (Lassey *et al.*, 2001). After 3 months the permeation rates become too uncertain, unless there is plan for tube recovery and re-calibration. Moate *et al.* (2013a) has proposed a method to predict release rate of SF_6 from permeation tubes with long deployment.
- Rumen-fistulated animals should not be used due to the risk of leakage at the cannula affecting emission estimates.

Animals, harnesses and sample collection gear

- Experimental animals should be representative of the population. Selecting submissive animals only will yield biased emission estimates. Obviously, balanced (e.g. by weight, milk yield, condition score, general behaviour) allocation of animals to treatments is necessary.
- Experimental animals need proper training to wear sample collection harness and canister. Untrained animals will have modified behaviour producing biased results and higher damage to collection gear.
- The harness for sample collection should be lightweight and comfortable for the animals, without interfering normal behaviour such as eating, drinking, ruminating and lying down and social interaction. Ruminants use a neck movement to get up from a lying position. Long and heavy yoke-shape canisters may prevent animals arising easily.
- The sample lines should be free of leakages, and the internal volume along the sample line should be as low as practical by using tubing of narrow internal diameter (e.g. 3 mm) without excessive length to achieve efficient sample collection.
- The inlet of sample line should be positioned just above the nostrils. Fitting a dual-entrance inlet (e.g. an inverted 'Y' tube) minimises blockage of sample line by water.
- The stainless steel capillary tubing (flow restrictor) should be well protected to prevent bending at the crimp point, and preferably insulated if condensation might be likely during cold periods of the day. Using alternative flow restrictors (e.g. ball bearing and orifice plate) is recommended. Fitting filters in the sample line prevents blockage of flow restrictor by dust, but not by water.
- The sample collection canisters can be made from PVC or stainless steel. Canister volume is relatively unimportant (range 0.5-2.5 l), but a nominal common volume is useful so that a common sampling rate and common residual vacuum apply to all canisters.
- Mean rate of sample collection is confirmed by meeting a target residual vacuum at the end of the collection period, which is about ½ of the full vacuum (~0.5 atm). Only collection canisters with proven vacuum tightness should be used, which can be assured by evacuating following previous usage and checking the vacuum before re-usage. If samples are going to be analysed without prior dilution, residual vacuum in canister at end of collection can be set to ~0.6 atm in order to counteract effects of decline in sample flow rate.
- Samples should be collected over a 5-7 day period, aiming to account for day to day variation in CH_4 emission. At the end of collection (usually 24 h), samples not meeting the target residual vacuum should be disregarded and the collection gear (sampling line and harness) replaced and later checked for faults. It is advisable that several sets of replacement collection harnesses be available for use.
- Collection canisters should be flushed with air or N_2 before re-use. Residual gases will contribute to inaccuracies and variation of estimates.

Background air sampling

- Ideally, samples of air from the experimental site with no animals present should be collected before an experiment starts. The background concentrations of SF_6 and CH_4 should be <10 ppt and <5 ppm, respectively, and as homogeneous as possible. Potential

non-experimental sources of SF_6 and CH_4 should be explored and eliminated where possible.
- Background air samples should be collected from multiple fixed locations within a building or pen, reflecting the air to which the animals are exposed, and if possible, individualised background sample should be aimed (Deighton *et al.*, 2014). In grazing experiments, extra animals can be used as background sample collectors ('mobile'), with inlet away from the mouth and nostrils. Back-up fixed background samples should be collected from fixed points near the experimental perimeter, including upwind.
- Background samples should be treated and managed in a similar way as the samples collected from animals.

Sample management
- Sample management should take account of the need to re-use canisters (depending on the canister turnaround required by the experiment).
- Samples can be analysed immediately after arrival at the laboratory (prior dilution with N_2 or not) or sub-sampled for later analysis. The choice of procedures will depend on the number of samples to be analysed, the need to reuse the collection canisters, and whether the analytical laboratory is remote from the experimental site. Samples or sub-samples that have to be stored for later analysis should either be over-pressurised so that any leaks are outward or stored in an environment free of contamination from CH_4 and SF_6.
- If dilution is made, it should stabilise for at least two hours before analyses.
- Sample analysis by gas chromatography should be conducted in duplicate or triplicate, and using appropriate gas standards that are run at the start and end of the day as well as at regular intervals (e.g. every 10 samples).

Allied measurements/estimations
- Methane emission data *per se* have little use for interpretation within the context of the experiment if no other data are also simultaneously collected. Important data to be collected in parallel with CH_4 emissions should include at least animal characteristics (e.g. breed, age, liveweight, production, liveweight gain, etc.), feed intake (and if possible feed digestibility) and feed composition.
- The experimental setting may also offer a possibility for sampling of rumen contents to gain insights into fermentation profiles associated with CH_4 emission. Rumen contents can be sampled at the end of the experimental period using 'stomach tubing'.

Scrutinizing for anomalous data
- Once the samples are analysed, data should be thoroughly scrutinised for anomalies and outliers. The concentration of both SF_6 and CH_4 in a sample should ideally be at least 10 times the background levels (Lassey, 2013). Therefore any concentration below 70 ppt SF_6 or 20 ppm CH_4 should be carefully examined.
- Exceptionally windy weather may lead to low concentrations of both gases, with little impact on CH_4/SF_6 ratio, but extreme weather will affect animal behaviour and feed intake,

hence yielding outlier data. Recording weather conditions should be standard practice during grazing experiments.

- Unusual data belonging to a particular animal should be assessed taking into account emission estimates across the days of sample collection. Particularly high or low ratios (CH_4/SF_6) lead to extreme high or low emission estimates that may be biologically unrealistic. The cause may be in the sampling (inefficient breath collection; ill-sited background samplers), in the sample collection apparatus (leaks before the flow restrictor), in the animal (not feeding adequately), or due to a faulty permeation. The last of these may not be determinable without post-mortem recovery of tube.
- Statistical tools should be used to identify 'outlier' or 'anomalous' data. For example, the 'modified Z- score' test, which is based on median of data and the median of the absolute deviation of the median, can be applied to the net concentrations of gases ($[SF_6]$ and $[CH_4]$, the ratio of $[CH_4]/[SF_6]$, the ratio $[SF_6]$/permeation rate of SF_6, estimated CH_4 emission, and CH_4 yield (emissions per unit of feed intake) (see Pinares-Patiño *et al.*, 2014).

10.9 Concluding remarks

The SF_6 fulfils most of the requirements for an ideal tracer gas to trace emissions of enteric CH_4 from ruminants. Nevertheless, pathways of transfer between ruminal and hindgut pools and relative rates of absorption and excretions (as compared to CH_4) from these pools remain to be elucidated.

The SF_6 tracer technique is associated with more variability in CH_4 emission estimates compared to emissions measured in respiration chambers. Sources of variation specific to the tracer technique remain to be fully understood, but most likely are related to the experimental conditions and improper handling of the technique (e.g. inaccurate assessment of background SF_6 and CH_4 concentrations). More recently, sources of inaccuracy and imprecision have been ascribed to the technique itself (e.g. instability of the sample flow rate). The SF_6 tracer technique remains the method of choice when CH_4 emission measurements are required from individual ruminant species under near production conditions. However, quality assurance and control measures should be implemented when applying the technique.

References

Armsby, H.P., 1904. The respiration calorimeter at the Pennsylvania Experiment Station. Agricultural Experiment Station Records 15: 1037-1050.

Bayat, A.R., Kairenius, P., Stefański, T., Leskinen, H., Comtet-Marre, S., Forano, E., Chaucheyras-Durand, F. and Shingfield, K.J., 2015. Effect of camelina oil or live yeasts (*Saccharomyces cerevisiae*) on ruminal methane production, rumen fermentation, and milk fatty acid composition in lactating cows fed grass silage diets. Journal of Dairy Science 98: 3166-3181.

Beauchemin, K.A., Coates, T., Farr, B. and McGinn, S.M., 2012. Technical note: can the sulfur hexafluoride tracer gas technique be used to accurately measure enteric methane production from ruminally cannulated cattle? Journal of Animal Science 90: 2727-2732.

Berends, H., Gerrits, W.J.J., France, J., Ellis, J.L., Van Zijderveld, S.M. and Dijkstra, J., 2014. Evaluation of the SF_6 tracer technique for estimating methane emission rates with reference to dairy cows using a mechanistic model. Journal of Theoretical Biology 353: 1-8.

Blaxter, K.L. and Clapperton, J.L., 1965. Prediction of the amount of methane produced by ruminants. British Journal of Nutrition 19: 511-522.

Boadi, D.A., Wittenberg, K.M. and Kennedy, A.D., 2002. Validation of the sulphur hexafluoride (SF_6) tracer gas technique for measurement of methane and carbon dioxide production by cattle. Canadian Journal of Animal Science 82: 125-131.

Clark, H., Pinares-Patiño, C.S. and De Klein, C., 2005. Methane and nitrous oxide emissions from grazed grasslands. In: McGilloway, D.A. (ed.) Grassland: a global resource. Wageningen Academic Publishers, Wageningen, the Netherlands, pp. 279-293.

Clemons, C.A., Coleman, A.I. and Saltzman, B.E., 1968. Concentration and ultrasensitive chromatographic determination of sulfur hexafluoride for application to meteorological tracing. Environmental Science and Technology 2: 551-556.

Colvin Jr., H.W., Wheat, J.D., Rhode, E.A. and Boda, J.M., 1957. Technique for measuring eructated gas in cattle. Journal of Dairy Science 40: 492-502.

Czerkawski, J.W., 1969. Methane production in the rumen and its significance. World Review of nutrition and Dietetics 11: 240-282.

Deighton, M.H., O'Loughlin, B.M., Williams, S.R.O., Moate, P.J., Kennedy, E., Boland, T.M. and Eckard, R.J., 2013a. Declining sulphur hexafluoride permeability of polytetrafluoroethylene membranes causes overestimation of calculated ruminant methane emissions using the tracer technique. Animal Feed Science and Technology 183: 86-95.

Deighton, M.H., Williams, S.R.O., Eckard, R.J., Boland, T.M. and Moate, P.J., 2013b. High concordance of CH_4 emissions is possible between the SF_6 tracer and respiration chamber techniques. Advances in Animal Biosciences 4: 411.

Deighton, M.H., Williams, S.R.O., Lassey, K.R., Eckard, R.J., Boland, T.M. and Moate, P.J., 2013c. Effect of temperature on the rate of sulphur hexafluoride release from permeation-tubes. Advances in Animal Biosciences 4: 416.

Deighton, M.H., Williams, S.R.O., Lassey, K.R., Eckard, R.J., Boland, T.M. and Moate, P.J., 2013d. Release rate of sulphur hexafluoride from permeation-tubes is not affected by submersion in water. Advances in Animal Biosciences 4: 269.

Deighton, M.H., Williams, S.R.O., Murray C.H., Eckard, R.J., Boland, T.M., Wales, W.J. and Moate, P.J., 2014. A modified sulphur hexafluoride tracer technique enables accurate determination of enteric methane emissions from ruminants. Animal Feed Science and Technology 197: 47-63.

Demeyer, D.I. and Van Nevel, C.J., 1975. Methanogenesis, an integrated part of carbohydrate fermentation, and its control. In: McDonald, I.M. and Warner, A.C.I. (eds.) Digestion and metabolism in the ruminant. The University of New England Publishing Unit, Armidale, Australia, pp. 366-382.

Derno, M., Hammon, H., Röntgen, M., Metges, C. and Kuhla, B., 2013. Periprandial methane production in late pregnant and early lactating German Holstein cows. Advances in Animal Biosciences 4: 374.

Dini, Y., Gere, J., Briano, C., Manetti, M., Juliarena, P., Picasso, V., Gratton, R. and Astigarraga, L., 2012. Methane emission and milk production of dairy cows grazing pastures rich in legumes or rich in grasses in Uruguay. Animals 2: 288-300.

Dougherty, R.W. and Cook, H.M., 1962. Routes of eructated gas expulsion in cattle – a quantitative study. American Journal of Veterinary Research 23: 997-1000.

Garnsworthy, P.C., Craigon, J., Hernandez-Medrano, J.H. and Daunders, N., 2012. On-farm methane measurements during milking correlate with total methane production by individual dairy cows. Journal of Dairy Science 95: 3166-3180.

Gere, J., Juliarena, P. and Gratton, R., 2014. Possible causes of high dispersion detected in methane emission measurements using the SF_6 tracer technique. In: Alfaro, M., Gonzales, S., Hube, S., Munoz, C., Pinares-Patiño, C. and Ungerfeld, E. (eds.) Libro de Resúmenes de la Primera Conferencia de Gases de Efecto Invernadero en Sistemas Agropecuarios de Latino America (in Spanish), Serie Actas INIA No. 54. INIA Remehue, Osorno, Chile, pp. 101-102.

Gere, J.I. and Gratton, R., 2010. Simple, low-cost flow controllers for time-averaged atmospheric sampling and other applications. Latin American Applied Research 40: 377-381.

Goopy, J.P., Woodgate, R., Donaldson, A., Robinson, D.L. and Hegarty, R.S., 2011. Validation of a short-term methane measurement using portable static chambers to estimate daily methane production in sheep. Animal Feed Science and Technology 166/167: 219-226.

Grainger, C., Clarke, T., McGinn, S.M., Auldist, M.J., Beauchemin, K.A., Hannah, M.C., Waghorn, G.C., Clark, H. and Eckard, R.J., 2007. Methane emissions from dairy cows measured using the sulfur hexafluoride (SF_6) tracer and chamber techniques. Journal of Dairy Science 90: 2755-2766.

Grainger, C., Williams, R., Clarke, T., Wright, A.D. and Eckard, R.J., 2010. Supplementation with whole cottonseed causes long-term reduction of methane emissions from lactating dairy cows offered a forage and cereal grain diet. Journal of Dairy Science 93: 2612-2619.

Hammond, K.J., Muetzel, S., Waghorn, G.C., Pinares-Patiño, C.S., Burke, J.L. and Hoskin, S.O., 2009. The variation in methane emissions from sheep and cattle is not explained by the chemical composition of ryegrass. Proceedings of the New Zealand Society of Animal Production 69: 174-178.

Hegarty, R.S., 2005. Development of methane-releasing permeation devices for intraruminal use. Recent Advances in Animal Nutrition 15: 3A.

Hoernicke, H., Williams, W.F., Waldo, D.R. and Flatt, W.P., 1965. Composition and absorption of rumen gases and their importance for the accuracy of respiration trials with tracheostomized ruminants. In: Blaxter, K.L. (ed.) Energy metabolism. Academic Press, London, UK, pp. 165-178.

Hofmeyr, H.S., Slabbert, N. and Pienaar, J.P., 1984. Partitioning of methane production between ruminal and hindgut fermentation sites in sheep. Canadian Journal of Animal Science 64 (Suppl.): 171-172.

Immig, I., 1996. The rumen and hindgut as source of ruminant methanogenesis. Environmental Monitoring and Assessment 42: 57-72.

Johnson, D.E., Ferrell, C.J. and Jenkings, T.G., 2003. The history of energetic efficiency research: where have we been and where are we going? Journal of Animal Science 81 (Suppl. 1): E27-E38.

Johnson, K.A. and Johnson, D.E., 1995. Methane emissions from cattle. Journal of Animal Science 73: 2483-2492.

Johnson, K.A., Huyler, M.T., Westberg, H.H., Lamb, B.K. and Zimmerman, P., 1994a. Measurement of methane emissions from ruminant livestock using a SF_6 tracer technique. Environmental Science and Technology 28: 359-362.

Johnson, K.A., Huyler, M.T., Westberg, H.H., Lamb, B.K. and Zimmerman, P., 1994b. Measurement of methane emissions from ruminant livestock using a sulphur hexafluoride tracer technique. In: Aguilera, J.F. (ed.) Energy metabolism of farm animals. EAAP publication no. 76. Servicio de Publicaciones, Consejo Superior de Investigaciones Cientificas, Granada, Spain, pp. 335-338.

Johnson, K.A., Westberg, H.H., Michal, J.J. and Cossalman, M.W., 2007. The SF_6 technique: methane measurement from ruminants. In: Makkar, H.P.S. and Vercoe, P.E. (eds.) Measuring methane production from ruminants. Springer, Dordrecht, the Netherlands, pp. 33-67.

Judd, M.J., Kelliher, F.M., Ulyatt, M.J., Lassey, K.R., Tate, K.R., Shelton, I.D., Harvey, M.J. and Walker, C.F., 1999. Net methane emissions from grazing sheep. Global Change Biology 5: 647-658.

Kellner, O., 1910. The scientific feeding of animals. The McMillan Company, NY, USA.

Lassey, K.R., 2007. Livestock methane emission: from the individual grazing animal through national inventories to the global methane cycle. Agriculture and Forest Meteorology 142: 120-132.

Lassey, K.R., 2013. On the importance of background sampling in applications of the SF_6 tracer technique to determine ruminant methane emissions. Animal Feed Science and Technology 180: 115-120.

Lassey, K.R., Martin, R.J., Williams, S.R.O., Berndt, A., Iwaasa, A.D., Hegarty, R.S., Moate, P.J., Deighton, M.H., Ribaux, B.E., 2014. Analysis of breath samples. In: Lambert, M.G. (ed.) Guidelines for use of sulphur hexafluoride (SF_6) tracer technique to measure enteric methane emissions from ruminants. New Zealand Agricultural Greenhouse Gas Research Centre, Palmerston North, New Zealand, pp. 89-113.

Lassey, K.R., Pinares-Patiño, C.S., Martin, R.J., Molano, G. and McMillan, A.M.S., 2011. Enteric methane emission rates determined by the SF_6 tracer technique: temporal patterns and averaging periods. Animal Feed Science and Technology 166-167: 183-191.

Lassey, K.R., Ulyatt, M.J., Martin, R.J., Walker, C.F. and Shelton, I.D., 1997. Methane emissions measured directly from grazing livestock in New Zealand. Atmospheric Environment 31: 2905-2914.

Lassey, K.R., Walker, C.F., McMillan, A.M.S. and Ulyatt, M.J., 2001. On the performance of SF_6 permeation tubes used in determining methane emission rates from grazing livestock. Chemosphere: Global Change Science 3: 367-376.

Laubach, J., Baib, M., Pinares-Patiño, C.S., Phillips, F.A., Naylor, T.A., Molano, G., Cárdenas Rocha, E.A. and Griffith, D.W.T., 2013a. Accuracy of micrometeorological techniques for detecting a change in methane emissions from a herd of cattle. Agriculture and Forest Meteorology 176: 50-63.

Laubach, J., Grover, S.P.P., Pinares-Patiño, C.S. and Molano, G., 2014. A micrometeorological technique for detecting small differences in methane emissions from two groups of cattle. Atmospheric Environment 98: 599-606.

Leuning, R., Baker, S.K., Jamie, I.M., Hsu, C.H., Klein, L., Denmead, O.T. and Griffith, D.W.T., 1999. Methane emission from free-ranging sheep: a comparison of two measurement methods. Atmospheric Environment 33: 1357-1365.

Levitt, M.D, Levitt, D.G. and Berggren, T., 1973. Use of inert gases to study the interaction of blood flow and diffusion during passive absorption from the gastrointestinal tract of the rat. Journal of Clinical Investigation 52: 1852-1862.

Machmüller, A. and Hegarty, R.S., 2006. Alternative tracer gases for the ERUCT technique to estimate methane emission from grazing animals. International Congress Series 1293: 50-53.

Madsen, J., Bjerg, B.S., Hvelplund, T., Weisbjerg, M.R. and Lund, P., 2010. Methane and carbon dioxide ratio in excreted air for quantification of the methane production from ruminants. Livestock Science 129: 223-227.

Martin, C., Koolaard, J., Rochette, Y., Clark, H., Jouany, J.P. and Pinares-Patio, C.S., 2012. Effect of release rate of the SF_6 tracer on methane emission estimates based on ruminal and breath gas samples. Animal 6: 518-525.

Martin, C., Morgavi, D.P. and Doreau, M., 2010. Methane mitigation in ruminants: from microbe to the farm scale. Animal 4: 351-365.

McCourt, A.R., Yan, T. and Mayne, S., 2008. Validation of the sulphur hexafluoride (SF_6) tracer technique as a method of predicting methane output from dairy cows. In: Proceedings of the British Society of Animal Science, Edinburgh, UK, p. 120.

McGinn, S.M., Beauchemin, K.A., Iwaasa, A.D. and McAllister, T.A., 2006. Assessment of the sulfur hexafluoride (SF_6) tracer technique for measuring enteric methane emissions from cattle. Journal of Environment Quality 35: 1686-1691.

Meyer, M., Tebbe, U. and Piiper, J., 1980. Solubility of inert gases in dog blood and skeletal muscle. Pflügers Archiv: European Journal of Physiology 384: 131-134.

Moate, P.J., Clarke, T., Davis, L.H. and Laby, R.H., 1997. Rumen gases and bloat in grazing dairy cows. The Journal of Agricultural Science 129: 459-469.

Moate, P.J., Williams, S.R.O., Deighton, M.H. and Hannah, M.C., 2013a. Rate of release of SF_6 from permeation tubes is described by Michaelis-Menten kinetics. Advances in Animal Biosciences 4: 421.

Moate, P.J., Williams, S.R.O., Deighton, M.H., Jacobs, J.L. and Wales, W.J., 2013b. Influence of rumen cannulation on feed intake, milk production, enteric methane production and composition of rumen headspace gas. Advances in Animal Biosciences 4: 423.

Morel, D.R., Schwieger, I., Hohn, L., Terrettaz, J., Llull, J.B., Cornioley, Y.A. and Schneider, M., 2000. Human pharmacokinetics and safety evaluation of SonoVue, a new contrast agent for ultrasound imaging. Investigative Radiology 35: 80-85.

Muñoz, C., Yan, T., Wills, D.A., Murray, S. and Gordon, A.W., 2012. Comparison of the sulfur hexafluoride tracer and respiration chamber techniques for estimating methane emissions and correction for rectum methane output from dairy cows. Journal of Dairy Science 95: 3139-3148.

Murray, P.J., Moss, A., Lockyer, D.R. and Jarvis, S.C., 1999. A comparison of systems for measuring methane emission from sheep. Journal of Agricultural Science 133: 439-444.

Murray, R.M., Bryant, A.M. and Leng, R.A., 1976. Rates of production of methane in the rumen and large intestine of sheep. British Journal of Nutrition 36: 1-14.

Oddy, V.H., de Haas, Y., Basarab, J., Cammack, K., Hayes, B., Hegarty, R.S., Lassen, J., McEwan, J.C., Miller, S., Pinares-Patiño, C.S., Shackell, G., Vercoe, P. and Pickering, N.K., 2014. Breeding ruminants that emit less methane – the role of international collaboration. In: Proceedings 10th World Congress of Genetics Applied to Livestock Production. Available at: http://tinyurl.com/qx3szfn.

Pinares-Patiño, C.S., 2000. Methane emission from forage-fed sheep, a study of variation between animals. PhD thesis, Massey University, Palmerston North, New Zealand. Available at: http://muir.massey.ac.nz/handle/10179/2454.

Pinares-Patiño, C.S. and Clark, H., 2008. Reliability of the sulfur hexafluoride tracer technique for methane emission measurement from individual animals: an overview. Australian Journal of Experimental Agriculture 48: 223-229.

Pinares-Patiño, C.S. and Waghorn, G. (eds.), 2014. Technical manual on respiration chambers designs. Manual commissioned by the New Zealand Government to support the goals and objectives of the Global Research Alliance on Agricultural Greenhouse Gases. Available at: http://tinyurl.com/nb87zlp.

Pinares-Patiño, C.S., Franco, F.E., Molano, G., MacLean, S., Kjestrup, H., Sandoval, E. and Laubach, J., 2013a. Methane emission from cattle grazed on pasture sprayed with canola oil. Advances in Animal Biosciences 4: 273.

Pinares-Patiño, C.S., Gere, J., Williams, K., Gratton, R., Juliarena, P., Molano, G., MacLean, S., Sandoval, E., Taylor, G. and Koolaard, J., 2012. Extending the collection duration of breath samples for enteric methane emission estimation using the SF_6 tracer technique. Animals 2: 275-287.

Pinares-Patiño, C.S., Hickey, S.M., Young, E.A., Dodds, K.G., MacLean, S., Molano, G., Sandoval, E. Kjestrup, H., Harland, R., Pickering, N.K. and McEwan, J.C., 2013b. Heritability estimates of methane emissions from sheep. Animal 7 (Suppl. 2): 316-321.

Pinares-Patiño, C.S., Holmes, C.W., Lassey, K.R. and Ulyatt, M.J., 2008a. Measurement of methane emission from sheep by the sulphur hexafluoride tracer technique and by the calorimetric chamber: failure and success. Animal 2: 141-148.

Pinares-Patiño, C.S., Koolaard, J.P., Williams, S.R.O., Martin, C., Swainson, N.M., Berndt, A. and Molano, G., 2014. Data quality assurance and quality control. In: Lambert, M.G. (ed.) Guidelines for use of sulphur hexafluoride (SF_6) tracer technique to measure enteric methane emissions from ruminants. New Zealand Agricultural Greenhouse Gas Research Centre, Palmerston North, New Zealand, pp. 134-145.

Pinares-Patiño, C.S., Lassey, K.R., Martin, R.J., Molano, G., Fernandez, M., MacLean, S., Sandoval, E., Luo, D. and Clark, H., 2011a. Assessment of the sulphur hexafluoride (SF_6) tracer technique using respiration chambers for estimation of methane emissions from sheep. Animal Feed Science and Technology 166/167: 201-209.

Pinares-Patiño, C.S., Machmüller, A., Molano, G., Smith, A., Vlaming, J.B. and Clark, H., 2008b. The SF_6 tracer technique for measurements of methane emission from cattle – effect of tracer permeation rate. Canadian Journal of Animal Science 88: 309-320.

Pinares-Patiño, C.S., McEwan, J.C., Dodds, K.G., Cárdenas, E.A., Hegarty, R.S., Koolaard, J.P. and Clark, H., 2011b. Repeatability of methane emissions from sheep. Animal Feed Science and Technology 166: 210-218.

Reynolds, C.K., Crompton, L.A., Barratt, C.E., Donovan, N. and Misselbrook, T., 2013. Arterial, rumen, and milk concentration and net absorption of methane into the portal vein of lactating dairy cattle. Advances in Animal Biosciences 4: 392.

Rochette, Y., Eugene, M., Doreau, M., Morgavi, D.P. and Martin, C., 2012. Determination of enteric methane emission by SF_6 tracer technique: permeation tubes must be calibrated after incubation in the rumen for an accurate quantification. In: Hassouna, M. and Guingand, N. (eds.) Proceedings of the International Symposium on Emission of Gas and Dust from Livestock. Saint-Malo, France, pp. 359-361.

Schimmel, C., Bernard, S.L., Anderson, J.C., Polissar, N.L., Lakshminarayan, S. and Hlastala, M.P., 2004. Soluble gas exchange in the pulmonary airways of sheep. Journal of Applied Physiology 97: 1702-1708.

Steinfeld, H., P. Gerber, T. Wassenaar, V. Castel, M. Rosales, and de Haan, C., 2006. Livestock's long shadow: environmental issues and options. Food and Agriculture Organization of the United Nations, Rome, Italy.

Swainson, N.M., Brookes, I.M., Hoskin, S.O. and Clark, H., 2011. Post-experiment correction for release rate in permeation tubes improves the accuracy of the sulphur hexafluoride (SF_6) tracer technique in deer. Animal Feed Science and Technology 166/167: 192-197.

Torrent, J. and Johnson, D.E., 1994. Methane production in the large intestine of sheep. In: Aguilera, J.F. (ed.) Energy metabolism of farm animals. EAAP publication no. 76. Servicio de Publicaciones, Consejo Superior de Investigaciones Cientificas, Granada, Spain, pp. 391-394.

Ulyatt, M.J., Baker, S.K., McCrabb, G.J. and Lassey, K.R., 1999. Accuracy of SF_6 tracer technology and alternatives for field measurements. Australian Journal of Agricultural Research 50: 1329-1334.

US Environmental Protection Agency (US EPA), 2002. Byproducts of sulfur hexafluoride (SF_6) use in the electric power industry. US EPA, Office of Air and Radiation, Global Programs Division, Washington, DC, USA.

Vlaming, J.B., 2008. Quantifying variation in estimated methane emission from ruminants using the SF_6 tracer technique. PhD Thesis, Massey University, Palmerston North, New Zealand. Available at: http://mro.massey.ac.nz/handle/10179/730.

Vlaming, J.B., Brookes, I.M., Hoskin, S.O., Pinares-Patiño, C.S. and Clark, H., 2007. The possible influence of intra-ruminal sulphur hexafluoride release rates on calculated methane emissions from cattle. Canadian Journal of Animal Science 87: 269-275.

Williams, S.R.O., Moate, P.J., Hannah, M.C., Ribaux, B.E., Wales, W.J. and Eckard, R.J., 2011. Background matters with the SF_6 tracer method for estimating enteric methane emissions from dairy cows: a critical evaluation of the SF_6 procedure. Animal Feed Science and Technology 170: 265-276.

Wolin, M.J., 1960. A theoretical rumen fermentation balance. Journal of Dairy Science 43: 1452-1459.

Zimmerman, P.R., 1993. System for measuring metabolic gas emissions from animals. US Patent and Trademark Office, United States Patent No. 5,265,618.

Chapter 11: Indirect calorimetry: assessing animal response to heat and cold stress

J.B. Gaughan[1], M.J.W. Heetkamp[2] and P. Hendriks[3]*
[1]School of Agriculture and Food Sciences, Animal & Wildlife Sciences Group, The University of Queensland, QLD 4343 Gatton, Australia; [2]Wageningen University, Adaptation Physiology Group, P.O. Box 338, 6700 AH, Wageningen, the Netherlands; [3]Wageningen University, Technical Services, P.O. Box 338, 6700 AH, Wageningen, the Netherlands; j.gaughan@uq.edu.au

Abstract

Calorimetric thermal stress studies where indirect calorimetry is used as a tool to estimate energy expenditure have been undertaken since this technique was developed. Some examples of these studies are presented in this chapter. The measurement of gas exchange by means of an open-circuit respiration chamber have been done in studies where climate needs to be constant but also in studies where animal reactions to differences in climate as part of the experimental design. In the past and in present, animal studies are done in chambers with and without integrated climate controls. Changing climate factors, e.g. temperature, relative humidity, and air speed, may change an animal's physiological responses. As climate can also affect energy metabolism it is necessary to have a well-designed, accurate and reliable air conditioning system inside a respiration chamber to rule out the unwanted effects of an uncontrollable climate. In this chapter important issues related to the design of climate control in respiration chambers are described and methods for improving accuracy are discussed. We conclude that the design of chambers with and without climate control should be undertaken by those that understand animal biology as well as technicians and engineers. Furthermore, the basic physics of heat measurement are discussed and we elaborate about possible biases from biological factors and physical factors and how to deal with them.

Keywords: climate control, respiration chamber, reducing error

11.1 Introduction

Indirect calorimetry based on respiratory exchange measurement has been successfully used from the beginning of the 20th century (e.g. Wesleyan University, circa 1904) to obtain an estimate of heat production (energy expenditure) in human subjects and animals. In some studies animals may have been exposed to variations in temperature and humidity, and exposed to various exercise regimes. Examples are given in this chapter. However, most have used animals in steady state, i.e. exposed to constant temperature/humidity and not exercising.

W.J.J. Gerrits and E. Labussière (eds.) **Indirect calorimetry**
DOI 10.3920/978-90-8686-261-0_11, © Wageningen Academic Publishers 2015

Animal studies can be undertaken in chambers with and without integrated climate controls. Systems without integrated controls usually depend on the removal of heat and water vapour via large ventilation or air turnover rates. Short term gas exchange accuracy is affected negatively because the ventilation part of gas exchange calculations will be enormous but gas concentrations errors will still have their effect. Long term accuracy strongly depends on the precision that ventilation rate can be measured.

In this chapter we provide information on the basic physics of heat measurement. Furthermore, important issues related to the design of climate control in respiration chambers are described and methods for improving accuracy are discussed. Advantages and disadvantages of indirect calorimetry will be evaluated and biasing factors, both biological and physical, will also be discussed. Finally, we discuss some examples of calorimetric thermal stress studies.

11.2 Basic physics of heat measurement

Determining heat flow from a living organism is complex. Whereas the thermal dynamics of a solid object (e.g. a solid black body often a globe) can be easily modelled using classical physics the same cannot be done with an animal. The heat balance of a living organism is affected by dynamic internal and external forces.

In physics mass and energy of for example a black solid body are conserved. A black body is an idealized physical body that absorbs all incident electromagnetic radiation, regardless of frequency or angle of incidence. A black body in thermal equilibrium (that is, at a constant temperature) emits electromagnetic radiation called black-body radiation. The radiation is emitted according to Planck's law ($E = hv$) where, h, is Planck's constant ($6.62606957 \times 10^{-34}$ J/s), E is energy of a charged atomic oscillator in the wall of a black body, and v is the frequency of its associated electromagnetic wave (http://en.wikipedia.org/wiki/Planck_constant).

A black body in thermal equilibrium has two notable properties:

- It is an ideal emitter: at every frequency, it emits as much energy as or more energy than any other body at the same temperature.
- It is a diffuse emitter: the energy is radiated isotropically, independent of direction (http://en.wikipedia.org/wiki/Black_body).

The energy entering into, the energy retained, and energy exiting the black body can be easily measured. Predictions of heat flow are mathematically testable and can be repeated often, with the same result. The heat flow into and out of a black body located in full sunlight can be calculated by measuring heat flow: the radiant heat gained from the sun (mainly infrared radiation) + radiation lost from the globes surface (emission of infrared radiation) + heat gained from or lost to the air (influenced by air speed and air moisture) + heat gained from or lost within the body. The answer will of course depend upon the material from which the black body is made and the mass of the body. Given the same environmental conditions and an identical black body then the outcome will be the same at all locations and at all times.

On the other hand heat balance in a living organism is much more complex and dynamic. The heat produced from an animal at a particular day may be very different to the heat produced from the same animal exposed to the same climatic conditions at a later date, or even a few minutes later in time. In animals fuel (food, stored fat, glucose) + O_2 = ΔH (change in enthalpy) + energy in excreted waste (urine and faeces). However animals are not static so energy is also produced by work (digestion of food, walking, metabolic activity, stress), and energy can be retained as an animal grows (muscle, fat), or if pregnant energy can be retained in the developing foetus. To further complicate matters as environmental conditions change animal responses will also change. There may be subtle or major metabolic shifts which need to be considered when assessing heat production. As heat load changes animals will respond with changes in heart rate, respiration rate, sweating rate, metabolic rate, as well as a number of behavioural changes such as shade seeking. Oxygen consumption, body temperature and thermal conductance of striped hamsters that were acclimated to 5 °C (cold), 21 °C (warm) and 31 °C (hot) and then exposed to a temperature range of 5 to 37.5 °C (Figure 1). This study demonstrates that animals acclimated to different climatic conditions will have differences in their thermal balance that is independent of ambient conditions.

At higher temperatures evaporative heat loss (sweating, wetting of skin, panting) becomes more important because radiation, conduction and convection lose their heat flow capacity when the difference between subject's temperature and its environment is getting smaller

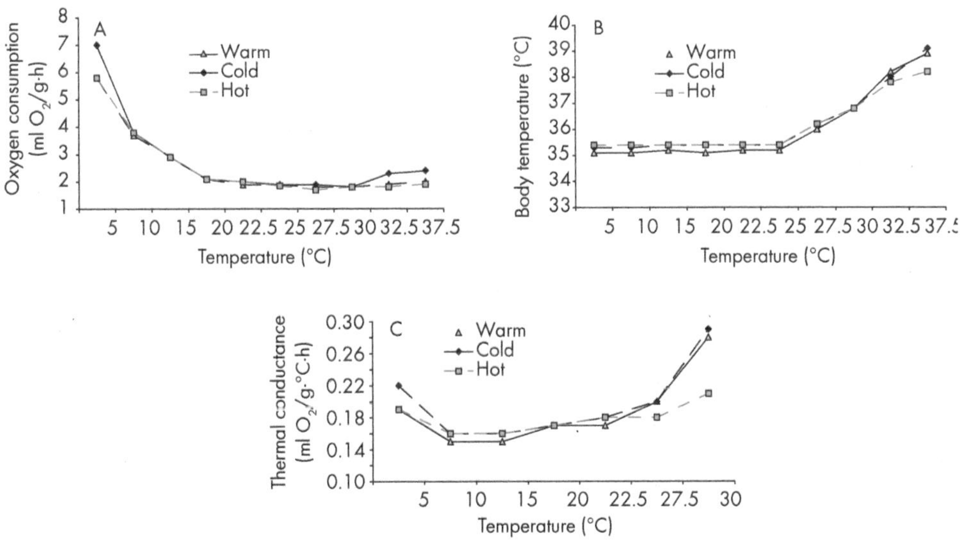

Figure 1. (A) Oxygen consumption, (B) body temperature and (C) thermal conductance of striped hamsters over a range of temperatures (Zhao et al., *2014).*

(Figure 2). Huynh *et al.* (2005), in a study with three levels of relative humidity and daily increasing ambient temperature inside a climate respiration chamber, concluded that thermal behaviour of growing pigs at high humidity changes at lower temperatures. Therefore, interactions between temperature and air humidity may affect behaviour and subsequently energy metabolism. This indicates that it is important to control temperature as well as relative humidity at a chosen level of steady state to minimize unforeseen effects on animals. The question here is: What would the animals reaction have been if temperature was increased daily but relative humidity was lowered to keep enthalpy steady? Temperature cannot be considered in isolation. Animals respond to the combined effects of temperature and humidity, and as such both need to be controlled.

11.3 Climate control

In studies where a constant ambient temperature and relative humidity is required but also if effects of climate change need to be studied one should be aware and try to measure what any living subject actually feels. Temperature sensors therefore should be placed at multiple positions near but out of reach of the subject. In climate (respiration) chambers used for animal studies, internally circulated air coming from the animal area should be mixed well inside the air conditioning unit(s) with the incoming ventilation air preferably before it comes to the heat exchangers. Air leaving the air-conditioning unit has to be very well distributed to give time to mix with chamber air before it reaches the animal(s). This will create a homogenous environment and the occurrence of draughts will be limited. To prevent draught at the animal level the maximum difference in temperature between air coming from an air-conditioning unit and the temperature near the measured subject(s) should not exceed approximately 3 °C while cooling and 7 °C while heating. In combination with a well-designed air distribution

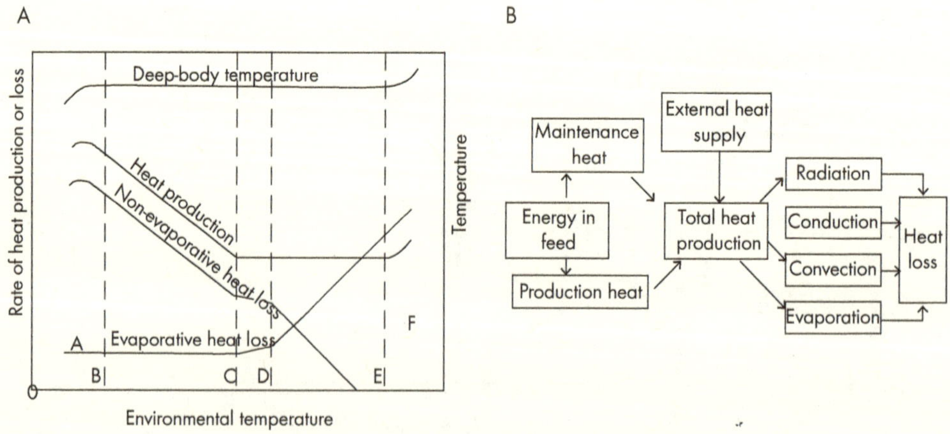

Figure 2. (A) Relation between air temperature and the components of heat loss. (B) Principle of heat balance (Mount, 1974)

system this will minimize the effect of blown-in colder air falling down before it is mixed with the present air. Increasing the amount of circulating air from animal room to air conditioning units and back will help to remove the heat load and at the same time reduce the temperature difference. When exchanging air within a chamber the airspeed at the animal level should not exceed 0.2 m/sec (Verstegen *et al.*, 1987). To prevent local high airspeed or even local draught ceilings with perforated plates or air socks can be used to distribute the air blown into an animal room before it reaches the animal(s).

Well-designed climate chambers with adequate air conditioning have little temperature variation in place and in time. To get to know this variation, temperature should be monitored within a climate room at various positions around (an) animal(s) during a time span where also changes in animal behaviour can render its effects on climate control. The number and position of sensors should be considered. To have a good measurement of ambient temperature around (an) animal(s) the variation should also be monitored continuously and therefore it is best to use at least three or more temperature sensors. Effective short term climate change is dependent on the chamber controls being able to provide enough power in heating, cooling, humidifying and dehumidifying the air. In animal studies the change in climate set-points for short- vs long-term environmental changes will differ. Where short term stress usually has a time span of a few hours to maximum a day, long term stress normally goes on from at least one day onwards. Also short term climate stress often means a sudden and larger change in temperature set-point cv longer-term climate stress. In all situations climate control software should be designed and tuned to prevent huge under- and over-shooting, at all changes in set points.

If climate control hard- and software is tuned well normally a sine shaped deviation from temperature set point in time occurs. This variation of temperature in time can be reduced by the correct choice and tuning of the algorithms that adjust the settings of the climate control hardware. At least in bigger chambers it seems best to measure temperature at various points close to a (group of) animal(s) and use the minimum, maximum, mean or median value as measured value in a cascade control. In cascade (= master/slave) control, one control loop applies control algorithms to a measured variable (= master) against a set-point, but then provides a varying set-point (= slave) to another control loop rather than affecting process variables directly (http://en.wikipedia.org/wiki/Control_system). It is safest to use a median calculated master value to prevent outlying values for example from sensor measured extremes or errors to have any effect. In climate control the set point of the slave sensor that measures the temperature of the air that is coming straight from the climate control unit(s) back into the animal room is changed, in a temporized way, according to the difference between master sensors median temperature and desired temperature. The slave sensor measured values are then used, in a direct control, to keep the blown in air at the slave set point temperature. In this master/slave control the delaying effect because of a large air buffer including the heat capacity of walls and inside materials is avoided.

Air humidity control is probably the most challenging aspect of climate control in respiration chambers. It can be recommended to understand the theory of the behaviour of water vapour in air when using climate control equipment. Mollier (1923) developed a h-x diagram for humid air from which a lot of derivatives are made like the one presented in Figure 3 from the former TFDL institute in Wageningen, the Netherlands.

It gives the relation between the (dry bulb) air temperature (Y-axis in °C), the absolute water content (X-axis in grams H_20/ kg dry air), the relative humidity (RH; the curved lines where the lowest line represents 100% saturation; φ=1.0), the enthalpy (kJ/kg) and the air water pressure (mbar). RH can be described as the ratio of the vapour pressure of moisture in air to the saturation pressure at the dry bulb temperature of the air. Actually the process of climate humidity control can be outlined in this diagram. For example, following the black arrows, one can see that coming from the 30 °C/60%RH point and cooling down the air will give 100% RH at approximately 21 °C. At 100% RH a temperature is called the dew point temperature. This is the saturation temperature of the moisture present in air. It can also be defined as the temperature at which the vapour changes into liquid (condensation). A simplified definition is the temperature at which the water vapour turns into 'dew'. Cooling down more until 17.5 °C will cause water condensation and loss of water vapour and dew point temperature then also goes to this 17.5 °C. Heating up air to again 30 °C will then cause about a 45%RH air, but dew point temperature stays at 17.5 °C. It is important to also understand from the Mollier diagram (Figure 3) (Mollier, 1923) that at higher temperatures the water vapour content in air can increase significantly. For example, following the curved φ=1.0 line: at 15 °C the maximum absolute water content in air is about 11 g/kg dry air, at 26 °C it is a factor of two more and at 40 °C it is a factor of four greater.

When animals or humans are present in a room the water vapour pressure in their expired air will have an effect on overall room humidity. For most species (an exception being camels) all air coming from lungs is at body temperature and 100% saturated with water vapour. You can easily measure this yourself! At higher ambient temperatures and humidity's, because of a higher water content in inspired air, the amount of lung water evaporation and therefore evaporative heat loss will decrease per breath stroke. Maintaining or even increasing evaporative heat loss then will only be possible by increased respiration rate (think of panting).

In respiration chambers, higher air humidity's in combination with a high air temperature can effect measurement of ventilation rate, air sampling and gas analysis. Because of increased water content in the air the risk of condensation of water in cold spots in the ventilation channels have to be considered. Measures should be taken to prevent water vapour to condensate effecting ventilation, air flow measurements, air-sampling and gas conditioning before dehumidification and gas analysis. The easiest way to prevent condensation of air water vapour is to keep systems in a room were temperature is never going lower than dew point temperature of the transported air. Also isolation can be considered for example in combination with heating of the sample air lines towards an analysis system. A disadvantage of

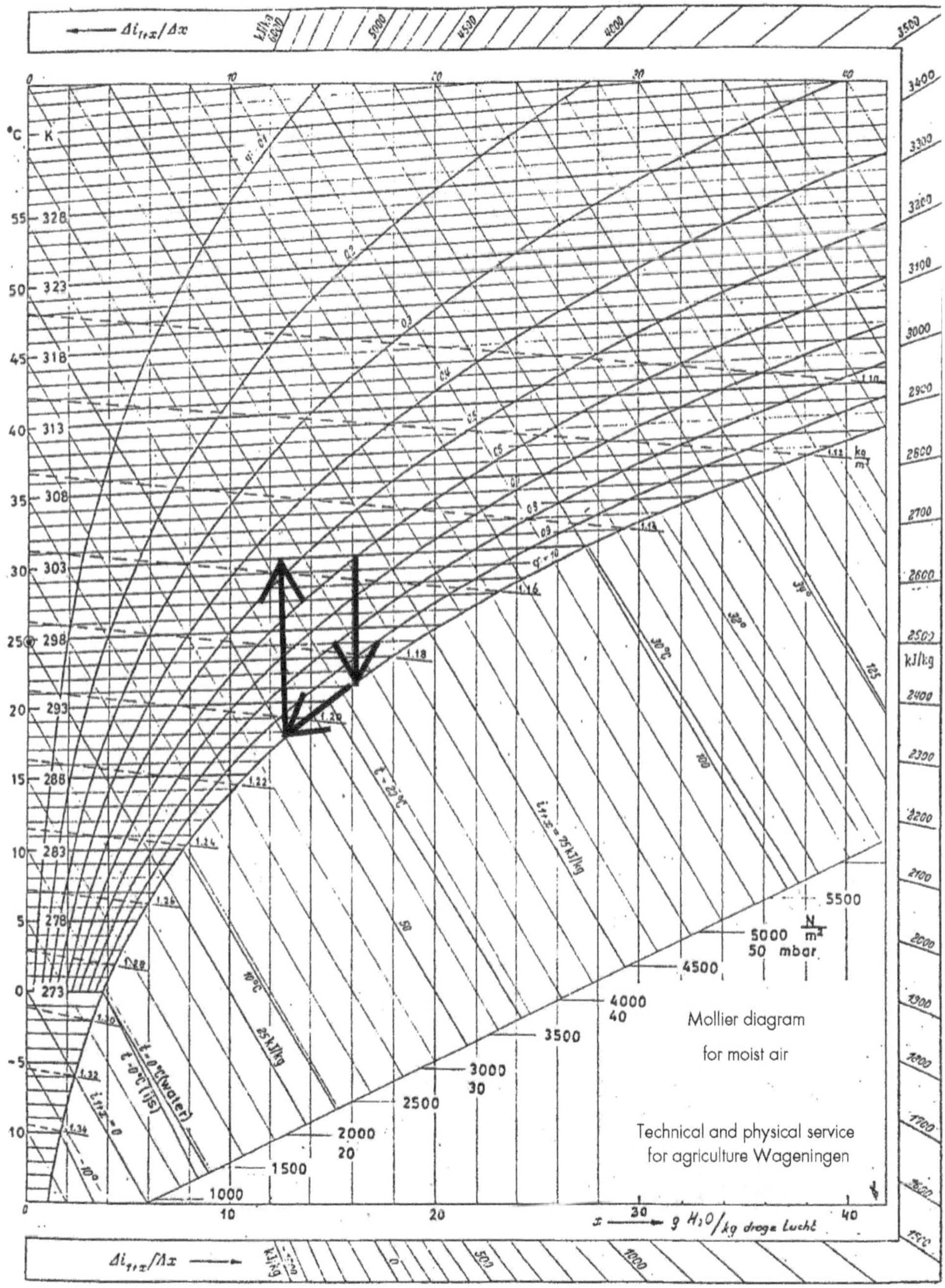

Figure 3. Mollier h-x diagram for humid air from the former TFDL Institute in Wageningen, the Netherlands.

putting isolation layers on top of ventilation channels with measuring and control equipment is that this will complicate detection when checking for leaks.

Surplus water vapour, coming from the breath of animals and wet surfaces in a chamber, or even from the incoming ventilation air, can be removed by the same heat exchanger (cooling device) in which the surplus heat load is removed from air. Water vapour will partially be condensed on a heat exchanger were liquid cooling water is streamed through. Because the same heat exchanger is used for cooling and for de-humidification special care must be taken to get the control right because valves und pumps should be assembled in a different way than is standard in climate control. As shown in Figure 4 the 3-way control valve (K2) in the cooling system (G-N) is differently placed then valve (K1) in the heating system (F-M). In this way the cooling system is not flushed continuously at a maximum level but only the amount that valve (K2) is controlling. For an accurate humidity control this setup is necessary. Actually a cold spot is established on heat exchanger G (starting at the lower left corner in Figure 4), where temperature comes below dew point of the air. The size of this cold spot depends on the amount of cold water controlled by valve K2 as a result of the water vapour produced inside room A and the desired RH. The size of this cold spot and the dew point temperature to which a part

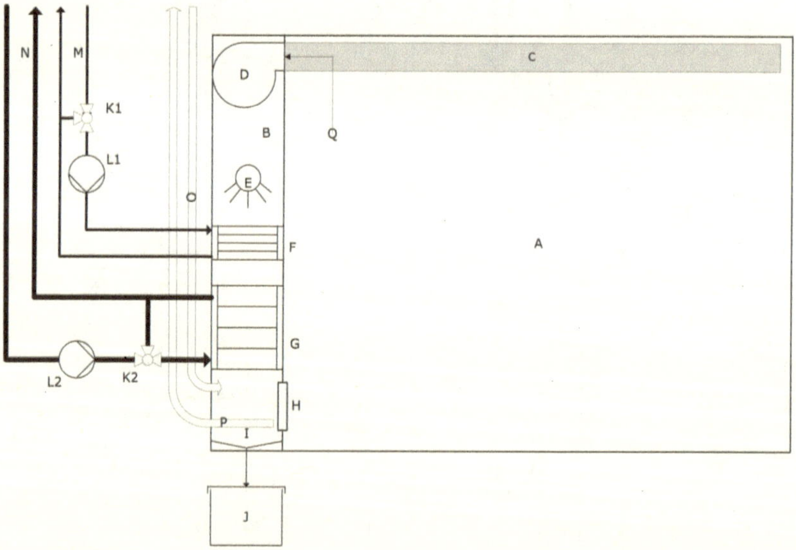

Figure 4. Layout of an air-conditioning unit from climate respiration chambers where: (A) animal room; (B) Air-conditioning unit; (C) air sock; (D) fan for re-circulating air; (E) humidifier in combination with an air-washer; (F) heat exchanger for heating; (G) heat exchanger for cooling and de-humidification; (H) dust filter; (I) condense water funnel; (J) condense water collection vessel; (K1 and K2) 3-way valves; (L1 and L2) liquid pumps; (M) warm water supply and return piping; (N) cold water supply and return piping; (O) incoming ventilation air pipe; (P) outgoing ventilation air pipe; (Q) Position of slave temperature sensor (Wageningen University, Wageningen, the Netherlands).

of the air is cooled down to defines the drying effect. If the pump L2 in the cooling system N-G would be placed as in the heating system M-F then one can imagine that if cold water is introduced it will spread out through the heat exchanger rather fast. There will be a cooling effect, but there will only be a de-humidification effect if enough cold water is introduced and dew point temperature is reached. Most likely the dew point then will be reached in a rather big part of the heat exchanger and the amount of de-humidification then suddenly is rather high. Probably then the RH is getting to low and controls will start humidification. Together with the humidity load of the animal(s) RH will increase soon and the drying process will start again. Because of continuous de-humidification followed by humidification the control stays out of balance.

Actually two independent controls, a temperature and a humidity control, need to calculate the position of valve K2 in the cooling system N-G depending on a recent temperature and humidity measurement, respectively. The highest calculated opening signal then should be send to the valve. If temperature at position Q is getting to low because the humidity control demands more cooling activity then the temperature control will gradually open valve K1 in the heating line M-F to inject warm water to heat exchanger F.

With de-humidifying by condensing water some of the ammonia gas will be washed out of the air and condensed water needs to be collected to estimate N-loss as part of a N-balance measurement. Animals like pigs, at higher temperatures, tend to pollute the floor they are living on (Aarnink *et al.*, 2006; Huynh *et al.*, 2005). A big problem then can be the development of ammonia gas which at levels >20 ppm becomes an animal and human welfare issue. A good way to get rid of ammonia is to 'wash' the air. In a system were water vapour surplus is already condensed on a cold heat exchanger, extra water in the form of big drops (like a shower) can be sprinkled on top of this exchanger. Ammonia then will be washed away very effectively and wash and condensed water can then be collected to estimate N-loss.

The capacity of heating, cooling, humidifying and dehumidifying systems should be calculated taking into account the desired climate range, mass and isolation characteristics of a chamber, estimated direct and evaporative heat load from animals, heat load from fans and lights, estimated water vapour load, the desired speed of change in climatic conditions, the amount of air circulation and the amount of ventilation.

As an example in the Wageningen chambers (Chapter 2, Heetkamp *et al.*, 2015) temperature can be set between 5 and 45 °C combined with a relative humidity of 50 to 90%. Temperature sensors with an accuracy of ≤± 0.1°C were chosen. In addition to 10 PT100 sensors (4-wire type 1/3DIN-B) per chamber there are three combined temperature/RH sensors (Novasina Hygrodat 100 with HIA-13/20M sensor and type CC-1 highly precise electrolytic measuring cell; ±2% RH accuracy), one for each possible room are used. The electrolytic measuring cell is also particularly suitable for humidity measurements in environments above 85% up to almost 100% RH.

Airspeed within a chamber, normally ≤0.2 m/sec, can be increased stepwise to approximately 1.0 m/sec. Another design characteristic of the Wageningen chamber is the selectable temperature control: either a master/slave-temperature control or direct temperature control. For this, 10 selectable 1/3 DIN-B PT100 temperature sensors are used per chamber at animal level. To reduce the possibility of a climate system becoming unstable, it is advised to place a 'slave' temperature sensor at a position directly downstream from the air-conditioning unit (Point Q in Figure 4). Temperature changes are then measured directly and heating and cooling valves can be changed accordingly. High circulating rates from air-conditioning units to animal rooms and back are desirable although airspeed at animal level should be observed. Therefore circulating rates can be fine-tuned resulting in a homogenous climate as well as homogenous gas concentrations inside a chamber. Dust can also be filtered out of the air but noise levels could be a point of concern.

In the Wageningen situation software and algorithms used were developed in-house. The software/user interface allows operators to make choices for all rooms, each of which has selectable air-conditioning units and sensors to measure and control the climate. The variations from set-point are tightly controlled. Temperature variation at one spot in time needs to be ≤0.5 °C and variation within a room at the same moment in time ≤1.0 °C. For relative humidity variation at one spot in time should be ≤5% and variation within a room at any time ≤10%.

11.4 Bias in the system – where does it come from, and how do we deal with it?

Calorimetric chambers are used to reduce the influence of external variables impacting on the animals being studied. By holding certain variables constant the impact of other variables can be assessed. While this is an experimental advantage it may also reduce the real world impact that the measured variables have on the species of interest. For example the effect of solar load cannot be adequately assessed in a chamber, and animals are rarely in a steady state environment.

When evaluating the advantages and disadvantages of indirect calorimetry there are biasing factors which need to be considered. Broadly stated there are biological factors and physical factors which can bias results. Essentially these biases are errors which the researchers need to account for. It is important that researchers understand where errors may enter the system, how to identify the errors and finally how to account for these errors when assessing the data during and at the end of the study. The sources of error within a study can be multi-faceted and are often interrelated. For example errors may result from: (1) the models used in the calculation of chamber conditions and associated assumptions; (2) the calorimetric factors used; (3) the duration of study; (4) the experimental design; (5) technical factors associated with the chamber and equipment used; (6) animal factors; and (7) human factors.

Animal factors which need to be considered include: (1) how quickly do animals 'adapt' or 'acclimate' to imposed conditions; (2) what are the thermal thresholds that need to be considered; and (3) animals are not static and will undergo metabolic change even during very short periods of time.

Animal factors are further influenced by:

- species;
- genotype, e.g. *Bos taurus* vs *Bos indicus* cattle;
- phenotype, e.g. animal variation within breed;
- size, e.g. rat vs elephant;
- age, e.g. young vs mature;
- nutritional intake, e.g. net energy balance;
- metabolic rate, e.g. heat load, exercise, time of day;
- metabolic state, e.g. pregnant, lactating;
- degree of adaptation/acclimation, e.g. seasonal effects;
- prior exposure to the tested environmental conditions.

Therefore, it is not possible to extrapolate except in very general terms the findings from one study of one species, genotype, metabolic state, etc. to another species, genotype or metabolic state.

The experimental design will also influence the results. Scientists need to be aware of how the experimental design may bias the result. As previously mentioned the outcomes from a particular calorimetry study (or any study for that matter) can be influenced by the design of the study. For example studies may be set up as a steady state study, e.g. a constant temperature is used or a variable state study where a cyclic temperature regimen is used. Studies may expose animals to chronic or acute stress – and again it is important that the scientists understand the differences between these. Outcomes will also be influenced by what is measured. For example body temperature in cattle can be measured as rectal temperature, vaginal temperature, tympanic temperature, rumen temperature and peritoneal cavity temperature. All measure body temperature but the values will not be the same, due to the location at which the temperature is measured, and the effects of animal activity and various metabolic processes may affect each location differently. Outcomes will be influenced by how often parameters are measured. Measurements taken of a particular parameter at 5 minute intervals are likely to present a different story to the same parameter being measure every 24 hour (this is especially important with many hormones). Sampling procedures are also important considerations. Does the sample procedure bias your result, e.g. restraining animals for blood collection may influence cortisol results. Housing of animals may also influence the outcome. Grouped housed animals are often behaviourally different to the same animals being housed individually.

11.5 Methods for improving accuracy

The accuracy and calibration of temperature and relative humidity sensors and also the accuracy of the measuring instrument where all sensors are actually connected to, as an interface towards the 'measuring and control' software on a computer, is important. Flexible positioning of sensors inside a chamber will make it possible to quickly check temperature sensors while putting them together anywhere to find out whether they give the same temperature. Sensors showing drift or variation from others will need to be checked, recalibrated or replaced. Temperature sensors can be calibrated by simply using melting ice water and water of for example 40 °C with an accurate lab thermometer available. In Wageningen interchangeable sensors are chosen and therefore standard plugs for ease of change are used. Because PT100 sensors are resistance temperature sensors and the electrical resistance in plugs can be changed, a 4-wired connection and measurement is chosen, meaning that variations in resistance in plugs and other connections will not have an effect on the measurements.

Relative humidity sensors can be calibrated by external companies, but they can also be tested and calibrated by the owner, although RH calibration is a complex and time-consuming process filled with potential pitfalls (Schellenberg, 2001). A RH sensor can be checked and calibrated with saturated salt solutions at different levels of relative humidity. For example, at 20 °C a solution of NaCl should produce a 75.4% RH and LiCl a 11.3% RH just above a saturated solution in a small sealed air compartment. Be sure that the sensor head is held just above but not in the saturated salt solution! If electrolytic sensors with an accuracy of ±2% RH are used the saturated salt calibration technique will give a total accuracy of about ±3% RH (Greenspan, 1976). Ambient temperature however should be well monitored and corrected for.

In Wageningen, the RH sensors are calibrated yearly on site by five different salt solutions (Novasina Sensorcheckers) which produce a relative humidity around a sensor of 11%, 33%, 53%, 75% or 90% respectively. These sensor checkers can be used easily on site without taking all sensors and wiring apart because it can be connected in any position on a free hanging sensor. Another example of a salt bath device is the Vaisala Humidity Calibrator HMK15 although somewhat more difficult to use on site because it has to stand straight up on a small table. One of the advantages of a multi-point field calibration is the ability to determine whether the field instrument operates correctly only in the range of interest. Because also devices like multiplexing multimeters and electrical connections from sensors to these meters can cause standard errors it is probably best to correct errors with a calibration line/curve in the data acquisition software. Then the measured values can be corrected easily before storage of data.

Improving climate control hardware and software is a challenge and troubled by the dynamics and variations in place and time, the accuracy of sensors, limitation of controlling systems used and relevant expertise of those operating the system. For example where there is the likelihood of sudden deviations in temperature and humidity set points, the risk of a significant temperature overshoots and/or undershoots is high. There is an on-going need to develop software to be able to tune climate control to prevent this.

If climate data is logged and stored frequently, effects of temperature and relative humidity changes on calculations of gas exchange can be taken care of as laws of Boyle and Gay-Lussac are used in calculations. And last but not least full system validation, like a CO_2 recovery test, under steady state as well as during changing climate conditions should prove the reliability of hard and software used while measuring and calculating gas exchange.

11.6 Interpretation of calorimetric studies – what we need to know

There are numerous examples of studies that have used calorimetric chambers to investigate heat stress responses in animals. Thermal stress studies may be defined as short-term, medium-term or long-term. Short-term studies are usually defined as those running for a few hours to a few days, medium-term studies run for 20 to 30 days and long-term studies are defined as those running for more than 30 days.

When reviewing published studies there are a number of factors which need to be considered, these can be considered as the *gold standards* for calorimetric studies (Table 1). It is important that we ascertain the duration that an animal is exposed to a thermal stress i.e. whether the study is investigating acute or chronic responses. If exposure times are not given then it is difficult to interpret the results. The duration of a study (minutes, hours, and days) is as important as the climatic conditions to which the animals are exposed. The sex, age, liveweight, health status, metabolic status of the animals used also needs to be determined. An understanding of animal adaptation or acclimation to the thermal conditions to which they have been exposed is also important. Nutrition, not just how much is eaten, but when it is eaten and the nutrient content of the diet is important. In ruminant studies, diets consisting of mostly grain will have a different impact on body temperature (due to fermentation) than will a grass based diet. It is also essential to determine if the equipment used was calibrated before the commencement of the study. Unfortunately this information is lacking in many publications.

The general responses of animals to thermal (heat) stress have been well documented. These include: reduced feed intake, an increase in water intake, elevated respiration rate, panting, increased sweating, wetting of body surfaces (e.g. licking legs), behavioural changes (e.g. spend more time standing), reduced reproductive performance and potential immune dysfunction. Under extreme conditions gut integrity may also be compromised which exposes the animal to toxins, which may lead to elevate body temperature, independent of ambient heat load. Changes in any of the aforementioned factors will likely affect body temperature, and again if these factors are not accounted for it is difficult to know if the animal response is being driven by ambient conditions or other factors. Sample collection period is important if changes in animal responses are to be determined. There are numerous papers where body temperature was measured once per day, and some where body temperature is only measured once per week. These data are of little value if we are trying to understand animal responses to a thermal challenge. A key question is how often should samples be collected? Current technology allows body temperature to be collected at a frequency of 1 second, although this is probably too often.

Table 1. Information required in calorimetric studies.

Animal data	Nutrition	Climate	Experiment
Species	Dietary ingredients	Dry bulb temperature	Duration
Breed	Time of feeding	Wet bulb temperature	Housing – full dimensions of pens
Age	Amount eaten	Relative humidity	Data collected from the animal
Liveweight	When eaten, e.g. how many meals	Dew point	Data collection sample period
Growth rate (if growing)	Nutritional status e.g. energy balance	Air speed (is air blowing on animals)	Full description of equipment used
Metabolic status[1]		Air turnover	Description of calibration methods used
Productivity[2]		Measurement of key gases, e.g. CO_2, NH_3, O_2	Analytical methods used, e.g. to determine blood metabolites
Adaptive status[3]		Location of sensors relative to the animals	Statistical analysis – models used
Health status			Fully explain outcomes, potential bias, and any limitations of the study
Behaviour related to thermal balance, e.g. wallowing, aestivation			

[1] Pregnant, lactating, activity.

[2] Dairy cow producing 45 l milk/d vs 22 l milk/d.

[3] Previous exposure to experimental conditions.

The frequency of data collection is to some extent study specific. However it would be useful to standardise data collection frequency across studies, e.g. collection of body temperature at 10 minute intervals. Doing so would allow better comparison across studies.

Researchers also need to be aware of the strategies that the animals they are studying use for thermal regulation and the implications if the research protocols do not allow these to happen. For example if the preferred strategy of the species being investigated is wallowing and animals are not able to do this then how reliable is the data produced?

As previously mentioned numerous studies have used calorimetric chambers to study the effects of thermal stress on animals. The following provides a brief overview of some of these studies. We will look at what was done, and determine how these studies have met the '*gold standard*'.

11.7 Some examples

In this section we will look at a number of published studies and determine if they have met the 'gold standard' as outlined above. We encourage the reader to look at these studies and others to gain a better understanding of what has been done, and how improvements could be made. A fundamental outcome of a good study is that you as a researcher should be provided with enough detail so that you can replicate the study.

Respiratory activity of calves was investigated in a classic early study by Findlay (1957). In this study calves were exposed to ambient temperatures of 20, 30 and 40 °C for 3 hours with an absolute humidity of 17 mg/l, and they also were exposed to 40 °C with an absolute humidity of 45 mg/l. Respiration rate and rectal temperature measurements were made every 10 to 15 minutes. Respiratory activity was further assessed by the measurement of tidal volume and minute volume. All equipment was calibrated.

This short-term study measures an acute response to heat stress and the findings cannot be extrapolated to long-term or chronic responses. The nutritional status was not known. The experimental outline is fully described.

Conley and Porter (1986) investigated heat loss in deer mice (*Peromyscus*). In this short term study 42 deer mice (body weight ranged from 11.3 to 20.2 g) were used. Each mouse was surgically fitted with 36 gauge thermocouples at four locations, and a thermocouple was placed in the rectum. Each mouse was fitted with a mask (to assess respiratory dynamics) and placed in a small respiratory chamber (Figure 5). The mice were exposed to 1 to 2 hours of experimental conditions. Studies were conducted at 15 and 27 °C. All equipment was calibrated.

The duration of exposure time (1 to 2 h) for the mice is possibly due to inter animal variation although this was not discussed, and differences in exposure time may bias the outcomes. Overall this study met most of the requirements as outlined in Table 1.

Blix *et al.* (2011) used winter acclimatized reindeer (*Rangifer tarandus*) to study the regulation of brain temperature when the animals were exposed to heat stress in a chamber. In this study they used 9 female reindeer. Each animal was surgically implanted with 0.2 mm diameter copper-constantan thermocouples. The thermocouples were placed in the brain, carotid artery and the jugular vein. A 200 mm thermocouple was also placed in the rectum to measure rectal temperature. Respiratory frequency was also measured. All equipment was calibrated and the procedures are clearly described in the paper. Reindeer were placed in a chamber that ranged from 0 to 50 °C and animal responses were observed.

It is not clear in the paper however how long the animals were maintained at any given temperature.

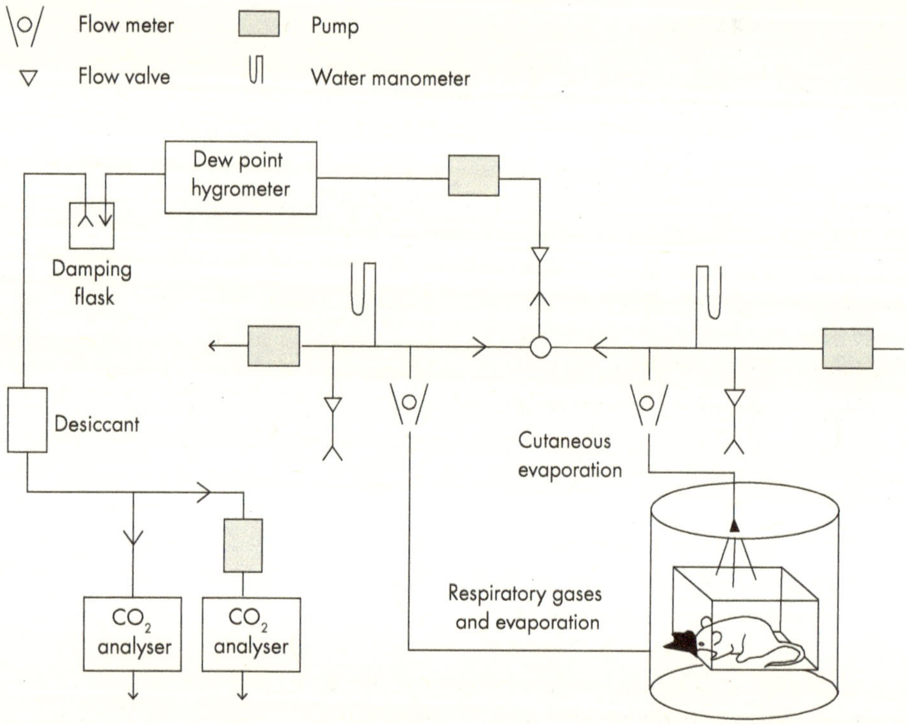

Figure 5. Diagram of the two-channel gas analysis system (Conley and Porter, 1986).

A good example of a study into the effects of a medium term heat stress is the work of Huynh *et al.* (2007). Thermal behaviour of growing pigs in response to high temperature and humidity was investigated. They calculated evaporative water and heat loss. In the study 62 kg pigs were used. The pigs were kept in groups of 10 on a concrete/slatted floor inside climate respiration chambers. Ambient temperature was increased 2 °C per day over 9 days from 16 to 32 °C and relative humidity was fixed at one of three levels: 50, 65 and 80%. Heat production, feed intake, rectal temperature, breathing frequency and behaviour was measured continuously to estimate the upper critical temperature. Evaporative water and heat loss was estimated after measuring and calculating all water and water vapour flows. This study highlights the need to understand the animals behaviour in relation to thermal stress.

Pigs don't sweat so rely on evaporative cooling of the skin. Pigs will often urinate and defecate on the floor to create a 'wallow'; they will then lie in this to get themselves wet which will result in water evaporating from the skin. However the rate of evaporation, and hence its efficiency in cooling the animal is dependent upon the water holding capacity of the air, and air speed over the animal. A failure to take into account animal behaviour may bias results.

A 12 day study was undertaken to examine cold stress in sheep (Dabiri *et al.*, 1995). In this study 16 sheep were used to investigate shearing techniques and resistance to cold stress.

Open-circuit calorimeter chambers were used. Oxygen consumption was measured over 6 days, 2 days before shearing and on 3 days post shearing. Sheep were exposed to a 7 °C ambient temperature +7 km/h wind for 18.5 h. This was followed by 4.5. h of 10 °C + wind + rain (overhead sprinklers; 25 mm/h).

The study was able to successfully measure heat production from the sheep. However the calibration of equipment was not mentioned and there are no details of the chamber. Nevertheless this is a good example of the use of open-circuit calorimetry.

The final example presented is a cold stress study using rats from 1954 (Cottle and Carlson, 1954). The study was designed to investigate adaptive processes to cold stress. Twenty-six 9 week old rats were used in the study. Heat loss (O_2 consumption) from the rats was measured using a closed circuit system. Rats were acclimated to 5 or 28 °C. After 60 days of acclimation the rats were exposed to 5, 15 and 25 °C and O_2 consumption was determined.

This study would be classified as a long term study. Feed intake and change in body condition, body weight and hair coat need to be accounted for. The study indicates that there is an acute response over the first 5 to 7 days of exposure to 5 °C. This highlights the importance of looking at data over time, and not just averaging results.

11.8 Conclusions

Since the early 1900's studies have investigated the effects of heat and cold stress across a range of conditions and species. The reliability of the data arising from these studies is dependent upon the accuracy of the measured parameters. When assessing studies the reader needs to critically examine the materials and methods to ensure that errors or biases have been accounted for.

On order to obtain reliable data potential biases or errors in the system must be accounted for. Respiration chambers must have accurate and reliable climate control systems when studies require a constant ambient temperature and/or relative humidity or where a programed change (e.g. a diurnal pattern) in these parameters is required as part of the experimental design.

Airspeed and drafts at the animal level must also be known and controllable because airflow over the animal may change its physiological response to a given climate variable. The design of chambers with and without climate control should be undertaken by those that understand animal biology as well as technicians and engineers.

Better technology is allowing us to gain a better understanding of the animal in response to heat and cold stress. However, it is important to remember that the use of chambers is the start. One of the key deficiencies in chamber studies is the lack of solar load on the animals. Animals also need to be assessed under 'natural' conditions if we are going to fully understand biological responses.

References

Aarnink, A.J.A., Schrama, J.W., Heetkamp, M.J.W., Stefanowska, J. and Huynh, T.T., 2006. Temperature and body weight affect fouling of pig pens. Journal of Animal Science 84: 2224-2231.

Blix, A.S., Walløe, L. and Folkow, L.P., 2011. Regulation of brain temperature in winter-acclimated reindeer under heat stress. Journal of Experimental Biology 214: 3850-3856.

Conley, K.E. and Porter, W.P., 1986. Heat loss from deer mice (*Peromyscus*) evaluation of seasonal limits to thermoregulation. Journal of Experimental Biology 126: 249-269.

Cottle, W. and Calson, L.D., 1954. Adaptive changes in rats exposed to cold. Caloric exchange. American Journal of Physiology 178: 305-308.

Dabiri, N., Holmes, C.W., McCutcheon, S.N., Parker, W.J. and Morris, S.T., 1995. Resistance to cold stress in sheep shorn by cover comb or standard comb. Animal Science 60: 451-456.

Findlay, J.D., 1957. The respiratory activity of calves subjected to thermal stress. Journal of Physiology 136: 300-309.

Greenspan, L., 1976. Humidity fixed points of binary saturated aqueous solutions. Journal of Research of the National Bureau of Standards A. Physics and Chemistry 81A-1: 89-96.

Heetkamp, M.J.W., Alferink, S.J.J., Zandstra, T., Hendriks, P., Van den Brand, H. and Gerrits, W.J.J., 2015. Design of climate respiration chambers, adjustable to the metabolic mass of subjects. In: Gerrits, W.J.J. and Labussière, E. (eds.) Indirect calorimetry. Techniques, computations and applications. Wageningen Academic Publishers, Wageningen, the Netherlands, pp. 35-56.

Huynh, T.T.T., Aarnink, A.J.A., Gerrits, W.J.J., Heetkamp, M.J.W., Canh, T.T., Spoolder, H.A.M., Kemp, B. and Verstegen, M.W.A., 2005. Thermal behaviour of growing pigs in response to high temperature and humidity. Applied Animal Behaviour Science 91: 1-16.

Huynh, T.T.T., Aarnink, A.J.A., Heetkamp, M.J.W., Verstegen, M.W.A. and Kemp, B., 2007. Evaporative heat loss from group-housed growing pigs at high ambient temperatures. Journal of Thermal Biology 32: 293-299.

Mollier, R., 1923. Ein neues diagram für dampfluftgemische. ZVDI 67: 9.

Mount, L.E., 1974. The concept of thermal neutrality. In: Monteith, J.L. and Mount, L.E. (eds.) Heat loss from animals and man. Assessment and control. Butterworths, London, UK, pp. 425-435.

Schellenberg, R., 2001. How hard could that be? Practical humidity calibration experiences. Cal Lab 2001 (January February March): 40-42. Available at: http://tinyurl.com/pvl3fqo.

Verstegen, M.W.A., Van der Hel, W., Brandsma, H.A., Henken, A.M. and Brandsen, A.M., 1987. The Wageningen respiration unit for animal production research: a description of the equipment and its possibilities. In: Verstegen, M.W.A. and Henken, A.M. (eds.) Energy metabolism in farm animals. Martinus Nijhoff Publishers, Dordrecht, the Netherlands.

Zhao, Z.J., Chi, Q.S., Liu, Q.S., Zheng, W.H., Lie, J.S. and Wang, D.H., 2014. The shift of thermoneutral zone in stripped hamster acclimated to different temperatures. PLoS ONE 9: e84396.

Chapter 12: Indirect calorimetry during incubation of hatching eggs

*H. van den Brand**, *M.J.W. Heetkamp and B. Kemp*
Adaptation Physiology Group, Department of Animal Sciences, Wageningen University, P.O. Box 338, 6700 AH, Wageningen, the Netherlands; henry.vandenbrand@wur.nl

Abstract

Indirect calorimetry can be used during incubation of avian eggs to monitor the quality of the incubation process, the development of the embryo and the utilization of nutrients. Indirect calorimetry has several benefits above direct calorimetry, particularly in hatching eggs. However, to obtain reliable measurements, some aspects of indirect calorimetry need to be taken into account. First, some attention is given to the use of respiration chambers for only one single egg or more eggs at the same time, with the advantages and disadvantages. Furthermore, attention is paid to the importance of using embryo or eggshell temperature as a tool to adjust the temperature in the respiration chamber. Since there is a strong relationship between embryo temperature and heat production (Q), the incubator temperature should be controlled depending on the desired eggshell temperature. Additionally, it is important to consider oxygen and carbon dioxide levels in the chamber, as low partial O_2 pressure is related to incubation at high altitude and high CO_2 levels can result from insufficient ventilation. Besides abiotic factors affecting Q, such as temperature, biotic factors, like egg weight and age of the breeder, affect Q as well. Finally, the use of stable isotopes during incubation to determine oxidation of different nutrients during the incubation process will be addressed. We conclude that indirect calorimetry during incubation of hatching eggs can give insight in the development of the embryo and the utilization of egg nutrients throughout incubation. However, to prevent misinterpretation of the results, some incubation factors (like temperature, O_2 and CO_2 concentration) and egg characteristics (like weight, breed and breeder age) need to be monitored and controlled.

Keywords: chickens, stable isotopes, oxygen, carbon dioxide, eggshell temperature

12.1 Introduction

During the incubation of eggs, nutrients are used to support growth and development of an embryo that gradually develops into a chicken. During the incubation process, oxygen is consumed and carbon dioxide, water and heat are produced by the embryo. Because of the consumption of oxygen and production of carbon dioxide, indirect calorimetry can be used as an indicator of the quality of the incubation process, the development of the embryo and the utilization of nutrients by the embryo (e.g. Janke *et al.*, 2004; Lourens *et al.*, 2006a; Molenaar *et al.*, 2010).

DOI 10.3920/978-90-8686-261-0_12, © Wageningen Academic Publishers 2015

Indirect calorimetry during the embryonic phase of birds was firstly used by Barrott in 1937. From that moment onward, numerous studies were performed to improve the quality of measurements. Romijn and Lokhorst (1960) compared the heat production (Q) determined by indirect calorimetry with the Q determined by direct calorimetry and they concluded that the relationship between both was very strong. Based on their work and that of others they calculated in 1961 the formulae for Q that still is in use for calculating Q of avian embryos (Romijn and Lokhorst, 1961):

$$Q\ (kJ) = 16.19 \times VO_2\ (l) + 5.00 \times VCO_2\ (l)$$

The advantages of indirect calorimetry above direct calorimetry for hatching eggs are evident: (1) indirect calorimetry is convenient and relatively easy to use in comparison to direct calorimetry; (2) corrections do not have to be made for heat loss by convection, conduction, evaporation, and radiation, as would have to be done in direct calorimetry; (3) the dynamics of respiration quotient (RQ) and Q within a day can be followed by indirect calorimetry, whereas this is considerably more difficult with direct calorimetry; (4) for hatching eggs manipulations in the internal and external environment of the egg can be investigated (e.g. incubation temperature), without affecting the methodology of measurement, which is more complicated with direct calorimetry.

Principally, the technique of indirect calorimetry has not changed for the last 50 years. Although VO_2 and VCO_2 are still determined, the accuracy and frequency of measurements have been improved (Frankenfield, 2010). Oxygen is used for the oxidation of nutrients from the egg, which can be converted in embryonic tissue, mainly via aerobic β-oxidation of fatty acids. Carbon dioxide is produced as a by-product of this oxidation.

However, to obtain reliable data from indirect calorimetry during incubation of hatching eggs, several aspects need to be considered. Firstly, the effects of different types of chambers used to determine gaseous exchange will be discussed. Secondly, effects of incubation temperature, oxygen and carbon dioxide concentrations on Q will be considered. Thirdly, egg related aspects like egg weight, breed and breeder age on Q will be discussed. Finally, attention will be paid to the use of stable isotopes to determine oxidation patterns of nutrients.

The aim of this paper is to discuss why and how incubation temperature, O_2 and CO_2 concentrations and some egg characteristics are important to obtain reliable results using indirect calorimetry during incubation of hatching eggs.

12.2 Oxygen consumption, carbon dioxide production and heat production

Using indirect calorimetry to determine metabolic rate of embryos always consists of the determination of VO_2 or VCO_2 or both. Based on these measurements Q can be calculated.

In case both O_2 and CO_2 are determined the formulae of Romijn and Lokhorst (1961) can be used (see above). When only VO_2 (e.g. Druyan, 2010; Janke *et al.*, 2004; Mortola and Awan, 2010; Mortola and Labbè, 2005; Mortola *et al.*, 2010) or VCO_2 (e.g. O'Dea *et al.*, 2004; Visschedijk, 1985) are determined, Q calculations are based on an assumed fixed RQ. However, it is important to realise that the RQ is not a constant value (Figure 1). During early incubation, CO_2 is liberated from the albumen due to liquefaction, resulting in a RQ exceeding unity for the first 4 days of incubation. Thereafter, hardly any carbon dioxide is released from the albumen and the oxidation of nutrients dominates the VO_2 and VCO_2. The final RQ value is a combination of amino acid oxidation (RQ=0.82), carbohydrate oxidation (RQ=1.00) and fat oxidation (RQ=0.71). RQ for mixed substrates is often calculated as 0.85 (Matarese, 1997), but this depends on which substrate is dominantly used. Because fat from the yolk is the dominant substrate for the embryo during incubation, the RQ decreases to approximately 0.68-0.72 from day 14 of incubation (E14) onwards (Chwalibog *et al.*, 2007; Decuypere, 1984; Molenaar *et al.*, 2010; Romijn and Lokhorst, 1960; Sato *et al.*, 2006; Figure 1).

Average RQ values used in literature for the whole incubation period are estimated to be 0.73 (Romijn and Lokhorst, 1961), 0.78 (Romanoff, 1967), 0.78 (Wangensteen and Rahn, 1970/1971), or 0.83 (Rahn *et al.*, 1974). Differences among studies in the observed RQ can probably be explained by the amount of yolk used, which in turn is related to e.g. eggshell temperature (EST, Molenaar *et al.*, 2010; Figure 1). Yolk is the most important supplier of energy during the incubation process and consists mainly of fat (Lourens *et al.*, 2006a). In case more yolk is oxidized during the incubation process, the average RQ will be lower. Furthermore, the RQ is affected by partial O_2 and CO_2 pressure in the environment (Rahn *et al.*, 1974). A reduced O_2 availability (due to low partial O_2 or high partial CO_2 pressure) hampers the oxidative metabolism of the embryo, thereby leading to a reduced contribution of fatty acid oxidation to the total oxidative metabolism. Hence a higher RQ will be observed (Figure 1).

12.3 Incubator size

In current indirect calorimetry, two approaches are used; incubation of one single egg in a small chamber or incubation of multiple eggs in a larger chamber. Barrott (1937) used a chamber with 89 eggs and when Romijn and Lokhorst (1960) referred to this study, they indicated clearly that the disadvantage of using a respiration chamber with more than one egg was the possibility of getting wrong results due to embryonic mortality during incubation. Indeed, when using more eggs in the same chamber, the moment of embryonic mortality needs to be known before Q can be calculated. Lourens *et al.* (2006a) developed a classification table to determine moment of death, based on morphological characteristics and size of the dead embryo.

On the one hand, in the Netherlands (Lourens *et al.*, 2006a, 2006b, 2007; Molenaar *et al.*, 2010; Nangsuay *et al.*, 2013) and in Belgium (De Smit *et al.*, 2005; Tona *et al.*, 2010) chambers with multiple eggs are used. The advantage of determination of Q per group of eggs is the reduced

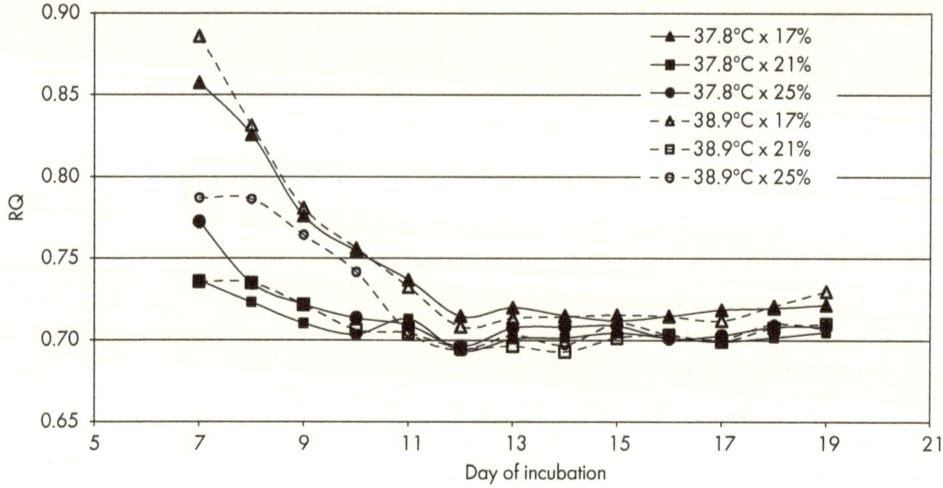

Figure 1. Respiration quotient (RQ) of broiler embryos incubated at two eggshell temperatures (37.8 or 38.9 °C) and three O_2 concentrations (17, 21 or 25%) from day 7 to 19 of incubation (based on Molenaar et al.*, 2010).*

variation among measurements in case gas exchange is determined frequently (e.g. each 9 min.). In case only one single egg is incubated, short term variation can be higher if lowering ventilation level is limited because of the minimum gas analysis flow. Small differences between gas concentrations in incoming and outgoing air may have large effects on VCO_2 and VO_2, resulting in more variation among measurements. The magnitude of this 'noise' effect mainly depends on the size of a chamber in relation to the ventilation rate, the so called washout effect (Chapter 2, Heetkamp *et al.*, 2015) or time constant (Chapter 3, Lighton, 2015). With more eggs, the metabolic mass in a chamber is higher and consequently, ventilation rate can be increased according to a minimal desired difference in incoming and outgoing gas concentrations. Increasing internal air circulation rate is of importance for a thorough internal gas mixing and for enough air speed across the eggs to maintain an uniform EST (Lourens *et al.*, 2011).

On the other hand, in Canada (Hamidu *et al.*, 2007; Menna and Mortola, 2002, 2003; O'Dea *et al.*, 2004; Segura *et al.*, 2006), Denmark (Chwalibog *et al.*, 2007), Germany (Janke *et al.*, 2002, 2004), Israel (Druyan, 2010; Druyan *et al.*, 2012) and Japan (Sato *et al.*, 2006), chambers for individual eggs are used. The advantages of individual incubation are evident: no miscalculations due to embryonic deaths and insight in variation among individual eggs, meaning that optimal EST can be realised for each individual egg. The disadvantage is evident as well: more variation among individual eggs and consequently more repetitions are needed.

12.4 Incubation temperature

Within an incubator, different temperatures need to be distinguished: incubator (air, chamber or room) temperature, embryo temperature and EST. Historically, mostly incubator temperature is used, which is set between 36 and 38 °C (Decuypere *et al.*, 2001). However, during incubation the embryo is losing heat continuously by evaporation (Rahn *et al.*, 1977), hence reducing the embryo temperature. Furthermore, Q increases exponentially after day 9 of incubation (E9) in chicken embryos (Chwalibog *et al.*, 2007; Lourens *et al.*, 2007; Molenaar *et al.*, 2010; Nangsuay *et al.*, 2013; Romijn and Lokhorst, 1960). This means that after day E9, embryo temperature will rise in case heat transfer from the egg to the environment (e.g. due to low air circulation) is insufficient or in case incubator temperature is not adjusted to maintain embryo temperature. Embryo temperature has been evaluated by measuring internal egg temperature at the allantoic fluid (Janke *et al.*, 2004), the yolk sac (Renema *et al.*, 2006) or the air cell (Pulikanti *et al.*, 2011). Recently, EST has been more and more used as a good indicator of embryo temperature. Indeed, for determination of embryo or internal egg temperature, the egg needs to be penetrated, which is more laborious and possibly detrimental for the embryo. It has been shown that EST deviates no more than 0.1 to 0.2 °C from embryo temperature (French, 1997; Meijerhof and Van Beek, 1993), indicating that EST can be used as a good alternative for embryo temperature.

In case incubator temperature is fixed, embryo temperature or EST will be 0.2-0.3 °C lower than incubator temperature during the first week of incubation due to evaporation and can be considerably higher during the remaining part of the incubation period (Figure 2, for schematic overview).

Monitoring and controlling embryo temperature is in our view of large importance to obtain reliable results using indirect calorimetry during incubation of hatching eggs because of two aspects. Firstly, because chicken embryos act as poikilotherms, their temperature and

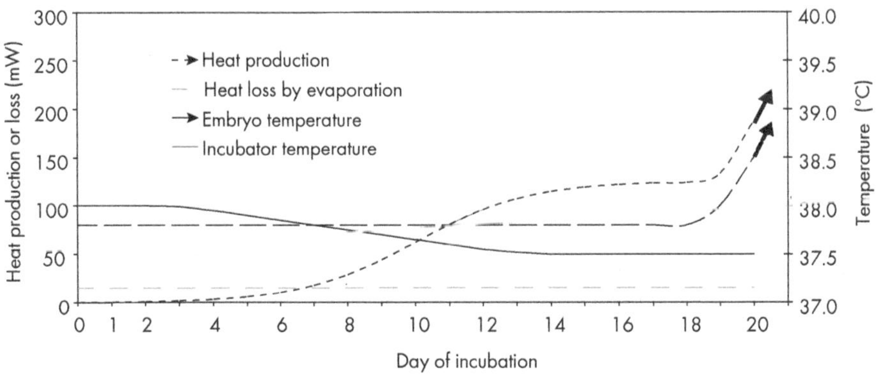

Figure 2. Schematic overview of the relationships between day of incubation and embryo temperature, incubator temperature, evaporation and heat production of chicken embryos.

consequently their metabolic rate depend on the incubator temperature (Janke *et al.*, 2002; Lourens *et al.*, 2006b). During the first 14 days of incubation, and in later incubation to a lesser extent (Willemsen, 2011), Q responds to short term variation in incubator temperature. This means that indirect calorimetry can help to evaluate the dynamics of embryonic development in relationship to variation in EST at different time points of incubation. Secondly, because incubation temperature affects the development of the embryos and consequently the metabolic rate, Q is affected by incubation temperature. In contrast to the first 14 days of incubation, during which a more or less linear relationship between incubator temperature and Q was found (Lourens *et al.*, 2006b), it has been demonstrated that this was not the case during the last week of incubation (Lourens *et al.*, 2006b). Several studies demonstrated an optimum EST throughout incubation. This optimum temperature has been shown to be between 37.5 and 38.0 °C EST, regardless of moment of incubation (Janke *et al.*, 2002; Joseph *et al.*, 2006; Leksrisompong *et al.*, 2007; Lourens *et al.*, 2005, 2007). In case the EST increased for a prolonged period, embryo development at the end of incubation was retarded, and consequently metabolic rate, as expressed by Q, was reduced (Lourens *et al.*, 2007; Willemsen, 2011).

It can be concluded that variation in EST can largely affect metabolic rate and thus Q. To prevent misinterpretation of obtained results from indirect calorimetry, EST need to be monitored and controlled.

12.5 Technique for controlling eggshell temperature

As an example, the technique used in our laboratory at Wageningen University, the Netherlands to control EST of a number of eggs, as sample of all eggs inside a chamber, is presented to allow the reader to use and apply information selectively. This technique is called a cascade (= master/slave) temperature control. In each respiration chamber (that is used as incubator), five accurate (≤0.1 °C) 4-wire connected 1/3DIN-B PT100 temperature sensors with a contact surface of about 1.5×1.5 mm are used. They are attached to the equator of five individual eggs using heat conduction paste (Dow Corning 340 Heat Sink Compound, Dow Corning GmbH, Wiesbaden, Germany) and a small piece of adhesive tape. Every measuring cycle (½ min), the median value of the five egg sensors is calculated (master temperature) and the machine sensor (slave temperature) is measured and both are stored. This median calculation of five egg temperatures is used to prevent extreme temperatures (like errors, due to e.g. infertile eggs) to have any effect on temperature control. The machine temperature sensor is positioned downstream the outlet of the air conditioning unit, preferably at a position were air is mixed by extra fans used to achieve a high air speed across the eggs. This slave sensor measures the temperature of the air that circulates around the eggs. The slave temperature is used for continuous direct control of the air conditioning unit. Every 5 minutes the last master value is used to change the set point of the slave sensor. This change is done in a temporized way, according to the difference between master temperature and set point egg temperature, to allow the eggs some time to react on the changed temperature of the air circulating in the chamber.

12.6 Oxygen

Effects of oxygen concentrations (both hypoxia and hyperoxia) on metabolic rate have been extensively studied, particularly because of the importance of reduced partial oxygen pressure at high altitudes (León-Velarde and Monge-C, 2004). In experimental settings, hypoxia is mostly induced by the introduction of N_2 gas (Azzam *et al.*, 2007; Lourens *et al.*, 2007; Molenaar *et al.*, 2010), whereas hyperoxia can be reached by the introduction of pure O_2 gas in a respiration chamber (Lourens *et al.*, 2007; Molenaar *et al.*, 2010).

One of the first studies into effects of oxygen availability during incubation on metabolic rate was performed by Barott (1937). It was shown that each deviation from normoxia (21%) resulted in a lower metabolic rate and Q, but that the effect of hypoxia (15 or 18% O_2) on Q was much larger than of hyperoxia (30, 40 or 50% O_2). More recent studies confirmed the negative effects of hypoxia on metabolic rate and Q. Visschedijk *et al.* (1980) found a small increase (8%) in VCO_2 when O_2 concentration was increased up to 39% for 3 hours during day E16 to E19. However, when O_2 concentration was lowered to 11% for only 3 hours on the same day, VCO_2 decreased with 70% compared to normoxia. Comparable results were found by Lokhorst and Romijn (1967) in a study where eggs were exposed to low levels of O_2 during the entire incubation. In more recent studies to hypoxia, lower levels of metabolic rate were found as well. Azzam *et al.* (2007), Mortola and Labbè (2005) and Mortola *et al.* (2010) found lower VO_2 when eggs were exposed to oxygen levels of 15% from day E5 onwards, 11% at day E11, E16 and E20 or 7% at only day E3, respectively. Moreover, Mortola *et al.* (2010) demonstrated at day E3 an increase in VO_2 with an increase in incubator temperature (31 to 39 °C), when eggs were exposed to normoxia, where this increase in VO_2 was absent at hypoxia (7% O_2). Mortola and Labbè (2005) demonstrated at normoxia that VO_2 increased with temperature (30, 33, 36, 39 °C) at day E11 and E16, whereas at hypoxia (11%) no effect of temperature on VO_2 was found. Consistently, Lourens *et al.* (2007) demonstrated that at 17% oxygen concentration a lower Q was found at 38.9 °C EST than at 37.8 °C EST, whereas the opposite was found at 25% oxygen concentration (Figure 3). Comparable results were found by Molenaar *et al.* (2010). However, in the latter two studies, eggs were exposed to different levels of O_2 from day E7 or E8 onward, whereas Mortola and Labbè (2005) exposed their eggs to different O_2 concentrations for only one day.

Because avian embryos act as poikilotherms, particularly during early incubation, a higher incubator temperature will increase the metabolic rate and consequently the oxygen requirements. In case of high incubation temperature combined with short term exposure to hypoxia (as in Mortola and Labbè, 2005), the embryo can cope with that, meaning that organ development is not retarded and metabolic rate can recover after return to normal temperature and normoxia. However, after prolonged exposure to high incubation temperature and hypoxia (as in Lourens *et al.*, 2007 and Molenaar *et al.*, 2010), the embryo is not able to cope anymore, meaning that metabolic rate will decrease with hypoxia and increase with hyperoxia.

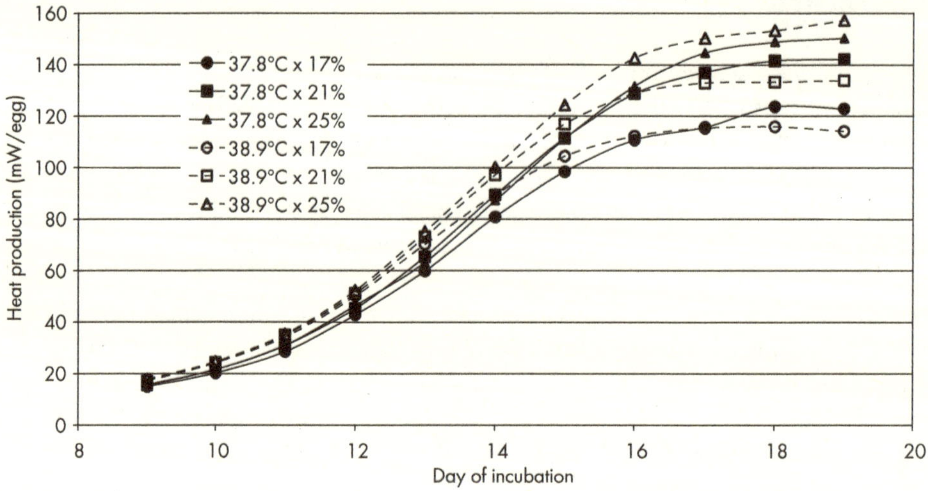

Figure 3. Heat production of broiler embryos incubated at two eggshell temperatures (37.8 or 38.9 °C) and three O_2 concentrations (17%, 21% or 25%) from day 9 to 19 of incubation (Lourens et al., *2007).*

12.7 Carbon dioxide

Effects of carbon dioxide level during incubation on the metabolic rate of the embryo have been hardly investigated. A lot of research was performed to study effects of CO_2 levels during incubation on embryo development, but almost never effects of CO_2 on embryonic metabolic rate or Q were considered. Barott (1937) found an approximately 5 and 20% lower Q during the plateau phase (day E16 to E18) when eggs were exposed to 2 or 4% CO_2, respectively throughout incubation compared to exposure to 0.5% CO_2. However, both higher CO_2 levels can be considered as extreme. It can be questioned whether embryos of modern breeds are sensitive to CO_2 levels below e.g. 1.0%; however, experimental data are currently lacking.

When increasing CO_2 levels by adding CO_2 to a respiration chamber, the level of O_2 will decrease because of dilution. Therefore, the effects of an increased CO_2 level may be confounded with decreased O_2 level. Even when the O_2 level is decreased with only e.g. 0.5% (20.95 to 20.45%, due to an increase of CO_2 level to 2.5%), these confounding effects cannot be ruled out. So, we advise in studies to effects of CO_2 levels during incubation using indirect calorimetry, to maintain the O_2 level at the normal ambient concentration of 20.95%. This implies that both CO_2 and O_2 should be injected in a chamber. Gas concentrations of incoming air and chamber air both need to be measured semi-continuously to make appropriate control of CO_2 and O_2 concentrations possible.

12.8 Techniques for controlling O_2 and CO_2

As an example, at Wageningen University, the Netherlands we use an incoming air buffer for each (of 2) chamber(s) of about ¼ the volume of a chamber. In this buffer incoming outside air is sucked by a chamber air pump (push type setting; Chapter 2, Heetkamp *et al.*, 2015) and, depending on desired CO_2 and/or O_2 concentration inside the chamber, additional CO_2 and/or O_2 and/or N_2 can be injected with different Mass-Flow-Controllers (MFC). The content of this buffer is mixed continuously and this mixed gas then is pumped into a chamber and analysed semi-continuously as incoming air for later VCO_2 and VO_2 calculations. The most appropriate method to control gas concentrations is probably by using extra gas concentration controlling modules within the gas analysis motoring software. Additional gas concentration control software modules can be developed, which in the end compute a digital signal towards a MFC to control the injection of O_2, CO_2 or N_2 in the incoming ventilation air stream. This digital signal can also be used in a digital-analogue converter, which can send an analogue electrical signal (e.g. 0-10V) to a MFC. To obtain reliable results, it is important to consider the requested range of gas injection flow needed during experiments. A narrow ranged MFC is preferred, because a wide range will reduce the accuracy of gas injection.

12.9 Effects of egg weight, breed and breeder age on heat production

Additional to temperature, egg related factors can also affect metabolic rate and consequently Q. The most important known egg related factors affecting Q are egg weight, breed and breeder age. Larger eggs contain more nutrients than smaller eggs (Lourens *et al.*, 2006a; Nangsuay *et al.*, 2011, 2013). It has been reported that due to this higher amount of nutrients during incubation, large eggs produce more heat than smaller eggs (Hoyt, 1987; Lourens *et al.*, 2006a; O'Dea *et al.*, 2004; Rahn *et al.*, 1974; Vleck and Vleck, 1987; Vleck *et al.*, 1980). Consequently, heat transfer from the egg to the environment (e.g. by changing air speed) or incubator temperature needs to be adjusted to prevent an increase in EST in larger eggs, because higher EST can affect the efficiency of energy utilization and finally Q (Lourens *et al.*, 2011; see above). Indirectly, this was also demonstrated by Elibol and Brake (2008), who showed that within a commercial incubator, larger eggs had a higher EST compared to smaller eggs, particularly when they were incubated far away from the fan, where the air speed and consequently the heat transfer was lower.

Egg size is partly confounded with breeder age. Older hens generally produce larger eggs than younger hens (O'Dea *et al.*, 2004; Suarez *et al.*, 1997), but even when eggs of young and older hens are selected on the same egg weight, eggs from older hens contain more nutrients, mostly due to a larger yolk size (Hamidu *et al.*, 2007; Nangsuay *et al.*, 2011). Consequently, eggs of similar weight from older hens, produce more heat than those of younger hens, as demonstrated by Hamidu *et al.* (2007) and Nangsuay *et al.* (2013; Figure 4).

The final egg related factor to be considered in relation to Q is the breed of the embryo. Due to selection in the last decades, growth and therefore Q post hatch is strongly increased, particularly in broilers, and it is hypothesised that this will also hold for the embryonic phase. However, literature in this perspective is ambiguous. On one hand, O'Dea *et al.* (2004) did not find any effect on Q during incubation among modern broiler strains and an unselected female breeder strain (breeds not defined). On the other hand, Druyan (2010) demonstrated a higher VO_2 (approximately 10-15%, expressed as ml/g egg weight/h) in broiler embryos (Ross 308; Aviagen, Huntsville, AL, USA and Cobb 500; Cobb-Vantress, Siloam Springs, AR, USA) than in layer embryos (Lohmann White Leghorn; Lohmann Tierzucht, Cuxhaven, Germany) from day E10 to E19. Consistently, Janke *et al.* (2004) found higher Q (approximately 15% at day E18, expressed as W/kg egg weight) in broiler embryos (Ross 308 and 508; Aviagen) compared to layer embryos (Lohmann White leghorn). However, Sato *et al.* (2006) found a tendency for a lower Q (expressed as J/min/g embryo weight) in broiler embryos (Cobb; Cobb-Vantress) than in layer embryos (Julia; not further defined).

These ambiguous results might be due to the following reasons:

- Eggs did not always have the same weight at start of incubation and as indicated above heavier eggs produce more heat. Even when Q was expressed per gram of egg weight, ratio

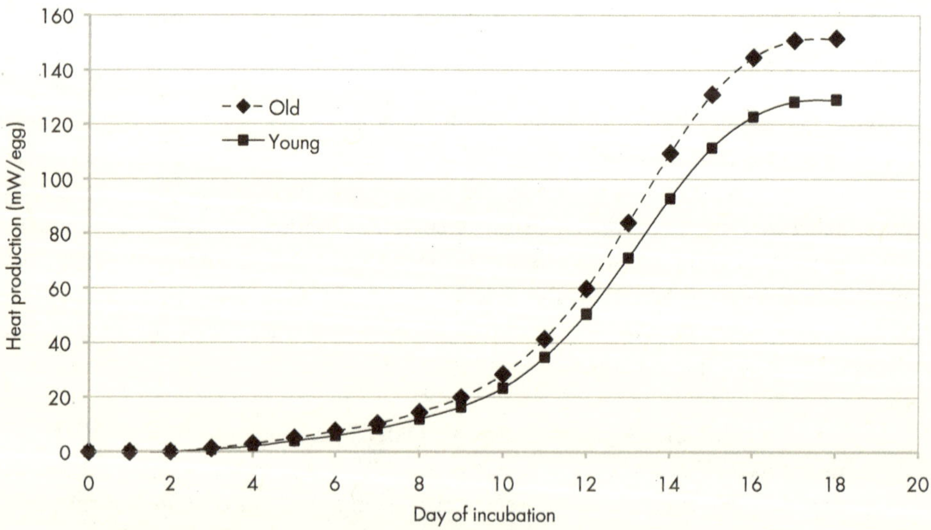

Figure 4. Heat production of broiler embryos during incubation originating from young (29 wk) or old (53 wk) breeder flocks and a same egg weight (58 to 61 g) (Nangsuay et al.*, 2013). Q between embryos of old and young breeder flocks differed from day 12 to 18 of incubation (P<0.05).*

between albumen and yolk differed between different egg sizes (Lourens *et al.* 2006a) and consequently nutrient density per gram of egg differed.

- The age of flocks from which eggs were obtained varied. As explained above, with variation in flock age, nutrient density varies and consequently variation in Q can be expected, meaning that obtained results are the confounded result of treatment and flock age.
- In most studies, eggs were incubated in the same incubator and in none of the studies eggs were incubated at controlled EST. In case eggs differ in energy content and are incubated at the same incubator temperature, they experience different EST (see above) and consequently Q will be affected (Lourens *et al.*, 2011).

Recent work of Nangsuay *et al.* (2015) covered all these three aspects in one study. Broiler (Ross 308) and layer (Lohmann Brown, Lohmann Tierzucht) eggs of the same weight and the same breeder age were incubated at the same EST (37.8 °C). They demonstrated that from day E12 to E18, broiler embryos had a higher Q than layer embryos.

Comparisons between broilers strains are also made. Hamidu *et al.* (2007) did not find any difference in Q between Ross (Aviagen) and Cobb 500 (Cobb-Vantress) embryos, which was confirmed by Druyan (2010) and Tona *et al.* (2010). Additionally, Janke *et al.* (2004), did not find differences in Q during incubation between Ross 308 and 508 (Aviagen) strains.

Based on these results, it appears most likely that the hypothesis that breeds with a higher Q post-hatch have a higher Q pre-hatch as well, can be confirmed. Consequently, layer and broiler breeds differ in their embryonic Q, but differences in embryonic Q among broiler strains appears to be limited.

12.10 Stable isotopes

Stable isotopes can be used to determine nutrient flows within the body of animals and man (Chapter 5, Van den Borne *et al.*, 2015). In eggs, particularly attention has been paid to carbohydrate metabolism, because of the limited carbohydrate content in the fresh egg. This necessitates gluconeogenesis and glycogen synthesis (Sunny and Bequette, 2010). With stable isotope traces, particularly the fate of carbohydrates can be studied more in detail. When performing indirect calorimetry, the production kinetics of $^{13}CO_2$ can be monitored as a measure of carbohydrate utilization (Chapter 5, Van den Borne *et al.*, 2015). Recently, Molenaar *et al.* (2013) did a first study on the glucose requirements of the developing chicken embryo as affected by EST. They showed by injection of [U-^{13}C] glucose in the allantoic fluid at several moments during the incubation period that the oxidation of glucose strongly increased with progressing incubation time (Figure 5). Furthermore, it was demonstrated that the oxidation of glucose was increased by a higher incubation temperature.

It can be concluded that the use of stable isotopes that can be expired in the air, is a useful tool in combination with indirect calorimetry. Particularly, when expired air can be analysed

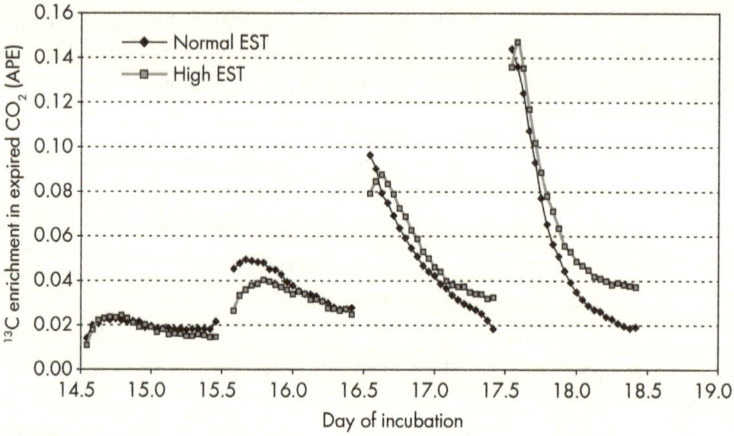

Figure 5. ^{13}C enrichment in expired CO_2 after injecting a single bolus of [U-^{13}C] glucose (0.73 mg in 250 μl sterile water) in the chorio-allantoic fluid of broiler embryos for each of four consecutive days from day 14.5 of incubation. Eggs were incubated at 37.8 (normal) or 38.9 °C (high) eggshell temperature (EST) from day 13 of incubation onward (Molenaar et al., *2013).*

continuously, variation in oxidation of specific nutrients can easily be traced, explaining physiological responses of embryos to variation in incubation conditions.

12.11 Conclusions

We conclude that indirect calorimetry during incubation of hatching eggs can give insight in the development of the embryo and the utilization of egg nutrients throughout incubation. However, to prevent misinterpretation of the results, some incubation factors (like temperature, O_2 and CO_2 concentration) and egg characteristics (like weight, breed and breeder age) need to be monitored and controlled.

References

Azzam, M.A., Szdzuy, K. and Mortola, J.P., 2007. Hypoxic incubation blunts the development of thermogenesis in chicken embryos and hatchlings. American Journal of Physiology – Regulatory and Comparative Physiology 292: R2373-R2379.

Barrott, H.G., 1937. Effect of temperature, humidity, and other factors on hatch of hens' eggs and on energy metabolism of chick embryos. Technical Bulletin U.S. Department of Agriculture 553: 1-45.

Chwalibog, A., Tauson, A.-H., Ali, A., Matthiesen, C., Thorhauge, K. and Thorbek, G., 2007. Gas exchange, heat production and oxidation of fat in chicken embryos from a fast or slow growing line. Comparative Biochemistry and Physiology Part A 146: 305-309.

De Smit,L., Tona, K., Bruggeman, V., Onagbesan, O., Hassanzadeh, M., Arckens, L. and Decuypere, E., 2005. Comparison of three lines of broilers differing in ascites susceptibility or growth rate. 2. Egg weight loss, gas pressures, embryonic heat production, and physiological hormone levels. Poultry Science 84: 1446-1452.

Decuypere, E., 1984. Incubation temperature in relation to post-natal performance in chickens. Archiv für Experimentelle Veterinärmedizin 38: 439-449.

Decuypere, E., Tona, K., Bruggemen, V. and Bamelis, F., 2001. The day-old chick: a crucial hinge between breeders and broilers. World Poultry Science Journal 57: 127-138.

Druyan, S., 2010. The effects of genetic line (broilers vs. layers) on embryo development. Poultry Science 89: 1457-1467.

Druyan, S., Levi, E., Shinder, D. and Stern, T., 2012. Reduced O_2 concentration during CAM development – Its effect on physiological parameters of broiler embryos. Poultry Science 91: 987-997.

Elibol, O. and Brake, J., 2008. Effect of egg weight and position relative to incubator fan on broiler hatchability and chick quality. Poultry Science 87:1913-1918.

Frankenfield, D.C., 2010. On heat, respiration, and calorimetry. Nutrition 26: 939-950.

French, N.A., 1997. Modelling incubation temperature: the effects of incubator design, embryonic development, and egg size. Poultry Science 76: 124-133.

Hamidu, J.A., Fasenko, G.M., Feddes, J.J.R., O'Dea, E.E., Ouellette, C.A., Wineland, M.J. and Christensen, V.L., 2007. The effect of broiler breeder genetic strain and parent flock age on eggshell conductance and embryonic metabolism. Poultry Science 86: 2420-2432.

Heetkamp, M.J.W., Alferink, S.J.J., Zandstra, T., Hendriks, P., Van den Brand, H. and Gerrits, W.J.J., 2015. Design of climate respiration chambers, adjustable to the metabolic mass of subjects. In: Gerrits, W.J.J. and Labussière, E. (eds.) Indirect calorimetry. Techniques, computations and applications. Wageningen Academic Publishers, Wageningen, the Netherlands, pp. 35-56.

Hoyt, D.F., 1987. A new model of avian embryonic metabolism. Journal of Experimental Zoology Suppl. 1: 127-138.

Janke, O., Tzschentke, B. and Boerjan, M., 2004. Comparative investigations of heat production and body temperature in embryos of modern chicken breeds. Avian and Poultry Biology Reviews 15: 191-196.

Janke, O., Tzschentke, B., Höchel, J. and Nichelmann, M., 2002. Metabolic responses of chicken and Muscovy duck embryos to high incubation temperatures. Comparative Biochemistry and Physiology Part A 131: 741-750.

Joseph, N.S., Lourens, A. and Moran Jr. E.T., 2006. The effects of suboptimal eggshell temperature during incubation on broiler chick quality, live performance, and further processing yield. Poultry Science 85: 932-938.

Leksrisompong, N. Romero-Sanchez, H., Plumstead, P.W., Brannan, K.E. and Brake, J. 2007. Broiler incubation. 1. Effect of elevated temperature during late incubation on body weight and organs of chicks. Poultry Science 86:2685-2691.

León-Velarde, F. and Monge-C, C., 2004. Avian embryos in hypoxic environments. Respiratory Physiology and Neurobiology 141: 331-343.

Lighton, J.R.B., 2015. Metabolic measurement techniques: baselining, mathematical correction of water vapour dilution and response correction. In: Gerrits, W.J.J. and Labussière, E. (eds.) Indirect calorimetry. Techniques, computations and applications. Wageningen Academic Publishers, Wageningen, the Netherlands, pp. 57-72.

Lokhorst, W. and Romijn, C., 1967. Energy metabolism of eggs incubated in air with a low oxygen content. In: Blaxter, K.L., Kielanowski, J. and Thorbeck, G. (eds.) Energy metabolism in farm animal, EAAP Publication no. 12. Oriel Press, Newcastle, UK, pp. 331-338.

Lourens A., Meijerhof, R., Kemp, B. and Van den Brand, H., 2011. Energy partitioning during incubation and consequences for embryo temperature: a theoretical approach. Poultry Science 90: 516-523.

Lourens, A., Molenaar, R., Van den Brand, H., Heetkamp, M.J.W., Meijerhof, R. and Kemp, B., 2006a. Effect of egg size on heat production and the transition of energy from egg to hatchling. Poultry Science 85: 770-776.

Lourens, A., Van den Brand, H., Heetkamp, M.J.W., Meijerhof, R. and Kemp, B., 2006b. Metabolic responses of chick embryos to short term temperature fluctuations. Poultry Science 85: 1081-1086.

Lourens, A., Van den Brand, H., Heetkamp, M.J.W., Meijerhof, R. and Kemp, B., 2007. Effects of eggshell temperature and oxygen concentration on embryo growth and metabolism during incubation. Poultry Science 86: 2194-2199.

Lourens, A., Van den Brand, H., Meijerhof, R. and Kemp, B., 2005. Effect of eggshell temperature during incubation on embryo development, hatchability and post-hatch development. Poultry Science 84: 914-920.

Matarese, L.E., 1997. Indirect calorimetry: technical aspects. Journal of the American Dietetic Association 97 (Suppl.): S154-S160.

Meijerhof, R. and Van Beek, G., 1993. Mathematical modelling of temperature and moisture loss of htaching eggs. Journal of Theoretical Biology 165: 27-41.

Menna, T.M. and Mortola, J.P., 2002. Metabolic control of pulmonary ventilation in the developing chick embryo. Respiratory Physiology and Neurobiology 130: 43-55.

Menna, T.M. and Mortola, J.P., 2003. Ventilatory chemosensitivity in the chick embryo. Respiratory Physiology and Neurobiology 137: 69-79.

Molenaar, R., Meijerhof, R., Van den Anker, I., Heetkamp, M.J.W., Van den Borne, J.J.G.C., Kemp, B. and Van den Brand, H., 2010. Effect of eggshell temperature and oxygen concentration on survival rate and nutrient utilization in chicken embryos. Poultry Science 89: 2010-2021.

Molenaar, R., Van den Borne, J.J.G.C., Hazejager, E., Kristensen, N.B., Heetkamp, M.J.W., Meijerhof, R., Kemp, B. and Van den Brand, H., 2013. High environmental temperature increases glucose requirement in the developing chicken embryo. Plos One 8: e59637.

Mortola, J.P. and Al Awan, K., 2010. Growth of the chicken embryo: implication of egg size. Comparative Biochemistry and Physiology Part A 156: 373-379.

Mortola, J.P. and Labbè, K., 2005. Oxygen consumption of the chicken embryo: interaction between temperature and oxygenation. Respiratory Physiology and Neurobiology 146: 97-106.

Mortola, J.P., Wills, K., Trippenbach, T. and Al Awam, K., 2010. Interactive effects of temperature and hypoxia on heart rate and oxygen consumption of the 3-day old chicken embryo. Comparative Biochemistry and Physiology Part A 155: 301-308.

Nangsuay, A., Meijerhof, R., Ruangpanit, Y., Kemp, B. and Van den Brand, H., 2013. Energy utilization and heat production of embryos from eggs originating from young and old broiler breeder flocks. Poultry Science 92: 474-482.

Nangsuay, A., Molenaar, R., Meijerhof, R., Van den Anker, I., Heetkamp, M.J.W., Kemp, B. and Van den Brand, H., 2015. Differences in egg nutrient availability, development, and nutrient metabolism of broiler and layer embryos. Poultry Science 94: 415-423.

Nangsuay, A., Ruangpanit, Y., Meijerhof, R. and Attamangkune, S., 2011. Yolk absorption and embryo development of small and large eggs originating from young and old breeder hens. Poultry Science 90: 2648-2655.

O'Dea, E.E., Fasenko, G.M., Feddes, J.J.R., Robinson, F.E., Segura, J.C., Ouellette, C.A. and Van Middelkoop, J.H., 2004. Investigating the eggshell conductance and embryonic metabolism of modern and unselected domestic avian genetic strains at two flock ages. Poultry Science 83: 2059-2070.

Pulikanti, R., Peebles, E.D. and Gerard, P.D., 2011. Physiological responses of broiler embryos to *in ovo* implantation of temperature transponders. Poultry Science 90: 308-313.

Rahn, H., Ackerman, R.A. and Paganelli, C.V., 1977. Humidity in the avian nest and egg water loss during incubation. Physiological Zoology 50: 269-283.

Rahn, H., Paganelli, C.V. and Ar, A., 1974. The avian egg: air-cell gas tension, metabolism and incubation time. Respiration Physiology 22: 297-309.

Renema, R.A., Feddes, J.J.R., Schmid, K.L., Ford, M.A. and Kolk A.R., 2006. Internal egg temperature in response to preincubation warming in broiler breeder and turkey eggs. Journal of Applied Poultry Research 15: 1-8.

Romanoff, A.L., 1967. Biochemistry of the avian embryo. John Wiley and Sons Inc., New York, NY, USA.

Romijn, C. and Lokhorst, W., 1960. Foetal heat production in the fowl. Journal of Physiology 150: 239-249.

Romijn, C. and Lokhorst, W., 1961. Some aspects of energy metabolism in birds. In: Brouwer, E. and Van Es, A.J.H. (eds.) Proceedings of the Second Symposium on Energy Metabolism. Methods and results of experiments with animals. EAAP Publication No. 10. EAAP, Rome, Italy, pp. 49-58.

Sato, M., Tachibana, T. and Furuse, M., 2006. Heat production and lipid metabolism in broiler and layer chickens during embryonic development. Comparative Biochemistry and Physiology Part A 143: 382-388.

Segura, J.C. Ouellette, C.A., Feddes, J.J.R., Fasenko, G.M. and Zuidhof, M.J., 2006. Development of a metabolic calorimeter system to measure heat production of domestic avian embryos during incubatin. Canadian Biosystems Engineering 48: 41-46.

Suarez, M.E., Wilson, H.R., Mather, F.B., Wilcox, C.J. and McPherson, B.N., 1997. Effect of strain and age of the broiler breeder female on incubation time and chick weight. Poultry Science 76: 1029-1036.

Sunny, N.E. and Bequette, B.J., 2010. Glucose and amino acid metabolism differ in developing chick-embryos derived from small compared to typical size broiler breeder eggs. Journal of Animal Science 88: 912-921.

Tona, K., Onagbesan, O.M., Kamers, B., Everaert, N., Bruggeman, V. and Decuypere E., 2010. Comparision of Cobb and Ross strains in embryo physiology and chick juvenile growth. Poultry Science 89: 1677-1683.

Van den Borne, J.J.G.C., Heetkamp, M.J.W., Alferink, S.J.J. and Gerrits, W.J.J., 2015. Moving from a complete energy balance towards substrate oxidation: use of stable isotopes. In: Gerrits, W.J.J. and Labussière, E. (eds.) Indirect calorimetry. Techniques, computations and applications. Wageningen Academic Publishers, Wageningen, the Netherlands, pp. 87-114.

Visschedijk, A.H.J., 1985. Gas exchange and hatchability of chicken eggs incubated at simulated high altitude. Journal of Applied Physiology 58: 416-418.

Visschedijk, A.H.J., Ar, A., Rahn, H. and Piiper, J., 1980. The independent effect of atmospheric pressure and oxygen partial pressure on gas exchange of the chicken embryo. Respiration Physiology 39: 33-44.

Vleck, C.M. and Vleck, D., 1987. Metabolism and energetics of avian embryos. Journal of Experimental Zoology Suppl. 1: 111-125.

Vleck, C.M., Vleck, D. and Hoyt, D.F., 1980. Patterns of metabolism and growth in avian embryos. American Zoologist 20: 405-416.

Wangensteen, O.D. and Rahn, H., 1970/1971. Respiratory exchange by the avian embryo. Respiratory Physiology 11: 31-45.

Willemsen, H., 2011. Effect of temperature manipulation during late incubation on metabolism and performance of broiler embryos and chickens. PhD thesis, Leuven University, Belgium.

Chapter 13: Conducting indirect calorimetry studies in companion animals

M.A. Gooding[1], L.L. McKnight[2], J.J. Ramsey[3] and A.K. Shoveller[1,2]*
[1]The Iams Company, 6574 State Route 503 North, Lewisburg, OH 45338, USA;[2]Centre for Nutrition Modelling, Department of Animal and Poultry Science, University of Guelph, 50 Stone Road East, Guelph, Ontario, N1G 2W1, Canada; [3]Department of Molecular Biosciences, School of Veterinary Medicine, University of California, One Shields Avenue, Davis, CA 95616, USA; akshoveller@gmail.com

Abstract

To optimize the health and well-being of companion animals, a robust understanding of energy and macronutrient metabolism in these species is necessary. Calorimetry is considered the gold standard for measuring energy expenditure (Q). This chapter examines the use of indirect calorimetry to quantify Q in companion animals and a summary of the published feline and canine literature is provided. As Q can be affected by a number of factors, especially environment, a discussion of the behavioural idiosyncrasies of cats and dogs are provided. To successfully achieve acclimation, behavioural idiosyncrasies of dogs and cats need to be considered and measures to adequately acclimate them to the equipment and handling associated with indirect calorimetry methodology should be utilized. Reports using calorimetry in cats have for the most part included details on behavioural acclimation, the feeding paradigm, and environmental factors and, in general, are in agreement. As a result, the range in published Q values is 124-250 kJ/kg body weight/d for fasting and post feeding and *ad libitum* food availability. In contrast, reports using calorimetry in dogs often lack any consideration of behavioural acclimation or understanding of breed differences, age, gender, body weight and condition, lean body mass, ambient and chamber temperature and relative humidity, feeding regime, and seasonal effects. As a result, the range in published Q values is considerable, 278-711 kJ/kg$^{0.75}$/d for fasted and post-prandial resting Q. Future literature should include descriptions of acclimation, monitoring of stress responses, and consideration of other measurements of energy expenditure in addition to environmental conditions and animal characteristics.

Keywords: dogs, cats, behaviour, acclimation, energy expenditure

13.1 Introduction

Adult dogs and cats require a continual supply of dietary energy to support basal metabolism, physical activity, maintenance of core body temperature and digestion, absorption and assimilation of nutrients (Figure 1). Gestation, lactation, growth, infection and disease require additional energy and protein beyond that required for basic maintenance of body weight in

DOI 10.3920/978-90-8686-261-0_13, © Wageningen Academic Publishers 2015

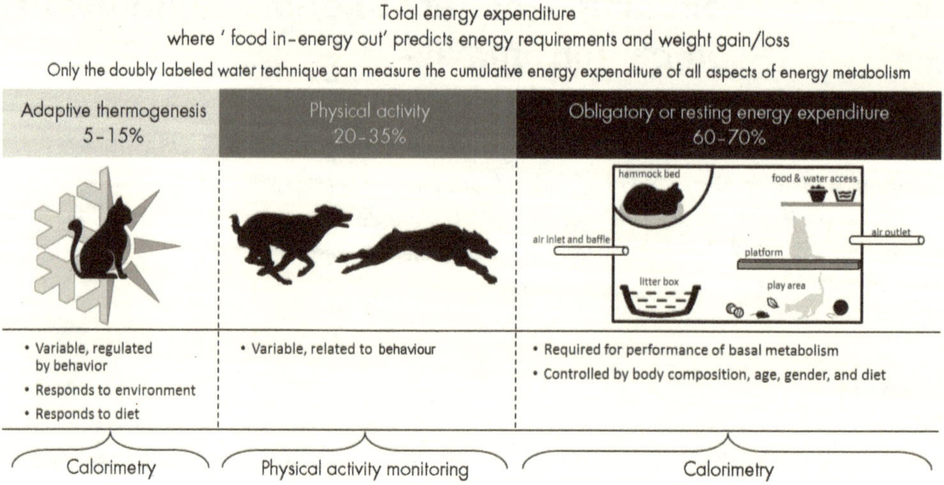

Figure 1. Components and their contributions to Q.

a healthy animal. This review will focus on investigations in adult dogs and cats using indirect calorimetry to quantify energy expenditure (Q) and metabolic fuel selection. Further, a review of the unique behavioural idiosyncrasies of dogs and cats that need to be considered when conducting indirect calorimetry studies is provided.

13.2 Factors influencing energy expenditure

The National Research Council (NRC) and the Association of American Feed Control Officials (AAFCO) outline maintenance energy requirements for adult dogs and cats. These energy requirements are based on data collected from healthy adult laboratory animals that are of typical body weight and condition, undergoing modest amounts of exercise and housed in a thermo-neutral environment to which they have been acclimated. Indirect respiration calorimetry measurements played an important role in establishing recommended maintenance energy requirements for both dogs and cats (NRC, 2006). In addition to providing information about Q, indirect respiration calorimetry studies increased current knowledge about the influence of dietary macronutrient composition on substrate oxidation in cats (Gooding *et al.*, 2014; Green *et al.*, 2008; Lester *et al.*, 1999; Nguyen *et al.*, 2002, 2004; Russell *et al.*, 2002; Wei *et al.*, 2011). The application of calorimetry has been limited primarily to healthy dogs and cats at adult maintenance. However, many factors affect Q, including illness, age, body size, breed, spay/neuter, and environment. The influence of these factors on Q could be addressed through the use of indirect calorimetry. Recommendations for energy requirements of injured or ill dogs and cats primarily rely on equations and guidelines derived from short-term measurements completed in healthy animals under resting conditions and may not be appropriate since physical activity is often decreased with disease states and increases in resting Q may over-estimate changes in daily Q (Burkholder, 1995). It is not entirely clear how accurately these equations and guidelines predict Q in companion animals. Calorimetry chambers can be constructed around many hospital cages to closely match actual housing conditions and provide measures of daily Q. Also, the combined use of calorimetry chamber and doubly labelled water approaches provides a way to determine the influence of injury or illness on resting and total daily Q (includes physical activity) in animals at home or during a hospital stay. Only a few feeding experiments, using a small number of animals, have measured energy requirements in older dogs (≥9 years) (Finke, 1991; Finke, 1994; NRC, 2006; Taylor *et al.*, 1995) and cats (>10 years) (Birmingham *et al.*, 2013; NRC, 2006; Taylor *et al.*, 1995). Along with age, environmental temperature can have a major impact on energy requirements and several studies have investigated the influence of ambient temperature on Q in dogs (reviewed in NRC, 2006). In contrast, relatively little information is available about the influence of environmental temperature on Q in cats and the thermoneutral zone has not been clearly defined in these animals. The influence of breed on Q is another area that has not been adequately studied. Although current NRC guidelines include a few breeds of dogs (NRC, 2006), there is relatively little information about Q in the majority of dog breeds and nearly a complete lack of data on Q in various breeds of cats. Overall there is a lack of scientific evidence defining energy requirements in dogs and cats with respect to age, disease, injury, environmental temperature, and breed. The application of calorimetry could be instrumental in assessing Q under these wide ranging physiological and environmental conditions.

13.3 The use of calorimetry in determining energy expenditure

Energy expenditure is comprised of adaptive thermogenesis (cold and dietary induced), physical activity and resting energy expenditure (Figure 1) and to date, it has generally been appreciated that the split in these functions is similar to other mammals. Indirect calorimetry can be utilized to measure resting Q in fasted animals, which accounts for approximately 60-70% of total daily Q in adult animals. Resting Q measured in the fasted state can be affected by age, gender and hormonal status (intact or neutered) and diet has not been shown to affect resting Q in dogs or cats (Table 1 and 2). When fasting Q is subtracted from post-prandial energy expenditure the heat increment of feeding (HI) can be calculated. HI is influenced by the energetic costs of digestion and assimilation that include bond breakage, nutrient absorption, and synthesis of digestive tract structural and export protein (i.e. enzymes), and storage of nutrients (Baldwin, 1995). When HI is subtracted from ME intake, net energy (maintenance and balance) can be determined. However, no studies in cats or dogs have sought to define the net energy for these animals. Calorimetry could be a novel method to assess the energy value of different dietary macronutrient profiles or individual ingredients. To be able to calculate absolute concentrations of substrates oxidized, complete urinary collection for quantification of nitrogen excretion is required. Collection of urine may require acclimation to the urine collection method in cats and dogs, unless a non-absorbent material is used within litter boxes for cats (Wichert *et al.*, 2007); however, this practice will not be described in this chapter. Stress free elimination in metabolic rates is difficult to achieve in dogs or cats and is rarely performed today, to the best of these authors' knowledge. Cats can be trained to urinate on nonabsorbent material but acclimation procedures should be considered (Wichert *et al.*, 2007). Indirect calorimetry allows for measurements of adaptive thermogenesis and obligatory or resting Q, but does not measure the typical daily energy expended during physical activity since the calorimetry chamber is often smaller than the normal housing environment for the animal. Additional capabilities must be combined to understand the effect of experimental treatment on typical daily physical activity. Such capabilities have been implemented in companion animal research (Belsito *et al.*, 2009) to demonstrate that spaying and food restriction decrease physical activity and may contribute to positive energy balance. Innate behaviour, physical environment (temperature, humidity, and partial pressure) and diet are known to affect adaptive thermogenesis and need to be considered when conducting indirect calorimetry in cats and dogs, although to date, no one has shown an effect of both environmental temperature and relative humidity on cat metabolism in a free living model. Because behaviour can affect adaptive thermogenesis it is important to ensure that a dog or cat that is not stressed is used for indirect calorimetry and we will describe the behavioural idiosyncrasies in dogs and cats and considerations for acclimation to calorimetry chambers to ensure measurements are minimally influenced by stress. It is important to point out that a non-stressed dog or cat may not actually represent the in-home animal in many cases, but regardless, stress is an important variable to control in research projects.

13.4 Regulatory guidance for temporary restriction of dogs and cats

The United States Department of Agriculture's Office of Animal Welfare defines that dogs and cats can spend up to 12 hrs in study areas (different than primary enclosures). For dogs and cats, crates and chambers for respiratory gas collection would be classified as a study area. However, the allowable space reported in all of the cat papers described in this chapter meets or exceeds that which is required for permanent housing of adult cats ($0.6 \times 0.6 \times 0.7$ m) and could, in fact, be used for permanent housing (USDA-NAL, undated.). The Council of Europe (2006) states that dogs can be single housed with reduced floor space (2 m^2 for dogs <20 kg and 4 m^2 for dogs >20 kg and minimum height of 2 m) for scientific purposes. Further, the period for which a dog should be constrained should be minimized and should not in any event exceed 4 h. Finally, the well-being of the animals should be assessed and additional enrichment (i.e. socialization, auditory, exercise) offered where possible. For cats that are accustomed to group housing, single housing is not recommended to exceed 24 h with a minimum provision of 1.5 m^2 floor space, 0.5 m^2 for shelves, and 2 m in height. Further, within cat enclosures feeding and litter trays should not be less than 0.5 m apart.

13.5 Behavioural idiosyncrasies in dogs and cats

13.5.1 Acute and chronic stress and the introduction of abnormal test subjects

To reduce animal variability, dogs and cats selected for nutritional studies should exhibit behaviour patterns within their 'normal' repertoire that requires the animals to be well-adapted to particular research practices and environments (Meunier, 2006). Variability in outcomes will be greater if an animal is subjected to a threatening environment. The animal will attempt to cope within the environment, modifying physiological parameters via activation of the sympatho-adrenal axis and the hypothalamo-pituitary-adrenal (HPA) axis releasing catecholamines and glucorticoids (Seematter *et al.*, 2005). Cumulatively, these physiological responses may modify metabolism by increasing: (1) Q; (2) glycolysis (Seematter *et al.*, 2000); (3) lipolysis (Wennlund *et al.*, 1994); (4) insulin insensitivity (Selye, 1971); (5) blood pressure (Vincent *et al.*, 1993); and (6) ventilation rate (Seematter *et al.*, 2005) all of which may impact indirect calorimetry data and outcomes.

The validity of physiological indicators of stress in kennelled dogs and laboratory/caged cats were recently investigated. The measurements included in the assessment for dogs were cortisol (cortisol: creatinine ratios), serotonin, oxidative stress and total antioxidant capacity (Hewison *et al.*, 2014; Part *et al.*, 2014; Passantino *et al.*, 2014; Titulaer *et al.*, 2013). For cats indicators of stress were adrenocorticotropic hormone, luteinizing hormone-releasing hormone, faecal glucorticoid, immunoglobulin A and cortisol: creatinine ratios (Carlstead *et al.*, 2006; Ellis *et al.*, 2013; Gourkow *et al.*, 2014a; Uetake *et al.*, 2013). While physiological indicators of stress may provide an objective assessment of stress they are limited as the physiological outcomes can lack specificity (Mason and Mendl, 1993) and be impacted by collection techniques (Hennessy *et al.*, 1998; Kobelt *et al.*, 2003). Further, the practicality of collecting physiological samples during calorimetry collections may be limited; therefore, the assessment of behavioural changes in response to exposure to novel research practices may offer a non-invasive assessment of stress of the dog or cat that is particularly important to monitor during indirect calorimetry studies.

Dogs are inclined to demonstrate overt behavioural signs of acute and chronic stress (Titulaer *et al.*, 2013). Previously described, acute signs of stress in the dog include: panting, vocalizing, paw lifting, snout licking, lower posture, salivation, yawning, shaking and urinating (Beerda *et al.*, 1997, 1999b, 2000). Some of the frequently observed signs used to describe chronic stress of dogs housed in kennel or laboratory environments include abnormal repetitive movements (pacing, tail chasing, jumping), excessive auto-grooming and oral manipulation of the environment (Beerda *et al.*, 1997, 1998, 2000; Hetts *et al.*, 1992; Hubrecht *et al.*, 1992). These behavioural indictors of stress in the dog have been correlated to physiological stress responses (increased heart rate, cortisol, etc.) in kennel or laboratory situations (Bodnariu, 2008; Hiby *et al.*, 2006; Part *et al.*, 2014; Rooney *et al.*, 2007) although the association is not consistently observed (Hewison *et al.*, 2014).

Cats are unusual in that they display a reduced or inhibited behavioural repertoire in situations of chronic stress (Rochlitz *et al.*, 1998). Behaviours that are often inhibited in response to stress in the domestic cat include self-maintenance and exploratory and play behaviours (McCune, 1992, 1994; Rochlitz *et al.*, 1998). Some abnormal behaviour exhibited by the cat in response to confinement includes pica, self-mutilation, increased aggression, fearfulness, and reduced reproductive success (Rochlitz, 2005). Understanding that cats may not overtly display a stress response warrants careful monitoring under situations that might be perceived as threatening. This fact is particularly important in research environments where novel practices are utilized. Short term stress has received much less attention than chronic stress responses in cats. Behavioural and postural changes describing extreme vigilance, piloerection, increased respiration and escape/hiding behaviour can be indicators of acute stress as they often decline after short term exposure to a novel environment (Kessler and Turner, 1997; McCune, 1992, 1994; Smith *et al.*, 1994).

Dogs and cats use different behavioural repertoires to cope with acute and chronic states of stress. Using behavioural indicators of stress is a useful and non-invasive tool to assess the effects of a particular environment or research practice. However, these tools need to be used with the understanding of their limitations (general lack of validation, animal variability, human scoring ability, timing of the assessment) and would be most effective if combined with a variety of measures of stress response that will have the least effect on animal welfare and research practices. As an example, measurements of respiration rate during indirect calorimetry acclimation may offer a non-invasive physical measure of stress to support behavioural observations during acclimation. The primary features of indirect calorimetry research methods and techniques that may specifically contribute (either individually or combined) to a stress response in companion animals include social restriction or isolation, spatial restriction and novelty of the chamber environment (or muzzle for the dog) and the environment associated with the collection methods.

13.5.2 Social restriction or isolation

Dogs have highly developed socio-cognitive capabilities and form complex social relationships (Miklosi *et al.*, 2003). Thus, social isolation from conspecifics can contribute to the initiation of physiological stress responses (Beerda *et al.*, 1999a) and demonstrations of abnormal behaviours such as altered activity patterns, stereotypies, vocalization and environmental/object manipulations in singly housed dogs (Hetts *et al.*, 1992; Mertens and Unshelm, 1996; Wells, 2004). The effect of social isolation becomes readily apparent when dogs previously housed in group situations are subjected to social/spatial restrictions where they demonstrate higher levels of over-excitability, aggression and uncertainty when challenged in a novel situation with fearful or startling stimuli (Beerda *et al.*, 1999b). Alternatively, the presence of familiar companions during exposure to novelty can often mitigate HPA responses (Hennesy, 2013). For the dog, interactions with humans in the form of exercise, soothing petting, training, playing and grooming, can reduce reactivity and stress response (HPA activation) to novel or stressful environments, such as kennel environments, and experimental procedures, such as taking a measure of heart rate (Bergamasco *et al.*, 2010; Coppola *et al.*, 2006; Hennesy *et al.*, 1998, 2002; Menor-Campos *et al.*, 2011; Shiverdecker *et al.*, 2013; Tuber *et al.*, 1996). If social isolation is required for research purposes, as is often the case for indirect calorimetry measures, providing predictable opportunities for positive human interaction or conspecific social interaction before or after testing may help reduce stress response to social isolation. In addition, allowing dogs to have the opportunity to have visual conspecific contact during indirect calorimetry measures may be mentally stimulating to dogs during testing (Wells and Hepper, 1998).

Cats are historically believed to be solitary species. However, sufficient evidence exists suggesting that feral cats may form large social groups called colonies when food resources are abundant (Crowell-Davis *et al.*, 2004). Alternatively, if food resources are limited cats can live solitary lives. For group housing situations to be successful, in shelters or laboratories, stable social groups need to be established and maintained. If not, cats may display greater signs of stress due to agonistic interactions in a dynamic group setting compared to cats that are singly housed (Durman, 1991; Bradshaw and Hall, 1999; Ottway and Hawkins, 2003; Smith *et al.*, 1994). There is some evidence suggesting no effect of single versus multi-cat housing on stress and that the features of the environment are likely more impactful on stress (reviewed in: Finka *et al.*, 2014; Gourkow and Fraser, 2006; Lichtsteiner and Turner, 2008; Ramos *et al.*, 2012; Uetake *et al.*, 2013). While there is a dearth of information investigating the effects of visual conspecific contact in cats, it can be hypothesized that limited visual interactions with unfamiliar conspecifics during research practices would be beneficial. However, efforts should be made to provide other opportunities for visual stimulation during chamber restriction including familiar cats and handlers (DeLuca and Kranda, 1992; Rochlitz, 2000). Overall, social restriction in suitably designed environments likely poses less of a risk for a stress response for the cat versus the dog. If respiration chambers are designed appropriately the social restriction should not significantly impact outcomes by contributing to a stress response.

13.5.3 Spatial environment of indirect calorimetry

Initial research investigating the effect of cage size on behaviour of laboratory dogs described the effects of spatial restriction on total activity or movement of singly housed dogs as an indicator of welfare. Most research groups concluded that there were no significant effects of cage size on outcome measures (Hite *et al.*, 1977; Hughes *et al.*, 1989; Neamand *et al.*, 1975). However, it was hypothesised that activity alone was an insufficient tool to assess welfare. Since there was movement towards pair/group housing for dogs, investigators sought to measure the effects of cage size within group housing situations (Bebak and Beck, 1993). Several research groups suggested that kennel size during group housing situations did not significantly influence stress response of the dogs (Pettijohn *et al.*, 1980) due to increased demonstrations of stress related behaviours during periods of social isolation versus spatial restriction (Hetts *et al.*, 1992; Hubrecht *et al.*, 1992). Lastly, dogs are naturally denning species (Hunt *et al.*, 1983). Temporary restriction within crates, which are similar in characteristics to indirect calorimetry chambers, have been successfully used to offer a place of security for the dog by providing opportunity for control of the environment and escape (Hubrecht *et al.*, 1992; Tuber *et al.*, 1999). With effective acclimation programs and predictable research and handling practices, respiration calorimetry chambers may have a similar benefit to a crate, offering dogs a place of comfort and security.

The characteristics of the environment that are most often cited as influencing cat behaviour include the quantity of space (size) and quality of space, specifically the social (reviewed above) and sensory characteristics of the environment (described in detail by Rochlitz, 2007). While meeting minimum housing requirements is vital, the vertical space of an enclosure is particularly important to support a cat's innate preference to utilise elevated structures and raised surfaces more often than lower or ground level beds/perches (Durman, 1991; Ellis *et al.*, 2014; Podberscek *et al.*, 1991; Rochlitz, 1999; Roy, 1992; Smith *et al.*, 1994; Uetake *et al.*, 2012). Also, sufficient space that allows for some separation of feeding, resting, elimination and playing areas is ideal (Rochlitz, 2007). In addition to offering the appropriate quantity of space for the cat, the characteristics of the environment are equally as important to consider during restriction (Finka *et al.*, 2014; Geret *et al.*, 2011; Gourkow and Fraser, 2006; Rochlitz, 2007). The features of an enriching environment include (reviewed in Rochlitz, 2000): providing elevated structures for the cats to climb and view surrounding from, an opportunity for hiding (Carlstead *et al.*, 1993), comfortable resting areas (Crouse *et al.*, 1995; Hawthorne *et al.*, 1995) and predictive positive human interactions (Gourkow and Fraser, 2006; Gourkow *et al.*, 2014b). We have previously demonstrated that step wise acclimation to temporary restriction in calorimetry chambers, designed to provide both sufficient quality and quantity of space, can take cats up to 11 weeks to fully acclimate and, as expected, cats that were not acclimated displayed significantly more behavioural responses associated with stress during temporary restriction (Gooding *et al.*, 2012).

13.5.4 Reactivity to novel research environments and practices and the benefit of acclimation

The novelty of an environment is a primary stressor for both cat and dog. Some of the stress of novelty can be mitigated through: (1) the design of chambers that have similar features (sounds, scent, temperature, humidity, air flow) to the home environment as this may reduce level of novelty; (2) the development of highly predictable testing and handling routines (Dess *et al.*, 1983; Muir and Pfister, 1986); and (3) acclimation to indirect calorimetry testing techniques via habituation through repeated exposure to a particular stimulus. With respect to acclimation, the response is decreased over time (Webster, 1994)) and desensitization (the process of minimizing/removing aversive stimuli associated to an event by providing a positive stimulus concurrently (Chance, 2003)). By allowing an increase in control of the environment and offering an opportunity to hide may act to reduce stress responses. However, these options are often not feasible during indirect calorimetry testing. It may be more practical to ensure that daily practices are consistent. Meunier (2006) reviewed and outlined certain strategies that may be used to acclimate dogs to certain research settings and practices. To our knowledge and despite the understanding that several features of indirect calorimetry may contribute to a stress response, there are no published studies measuring or validating the effects of differing acclimation programs for research practices that require crate, chamber or muzzle restriction for dogs. Such information would be highly valuable to ensure we support optimal well-being in research practices while reducing animal variability to allow for the robust comparison of research data across studies (Turner *et al.*, 2003). Several studies have outlined the adaptation of dogs to kennel environments or their behavioural and physiological response to certain stimuli (Beerda *et al.*, 1998, 1999 a,b; Coppola *et al.*, 2006; Hennesy, 2013; Hiby *et al.*, 2006; Rooney *et al.*, 2007; Stephen *et al.*, 2006; Turner *et al.*, 2003). This information may be used to drive the design and assessment of the success of acclimation programs to chamber restriction or muzzle acclimation for dogs.

Acclimation procedures to temporary restriction within calorimetry chambers in cat studies range from no acclimation to an incremental acclimation of 11 weeks, where measures of behavioural signs of stress (cat-stress-score, CSS; Kessler and Turner, 1997), food intake, response to novelty and urination/defecation, were determined throughout the study (Gooding et al, 2012). Based on previous research, the majority of acclimation procedures described in indirect calorimetry studies, where naïve cats are first exposed to respiration chambers, was insufficient to fully adapt cats to restriction within chambers. Kessler and Turner, 1997, demonstrated that two thirds (n=140) of the cats adjusted to a novel environment (boarding cattery) after 2 weeks of exposure. Furthermore, when cats were exposed to a novel quarantine situation or group/individual cage restriction, stress response (measured using behavioural scoring systems for stress) declined within the first 24 hours, continued to decline after several days of exposure and further declined with prolonged exposure periods up to 5 weeks when measurements ended (McCune, 1992; Rochlitz, 1997; Smith *et al.*, 1994). As suggested previously, behavioural assessment of stress is valuable. However, animals are highly individualized in response patterns and changes in behavioural styles may be representative of a change in coping strategy as the stressor may be unresolved. For instance, Ellis *et al.*, 2014, noted that CSS declined within the first weeks of exposure but a longer duration of time was required for faecal glucorticoids to stabilize. Combining behavioural and physiological outcomes to assess stress level during indirect calorimetry studies will likely provide the most robust data to assess welfare and determine the validity of the calorimetry data collected. However, to date, there has been no study that combines behavioural and physiological outcomes to assess acclimation to indirect calorimetry research practices. As such, indirect calorimetry results need to be interpreted with that limitation in mind.

Some of the stress reactions to research practices requiring temporary spatial and social restriction or exposure to novelty may be mitigated if the appropriate environment is provided and dedicated populations of research animals are used for similar studies, such as indirect calorimetry. Ideally, animals selected for indirect calorimetry studies should be acclimated to chambers and testing procedures prior to study collections (see Gooding *et al.*, 2012 for a detailed acclimation protocol for cats). Following acclimation, repeated exposure, in a predictive manner, will allow the animal to feel control over the environment and become more adapted to certain research practices (Meunier, 2006). Outside of indirect calorimetry testing, opportunities for positive socialization with familiar conspecifics (especially for the dog) and human handlers, exercise and play will help improve the welfare of the test population leading to more robust outcome measures (Bergamasco *et al.*, 2010; Coppola *et al.*, 2006). Recognition of individual behaviour and innate motivations or preferences for cats and dogs need to be considered during indirect calorimetry studies to contribute to the most robust collection and comparison of scientific data across research groups.

13.6 Energy expenditure estimates in cats

Meta-analysis of the literature from 1933-2009 suggests that mean daily energy requirement for cats, as measured during indirect calorimetry experiments, is equal to 852.3±34.3 kJ ME/d or 209.2±6.3 kJ ME/kg body weight (BW) (Birmingham *et al.*, 2010). Of the estimates reported by Birmingham and colleagues (2010) only estimates of resting energy metabolism in normal, healthy, adult cats were included. In addition to those estimates, we included estimates of Q from calorimetry where defining the energy requirement was not a key outcome (Appendix 13A).

The majority of authors described or referenced the cat demographics, chamber configuration, and environment. Most studies conducted experiments in adult cats at temperature ranges of ~20-24 °C, although the effect of ambient temperature on feline metabolism is not entirely clear. Measuring Q in a thermo-neutral environment for humans is the best representation of North America where the cats are largely housed entirely indoors, but might not fully represent cats that have both indoor and outdoor access. Of the reports, only Gooding *et al.* (2014), Shoveller *et al.* (2014) and Berendt (2014) measure Q in cats that, in their daily housing, have outdoor access outside of collection days although measurements were conducted at a constant room temperature and humidity. Although we did not statistically compare these results, the estimates of Q from these cats allowed outside access were similar to other estimates where cats were housed indoors only, suggesting changes in housing had little to no effect on the calorimetry measurements. It may be interesting to look at whether the time of year, specifically how environmental climate (seasonal changes) alters Q in cats with indoor/outdoor access. Furthermore, there is general inconsistency among the literature cited as to the level of enrichment and human socialization that the cats receive. Such information should be reported since these factors can contribute the level of stress that cats may experience.

The environment that the respiratory chambers provide is also not consistently reported among the literature. We believe that authors need to clearly define the environmental enrichment provided to the cats and what type of micro-environments that the chambers provide, namely the design/complexity of the chamber (i.e. the provision of separate spaces for feeding, sleeping, resting and elimination) and the impacts of the external environment such as lighting, feeding and management (noise) schedule. Furthermore, the literature from the majority of authors, with the exception of Nguyen (2002 and 2004), provides the dimensions of the calorimetry chambers, but this information should always be included in the description of the methods.

Nguyen (2002 and 2004) described an acclimation, but cats were confined for 5 days with the first being used as acclimation, although no measures of stress were completed and one day may not lead to behavioural acclimation to a space restricted environment (Gooding *et al.*, 2012; Kessler and Turner, 1997; McCune, 1992; Smith *et al.*, 1994). A more in depth description of what criteria is used to determine the level of stress and the exclusion criteria should also be included to enable the reader to determine the potential effects on the data. Only Wichert *et al.* (2007) described chambers as being positioned side by side to facilitate visual contact between familiar cats, and amount of visual contact should also be included in the description of the methods. In general, the awareness of the idiosyncrasies of the behaviour of cats is realized by the individuals that have conducted studies in this species.

The majority of the studies examine Q in cats when offered food; however, Nguyen (2002, 2004) measured least observed metabolism (LOM) by measuring Q after a long duration fast (20-24 hr) with minimal physical activity. We interpret LOM as a measure close to basal metabolism. The estimates of Nguyen are approximately 60-80% of the other estimates of Q. Given that a small amount of voluntary activity would be accounted for in these longer duration calorimetry exposures, it appears as though the basal metabolic rate in cats accounts for a similar proportion of total Q as in other mammals. The proportion of total Q that is accounted for by basal Q appears to be greater in the situation of obese cats and this is also to be expected since obese cats have lower levels of physical activity (Belsito *et al.*, 2009). Gooding *et al.* (2014) and Shoveller *et al.* (2014) measured Q for 15-22 h after a meal and these data could be used to measure the heat increment of food. The heat increment of food is particularly important to measure for dietary paradigms, such as high fat feeding, which are hypothesized to alter Q following meal consumption. Environmental temperature, and in particular, cold exposure also affects the thermic effect of feeding and could provide another means by which to examine the similarity of cats to other mammalian species. Separately, this information could lead to a better understanding of the thermal neutral zone of cats and allow better understanding of the effects of environment on energy metabolism.

13.7 Energy expenditure estimates in dogs

The application of calorimetry for estimating Q in dogs has been limited (Appendix 13B). In general, enclosed chamber systems have been used to estimate Q in healthy animals in response to a range of physiological conditions, including cold exposure, prolonged fasting, starvation, and dietary interventions (i.e. high fat vs low fat feeding). Pooling the available data from chamber studies yields crude means of 341 (range, 279-511) kJ/kg$^{0.75}$/d and 422 (range, 328-711) kJ/kg$^{0.75}$/d for fasted and post-prandial resting Q. Expressing Q of dogs body weight to the exponent 0.75 is most appropriate given the wide range of body conformations that exist across dogs breed. The wide range in reported Q values likely reflects the variation in breeds, body weights (8-41 kg), and ages (2-11 y) of the dogs, in addition to the variation in the duration of measurements (4-24 h) between studies. Indeed, between dog variability in a cohort of similar dogs accounts for at least 55% of the random effects variability associated with measurements of indirect calorimetry and chamber and random error account for 12 and 32%, respectively (unpublished data). It is therefore important to consider the experimental design and we recommend to use the same chamber with the same line of sight for repeated measures within dogs and to conduct measurements under similar external influences, such as day of the week and time of day.

A lack of consistent reporting with respect to the characteristics of the calorimetry system was noted. Five out of the seven chamber studies reported some degree of acclimation to chamber (1-3 months), but a detailed description of acclimation was not provided by any of the authors (Crist and Romsos, 1987; McKnight *et al.*, 2014a,b; Richey *et al.*, 2009; Yoo *et al.*, 2006). Furthermore, no reported efforts were made to quantify stress related behaviours. Of the two studies that did not report acclimation, one study (Larsson *et al.*, 2014) grouped animals based on their behaviours during measurements into 'resting' or 'active'. Larsson *et al.* (2014) reported that dogs defined as active had high Q (718 $kJ/kg^{0.75}/d$) and respiratory quotient (0.94) values. Such high values, may reflect stress associated with measurements and/or environment, especially since these were client owned animals. However, a limited description of 'active' behaviours was provided. Specifically, the authors noted that active animals were considered to be more active than resting animals. In contrast, the other report which did not mention acclimation (Pouteau *et al.*, 2002), reported Q values close to those observed in other studies.

Application of sealed face mask calorimetry systems to estimate resting Q appears to be limited to clinical settings. This finding is likely due to the portability of such systems making it ideal in hospital settings. Only the Q estimates from the healthy dogs are given in appendix 13B. As numerical data was not given in two of the studies (O'Toole *et al.*, 2001, 2004), estimates of Q were crudely obtained from visual interpretation of the illustrations. The range of Q estimates appeared to be from 288-336 $kJ/kg^{0.75}/d$, which is somewhat lower than the values obtained by chamber systems and may suggest incomplete capture of volume of CO_2 and O_2. In all studies, dogs were allowed 10-15 min to adapt to wearing the face mask. Mazzaferro *et al.* (2001) acclimated the dogs to the mask once weekly over 8 consecutive weeks. During measurements the dogs were noted to be lying down calmly.

In conclusion, very few studies have utilized indirect calorimetry for the determination of resting Q in dogs. In general, there is a lack of reporting in canine studies with respect to calorimetry systems and acclimation protocols. For those that provided a limited description of the acclimation protocols, the effectiveness and validity of the protocols were not discussed. Given the lack available evidence, the impact of prior acclimation to chambers on Q measurements remains unclear.

13.8 Conclusions

Indirect calorimetry is considered the gold standard in measuring Q in clinical environments since the technique provides a sensitive, relatively inexpensive and non-invasive means to monitor energy metabolism. Cats are highly sensitive to the characteristics of their external environment and a novel environment, such as a respiratory chamber, may trigger a stress response. Stress responses in the cat are not overt and pose an additional issue for researchers to accurately monitor welfare via behavioural changes indicative of stress. To reduce variability and improve validity of results cats selected for nutritional trials should be healthy, display normal behaviours, and be well-adapted to a particular environment to represent a physiological replica of the 'pet cat' characterized by a constricted range of physical, behavioural, and clinical factors (Meunier, 2006). Reports using calorimetry in dogs lack sufficient information on behavioural acclimation and there is relatively little information about the influence of breed differences, age, gender, body weight and condition, lean body mass, ambient and chamber temperature and relative humidity, feeding regime, time of year and the effects of environment on stress. Overall, it may be concluded, that space limitation during indirect calorimetry, if compliant with minimum space requirements, is unlikely to be the primary contributor of a stress response (Hetts *et al.*, 1992) especially if dogs are provided with opportunities to exercise, socialize with other dogs and humans in a positive and predictive manner after a bout of temporary restriction (Coppola *et al.*, 2006; Hubrecht, 1993; Wells, 2004). Future literature should include descriptions of acclimation, monitoring of stress responses, and consideration of other measurements of Q.

References

Association of American Feed Control Officials (AAFCO), 2013. Official Publication, 2013. AAFCO, Atlanta, GA, USA.

Baldwin, R.L., 1995. Modeling ruminant digestion and metabolism. Chapman and Hall, London, UK.

Bebak, J. and Beck, A.M., 1993. The effect of cage size on play and aggression between dogs in purpose-bred beagles. Laboratory Animal Science 43: 457-459.

Beerda, B., Schilder, M.B., Van Hooff, J.A. and De Vries, H.W., 1997. Manifestations of chronic and acute stress in dogs. Applied Animal Behaviour Science 52: 307-319.

Beerda, B., Schilder, M.B.H., Bernadina, W., Van Hooff, J.A., De Vries, H.W. and Mol, J.A., 1999a. Chronic stress in dogs subjected to social and spatial restriction. II. Hormonal and Immunological Responses. Physiology and Behaviour 66: 243-254.

Beerda, B., Schilder, M.B.H., Van Hooff, J.A., De Vries, H.W. and Mol, J.A., 1998. Behavioural, saliva cortisol and heart rate responses to different types of stimuli in dogs. Applied Animal Behaviour Science 58: 365-381.

Beerda, B., Schilder, M.B.H., Van Hooff, J.A., De Vries, H.W. and Mol, J.A.,1999b. Chronic stress in dogs subjected to social and spatial restriction. I. Behavioural responses. Physiology and Behaviour 66: 233-242.

Beerda, B., Schilder, M.B.H., Van Hooff, J.A., De Vries, H.W. and Mol, J.A., 2000. Behavioural and hormonal indicators of enduring environmental stress in dogs. Animal Welfare 9: 49-62.

Belsito, K.R., Vester, B.M., Keel, T., Graves, T.K. and Swanson, K.S., 2009. Impact of ovariohysterectomy and food intake on body composition, physical activity, and adipose gene expression in cats. Journal of Animal Science 87: 594-602.

Berendt, K., 2014. Starch: an alternative energy source for cats. Department of Agricultural, Food and Nutritional Science, University of Alberta, Canada.

Bergamasco, L., Osella, M.C., Savarino, P., Larosa, G., Ozella, L., Manassero, M., Badino, P., Odore, R., Barbero, R. and Re, G., 2010. Heart rate variability and saliva cortisol assessment in shelter dog: human-animal interaction effects. Applied Animal Behaviour Science 125: 56-68.

Birmingham, E.N., Thomas, D.G., Morris, P.J. and Hawthorne, A.J., 2010. Energy requirements of adult cats. British Journal of Nutrition 103: 1083-1093.

Birmingham, E.N., Weidgraaf, K., Hekman, M., Roy, N.C., Travendale, M.H. and Thomas, D.G., 2013. Seasonal and age effects on energy requirements in domestic short-hair cats (*Felis catus*) in a temperate environment. Journal of Animal Physiology and Animal Nutrition 97: 522-530.

Bodnariu, A., 2008. Indicators of stress and stress assessment in dogs. Lucrari Stiintifice Medicina Veterinara 41: 20-26.

Bradshaw, J.W. and Hall, S.L., 1999. Affiliative behaviour of related and unrelated pairs of cats in catteries: a preliminary report. Applied Animal Behaviour Science 63: 251-255.

Burkholder, W.J., 1995. Metabolic rates and nutrient requirements of sick dogs and cats. Journal of the American Veterinary Medical Association 206: 614-618.

Carlstead, K., Brown, J.L. and Strawn, W., 1993. Behavioural and physiological correlates of stress in laboratory cats. Applied Animal Behaviour Science 38: 143-158.

Center, S.A., Warner, K.L., Randolph, J.F., Sunvold, G.D. and Vickers, R.J., 2012. Influence of L-carnitine on metabolic rate, fatty acid oxidation, body condition, and weight loss in overweight cats. American Journal of Veterinary Research 72: 1002-1015.

Chance, P., 2003. Learning and behaviour, 5th ed. Wadsworth Publishing Co. Belmont, CA, USA.

Coppola, C.L., Grandin, T. and Enns, M., 2006. Human interaction and cortisol: can human contact reduce stress for shelter dogs? Physiology and Behaviour 87: 537-541.

Council of Europe, 2006. Appendix A: European Convention for the Protection of Vertebrate Animals used for Experimental and other Scientific Purposes. European Treaty Series 123. Council of Europe, Strasbourg, France. Available at: http://tinyurl.com/pg9drk3.

Crist, K.A. and Romsos, D.R., 1987. Evidence for cold-induced but not for diet-induced thermogenesis in adult dogs. Journal of Nutrition 117: 1280-1286.

Crouse, S.J., Atwill, E.R., Lagana, M. and Houpt, K.A., 1995. Soft surfaces: a factor in feline psychological well-being. Contemporary Topics in Laboratory Animal Science 34: 94-97.

Crowell-Davis, S.L., Curtis, T.M. and Knowles, R.J., 2004. Social organization in the cat: a modern understanding. Journal of Feline Medicine and Surgery 6: 19-28.

DeLuca, A.M. and Kranda, K.C., 1992. Environmental enrichment in a large animal facility. Laboratory Animals 21: 38-44.

Dess, N.K., Linwick, D., Patterson, J., Overmier, J.B. and Levine, S., 1983. Immediate and proactive effects of controllability and predictability on plasma cortisol responses to shocks in dogs. Behavioural Neuroscience 97: 1005-1016.

Durman, K.J., 1991. Behavioural indicators of stress. BSc. Thesis. University of Southampton, Southampton, UK.

Ellis, J.J., Protopapadaki, V., Stryhn, H., Spears, J. and Cockram, M.S., 2014. Behavioural and faecal glucocorticoid metabolite responses of single caging in six cats over 30 days. Veterinary Record Open 1: e000056.

Finka, L.R., Ellis, S.L. and Stavisky, J., 2014. A critically appraised topic (CAT) to compare the effects of single and multi-cat housing on physiological and behavioural measures of stress in domestic cats in confined environments. BMC Veterinary Research 10: 73.

Finke, M.D., 1991. Evaluation of the energy requirements of adult kennelled dogs. Journal of Nutrition121: S22-S28.

Finke, M.D., 1994. Energy requirements of adult female Beagles. Journal of Nutrition 124: 2605S-2608S.

Geret, P.C., Riond, B., Cattori, V., Meli, L.M., Hofmann-Lehmann, R. and Lutz, H., 2011. Housing and care of laboratory cats: from requirements to practice. Schweizer Archiv für Tierheilkunde 153: 157-164.

Gooding, M.A., Duncan, I.J.H., Atkinson, J.L. and Shoveller, A.K., 2012. Development and validation of a behavioural acclimation protocol for cats to respiration chambers used for indirect calorimetry studies. Journal of Applied Animal Science 15: 144-162.

Gooding, M.A., Duncan, I.J.H., Atkinson, J.L. and Shoveller, A.K., 2014. Effects of high-fat and high-carbohydrate diets on fat and carbohydrate oxidation and plasma metabolites in healthy cats. Journal of Animal Physiology and Animal Nutrition 98: 596-607.

Gourkow, N. and Fraser, D., 2006. The effect of housing and handling practices on the welfare, behaviour and selection of domestic cats (*Felis sylvestris catus*) by adopters in an animal shelter Animal Welfare 15: 371-377.

Gourkow, N., Hamon, S.C. and Phillips, C.J., 2014b. Effect of gentle stroking and vocalization on behaviour, mucosal immunity and upper respiratory disease in anxious shelter cats. Preventive Veterinary Medicine 117: 266-275.

Gourkow, N., LaVoy, A., Dean, G.A. and Phillips, C.J., 2014a. Associations of behaviour with secretory immunoglobulin A and cortisol in domestic cats during their first week in an animal shelter. Applied Animal Behaviour Science 150: 55-64.

Green, A.S., Ramsey, J.J., Villaverde, C., Asami, D.K., Wei, A. and Fascetti, A.J., 2008. Cats are able to adapt protein oxidation to protein intake provided their requirement for dietary protein is met. Journal of Nutrition 138: 1053-1060.

Hawthorne, A.J., Loveridge, G.G. and Horrocks, L.J., 1995. The behaviour of domestic cats in response to a variety of surface-textures. In: Holst, B. (ed.) Proceedings of the Second International Conference on Environmental Enrichment, Copenhagen Zoo, Copenhagen, Denmark, pp. 84-94.

Hennessy, M.B., 2013. Using hypothalamic-pituitary-adrenal measures for assessing and reducing the stress of dogs in shelters: a review. Applied Animal Behaviour Science 149: 1-12.

Hennessy, M.B., Voith, V.L., Hawke, J.L., Young, T.L., Centrone, J., McDowell, A., Linden, F. and Davenport, G.M., 2002. Diet and human interaction moderate hypothalamic-pituitary-adrenal responses of dogs in an animal shelter. Journal of the American Veterinary Medical Association 221: 65-71.

Hennessy, M.B., Williams, M.T., Miller, D.D., Douglas, C.W. and Voith, V.L., 1998. Influence of male and female petters on plasma cortisol and behaviour: Can human interaction reduce the stress of dogs in a public animal shelter? Applied Animal Behavioural Science 61: 63-77.

Hetts, S., Derrell Clark, J., Calpin, J.P., Arnold, C.E. and Mateo, J.M., 1992. Influence of housing conditions on beagle behaviour. Applied Animal Behaviour Science 34: 137-155.

Hewison, L.F., Wright, H.F., Zulch, H.E. and Ellis, S.L., 2014. Short term consequences of preventing visitor access to kennels on noise and the behaviour and physiology of dogs housed in a rescue shelter. Physiology and Behavior 133: 1-7.

Hiby, E.F., Rooney, N.J. and Bradshaw, J.W., 2006. Behavioural and physiological responses of dogs entering re-homing kennels. Physiology and Behaviour 89: 385-391.

Hite, M., Hanson, H.M., Bohidar, N.R., Conti, P.A. and Mattis, P.A., 1977. Effect of cage size on patterns of activity and health of beagle dogs. Laboratory Animal Science 27: 60-64.

Hubrecht, R.C., 1993. A comparison of social and environmental enrichment methods for laboratory housed dogs. Applied Animal Behaviour Science 37: 345-361.

Hubrecht, R.C., Serpell, J.A. and Poole, T.B., 1992. Correlates of pen size and housing conditions on the behaviour of kenneled dogs. Applied Animal Behaviour Science 34: 365-383.

Hughes, H.C., Campbell, S.A. and Kenney, C., 1989. The effects of cage size and pair housing on exercise of beagle dogs. Laboratory Animal Science 31: 302-305.

Hunt, R.M., Xiang-Xu, X. and Kaufman, J., 1983. Miocene burrows of extinct bear dogs: Indication of early denning behavior of large mammalian carnivores. Science 221: 364-366.

Kessler, M.R. and Turner, D.C., 1997. Stress and adaptation of cats (*Felis silvestris catus)* housed singly, in pairs and in groups in boarding catteries. Animal Welfare 6: 243-254.

Kobelt, A.J., Hemsworth, P.H., Barnett, J.L. and Butler, K.L., 2003. Sources of sampling variation in saliva cortisol in dogs. Research in Veterinary Science 75: 157-161.

Larsson, C., Jensen, R.B., Junghans, P. and Tauson, A.H., 2014. The oral ^{13}C-bicarbonate technique for estimation of energy expenditure in dogs: validation against indirect calorimetry. Archives of Animal Nutrition 68: 42-54.

Lester, T., Czarnecki-Maulden, G. and Lewis, D., 1999. Cats increase fatty acid oxidation when isocarorically fed meat-based diets with increasing fat content. American Journal of Physiology 277 (3 pt 2): R878-R886.

Lichtsteiner, M. and Turne, D.C., 2008. Influence of indoor-cat group size and dominance rank on urinary cortisol levels. Animal Welfare 17: 215-237.

Mason, G. and Mendl, M., 1993. Why is there no simple way of measuring animal welfare? Animal Welfare 2: 301-319.

Mazzaferro, E.M., Hackett, T.B., Stein, T.P., Ogilvie, G.K., Wingfield, W.E., Walton, J., Turner, A.S. and Fettman, M.J., 2001. Metabolic alterations in dogs with osteosarcoma. American Journal of Veterinary Research 62: 1234-1239.

McCune, S., 1992. Temperament and the welfare of caged cats. PhD thesis, University of Cambridge, UK.

McCune, S., 1994. Caged cats: avoiding problems and providing solutions. Newsletter of the Companion Animal Study Group 7: 1-9.

McKnight, L.L., Flickinger, E.A., Davenport, G.M., France, J. and Shoveller, A.K., 2014b. Dietary mannoheptulose has differential effects on fasting and post-prandial glucose oxidation in Labrador Retrievers. Journal of Applied Animal Research 43: 357-365.

McKnight, L.L., Flickinger, E.A., France, J., Davenport, G.M. and Shoveller, A.K., 2014a. Mannoheptulose has differential effects on fasting and post-prandial energy expenditure and respiratory quotient in adult Beagle dogs fed diets of different macronutrient contents. Journal of Nutritional Sciences 3: e17.

Menor-Campos, D.J., Molleda, J.M. and López-Rodriguez, R., 2011. Effects of exercise and human contact on animal welfare in a dog shelter. Veterinary Record 169: 388-391.

Mertens, P.A. and Unshelm, J., 1996. Effects of group and individual housing on the behaviour of kennelled dogs in animal shelters. Anthrozoös 9: 40-51.

Meunier, L.D., 2006. Selection, acclimation, training, and preparation of dogs for the research setting. Institute for Laboratory Animal Research 47: 326-347.

Miklósi, Á., Kubinyi, E., Topál, J., Gácsi, M., Virányi, Z. and Csányi, V., 2003. A simple reason for a big difference: wolves do not look back at humans, but dogs do. Current Biology 13: 763-766.

Muir, J.L. and Pfister, H., 1986. Corticosterone and prolactin responses to predictable and unpredictable novelty in rats. Physiology and Behaviour 37: 285-288.

National Research Council (NRC), 2006. Nutrient requirements of dogs and cats. National Academy Press, Washington, DC, USA.

Neamand, J., Sweeny, W.T., Creamer, A.A. and Conti, P.A., 1975. Cage activity in the laboratory beagle: a preliminary study to evaluate a method of comparing cage size to physical activity. Laboratory Animal Science 25: 180-183.

Nguyen, P., Dumon, H., Martin, L., Siliart, B., Ferrier, L., Humbert, B., Diez, M., Breul, S. and Biourge, V., 2002. Weight loss does not influence energy expenditure or leucine metabolism in obese cats. Journal of Nutrition 132: 1649S-1651S.

Nguyen, P., Leray, V., Dumon, H., Martin, L., Siliart, B., Diez, M. and Biourge, V., 2004. High protein intake affects lean body mass but not energy expenditure in nonobese cats. Journal of Nutrition 134: 2084S-2086S.

O'Toole, E., McDonell, W.N., Wilson, B.A., Mathews, K.A., Miller, C.W. and Sears, W.C., 2001. Evaluation of accuracy and reliability of indirect calorimetry for the measurement of resting energy expenditure in healthy dogs. American Journal of Veterinary Research 62: 1761-1767.

O'Toole, E., Miller, C.W., Wilson, B.A., Mathews, K.A., Davis, C. and Sears, W.C., 2004. Comparison of the standard predictive equation for calculation of resting energy expenditure with indirect calorimetry in hospitalized and healthy dogs. American Veterinary Medical Association 225: 58-64.

Ottway, D.S. and Hawkins, D.M., 2003. Cat housing in rescue shelters: A welfare comparison between communal and discrete-unit housing. Animal Welfare 12: 173-189.

Part, C.E., Kiddie, J.L., Hayes, W., Mills, D., Neville, R.F., Morton, D.B. and Collins, L.M., 2014. Physiological, physical and behavioural changes in dogs (*Canis familiaris*) when kennelled: testing the validity of stress parameters. Physiology and Behavior. 133: 260-271

Passantino, A., Quartarone, V., Pediliggeri, M.C., Rizzo, M. and Piccione, G., 2014. Possible application of oxidative stress parameters for the evaluation of animal welfare in sheltered dogs subjected to different environmental and health conditions. Journal of Veterinary Behavior: Clinical Applications and Research 9: 290-294.

Pettijohn, T.F., Davis, K.L. and Scott, J.P., 1980. Influence of living area space on agonistic interaction in telomian dogs. Neuroscience and Behavioural Biology 28: 343-349.

Podberscek, A.L., Blackshaw, J.K. and Beattie, A.W., 1991. The behaviour of laboratory colony cats and their reactions to a familiar and unfamiliar person. Applied Animal Behaviour Science 31: 119-130.

Pouteau, E.B., Mariot, S.M., Martin, L.J., Dumon, H.J., Mabon, F.J., Krempf, M.A., Robins, R.J., Darmaun, D.H., Naulet, N.A. and Nguyen, P.G., 2002. Rate of carbon dioxide production and energy expenditure in fed and food-deprived adult dogs determined by indirect calorimetry and isotopic methods. American Journal of Veterinary Research 63: 111-118.

Ramos, D., Arena, M.N., Reche-Junior, A., Daniel, A.G.T., Albino, M.V.C., Vasconcellos, A.S., Viau, P. and Oliveira, C.A., 2012. Factors affecting faecal glucocorticoid levels in domestic cats (*Felis catus*): a pilot study with single and large multi-cat households. Animal Welfare 21: 285-291.

Richey, J.M., Woolcott, O.O., Stefanovski, D., Harrison, L.N., Zheng, D., Lottati, M., Hsu, I.R., Kim, S.P., Kabir, M., Catalano, K.J., Chiu, J.D., Ionut, V., Kolka, C., Mooradian, V., Bergman, R.N., 2009. Rimonabant prevents additional accumulation of visceral and subcutaneous fat during high-fat feeding in dogs. American Journal of Physiology – Endocrinology and Metabolism 296: E1311-E1318.

Rochlitz, I., 1997. The welfare of cats kept in confined environments. PhD thesis, University of Cambridge, Cambridge, UK.

Rochlitz, I., 1999. Recommendations for the housing of cats in the home, in catteries and animal shelters, in laboratories and in veterinary surgeries. Journal of Feline Medicine and Surgery 1: 181-191.

Rochlitz, I., 2000. Recommendations for the housing and care of domestic cats in laboratories. Laboratory Animal 34: 1-9.

Rochlitz, I., 2005. A review of the housing requirements of domestic cats (*Felis silvestris catus*) kept in the home. Applied Animal Behaviour Science 93: 97-109.

Rochlitz, I., 2007. Housing and welfare. In: Rochlitz, I. (ed.) The welfare of cats. Springer, Dordrecht, the Netherlands, pp. 177-203.

Rochlitz, I., Podberscek, A.L. and Broom, D.M., 1998. The welfare of cats in a quarantine cattery. The Veterinary Record 143: 35-39.

Rooney, N.J., Gaines, S.A. and Bradshaw, J.W., 2007. Behavioural and glucocorticoid responses of dogs (*Canis familiaris*) to kennelling: investigating mitigation of stress by prior habituation. Physiology and Behaviour 92: 847-854.

Roy, D., 1992. Environmental enrichment for cats in rescue centres. BSc. thesis, University of Southampton, Southampton, UK.

Russell, K., Murgatroyd, P.R. and Batt, R.M., 2002. Net protein oxidation is adapted to dietary protein intake in domestic cats (*Felis silvestris catus*). Journal of Nutrition 132: 456-460.

Seematter, G., Binnert, C. and Tappy, L., 2005. Stress and metabolism. Metabolic Syndrome and Related Disorders 3: 8-13.

Seematter, G., Guenat, E., Schneiter, P., Cayeux, C., Jequier, E. and Tappy, L., 2000. Effects of mental stress on insulin-mediated glucose metabolism and energy expenditure in lean and obese women. American Journal of Physiology – Endocrinology and Metabolism 279: E799-E805.

Selye, H., 1971. Hormones and resistance. Parts 1 and 2. Journal of Pharmacological Science. 60: 1-28.

Shiverdecker, M., Schiml, P.A. and Hennessy, M.B., 2013. Human interaction moderates plasma cortisol and behavioural responses of dogs to shelter housing. Physiology and Behaviour 109: 75-79.

Shoveller, A.K., Minikhiem, D.L., Carnagey, K., Brewer, J., Westendorf, R., DiGennaro, J. and Gooding, M.A., 2014. Low level of supplemental dietary L-carnitine increases energy expenditure in overweight, but not lean, cats fed a moderate energy density diet to maintain body weight. International Journal of Applied Research in Veterinary Medicine 12: 33-43.

Smith, D.F.E., Durman, K.J., Roy, D.B. and Bradshaw, J.W.S., 1994. Behavioural aspects of the welfare of rescued cats. Journal of the Feline Advisory Bureau 31: 25-28.

Stephen, J.M. and Ledger, R.A., 2006. A longitudinal evaluation of urinary cortisol in kennelled dogs (*Canis familiaris*). Physiology and Behaviour 87: 911-916.

Taylor, E.J., Adams, C. and Neville, R., 1995. Some nutritional aspects of ageing in dogs and cats. Proceedings of the Nutrition Society 54: 645-656.

Titulaer, M., Blackwell, E.J., Mendl, M. and Casey, R.A., 2013. Cross sectional study comparing behavioural, cognitive and physiological indicators of welfare between short and long term kennelled domestic dogs. Applied Animal Behaviour Science 147: 149-158.

Tuber, D.S., Hennessy, M.B., Sanders, S., Miller, J.A., 1996. Behavioural and glucocorticoid responses of adult domestic dogs (*Canis familiaris*) to companionship and social separation. Journal of Comparative Psychology 110: 103-108.

Tuber, D.S., Miller, D.D., Caris, K.A., Halter, R., Linden, F., Hennessy, M.B., 1999. Dogs in animal shelters: problems, suggestions, and needed expertise. Psychological Science 10: 379-386.

Turner, P.A., Smiler, K.L., Hargaden, M. and Koch, M.A., 2003. Refinements in the care and use of animals in toxicology studies – regulation, validation, and progress. Contemporary Topics 42: 8-15.

Uetake, K., Goto, A., Koyama, R., Kikuchi, R. and Tanaka, T., 2013. Effects of single caging and cage size on behavior and stress level of domestic neutered cats housed in an animal shelter. Animal Science Journal 84: 272-274.

United States Department of Agriculture – National Agricultural Library (USDA-NAL), undated. Animal Welfare Information Center. Available at: http://tinyurl.com/nw7tzpb.

Villaverde, C., Ramsey, J.J., Green, A.S., Asami, D.K., Yoo, S. and Fascetti, A.J., 2008. Energy restriction results in a mass-adjusted decrease in energy expenditure in cats that is maintained after weight regain. Journal of Nutrition 138: 856-860.

Vincent, I.C., Michell, A.R. and Leahy, R.A., 1993. Non-invasive measurement of arterial blood pressure in dogs: a potential indicator for the identification of stress. Research in Veterinary Science 54: 195-201.

Webster, J., 1994. Animal welfare: a cool eye towards Eden. Blackwell Science, Oxford, UK.

Wei, A., Fascetti, A.J., Villaverde, C., Green, C., Manzanilla, E.G., Havel, P.J. and Ramsey, J.J., 2011. Influence of a high protein diet on energy balance in obese cats allowed *ad libitum* access to food. Journal of Animal Physiology and Nutrition 95: 359-367.

Wei, A., Fascetti. A,J., Kim, K., Lee, A., Graham, J.L., Havel, P.J. and Ramsey, J.J., 2014. Early effects of neutering on energy expenditure in adult male cats. PLoS One 9: e89557.

Wells, D.L. and Hepper, P.G., 1998. A note on the influence of visual conspecific contact on the behaviour of sheltered dogs. Applied Animal Behaviour Science 60: 83-88.

Wells, D.L., 2004. A review of environmental enrichment for kennelled dogs, *Canis familiaris*. Applied Animal Behaviour Science 85: 307-317.

Wennlund, A., Wahrenberg, H., Hagstromtoft, E., Bolinder, J. and Arner, P., 1994. Lipolytic and cardiac responses to various forms of stress in humans. International Journal of Sports Medicine 15: 408-413.

Wichert, B., Muller, L., Gebert, S., Wenk, C. and Wanner, M., 2007. Additional data on energy requirements of young adult cats measured by indirect calorimetry. Journal of Animal Physiology and Nutrition 91: 278-281.

Yoo, S., Ramsey, J.J., Havel, P.J., Jones, P.G. and Fascetti, A.J., 2006. Resting energy expenditure and body fat composition of Labrador Retrievers fed high fat and low fat diets. Journal of Animal Physiology and Nutrition 90: 185-191.

Appendix 13A.

Estimates of energy expenditure (Q) (as measured by indirect calorimetry and influenced by diet composition), description of population used, chamber design, behavioural acclimation to respiration chambers, and description of methods reported in the feline literature.

Study	Details[1,2]
No acclimation reported	
Nguyen *et al.* (2002)	
Population	7 neutered obese (1 female, 6 male); 6.2 ± 0.7 yr; 6.5 ± 0.3 kg and 4.4 BCS
Calorimetry system characteristics	energy expenditure and requirement assessed using three different undescribed methods
Adaptation details	no description of behavioural acclimation
Measurement and treatments	measurements taken for 2-4 hrs and extrapolated to whole day; not clear as to time spent fasting
Energy expenditure	daily Q; 124±4 kJ/kg/d; 218 kJ/kg fat free mass.
Outcomes	experimental HP (n=4) or Hills Feline r/d (n=3); no differences in Q between groups, and when expressed per kg increased as BW and body fat decreased
Nguyen *et al.* (2004)	
Population	16 neutered normal weight (11 female and 5 male); 4.5±0.1 yr; 4.83±0.13 kg and 3.4 BCS
Calorimetry system characteristics	analysers calibrated once hourly; 4 l/min flow
Adaptation details	no description of behavioural acclimation
Measurement and treatments	measurements taken after an overnight fast and for 14 h after to measure least observed metabolism; observed 30 min period
Energy expenditure	Q (described as least observed metabolism; start of study = 163±5 kJ/kg/d or 255±5 kJ/kg fat free mass; end of study = 165±3 kJ/kg/d or 240±4 kJ/kg fat free mass
Outcomes	high protein diets may promote an increased lean tissue mass in cats that may be helpful in treating adiposity
Acclimation reported	
Russel *et al.* (2002)	
Population	6 neutered cats (3 male and 3 female); 4.0±0.35 yr; 5.0±0.45 kg
Calorimetry system characteristics	1×0.6×0.6 m; ambient temperature was maintained at 19 °C; air flow was maintained at 10 l/min; food, water, a bed and litter were provided in the chamber
Adaptation details	before the study, cats were familiarised with the chamber, but there is no detail; cats urinated immediately upon entering the chamber
Measurement and treatments	18 hr collections; MP (77.6 g/kg or 35.3% of energy) and HP (96.5 g/kg); food available at all times
Energy expenditure	776.8±40.1 (MP) and; 745±24.7 kJ (HP) over 18 hr (~199-207 kJ/kg/d)
Outcomes	RQ of 0.787±0.002 vs 0.814±0.002; protein oxidation 20.4±0.8 (mg/min) vs 28.4±0.7; fat oxidation 9.0±0.7 vs 4.7±0.5; oxidation of protein over the 18 hr was significantly related with protein intake; oxidation of fat over the 18 hr was significantly related with fat intake

Study	Details[1,2]
Wichert *et al.* (2007)	
Population	8 adult cats; 16.5±3.5 months; 2.8±0.3 kg
Calorimetry system characteristics	two chambers side by side; no description of chambers; urine and faecal collection in chamber and non-absorbent material used to enable urine collection
Adaptation details	5 days in chambers, first day used for acclimation and day 2-5 used for data collection
Measurement and treatments	each day, cats were confined in chambers for 22.5 hr; food available at all times; no dietary treatments, a single diet was fed and contained 32.6% protein and 18.9% fat as fed
Energy expenditure	energy balance achieved at 239.6 kJ/kg BW/d, 331.2 kJ/kg $BW^{0.75}$/d, 365 kJ/kg $BW^{0.67}$/d
Outcomes	energy requirements of young active cats are greater than those of older cats and indirect calorimetry does not necessarily lead to lower estimates of EE as compared to other methods
Villaverde *et al.* (2008)	
Population	10 adult neutered cats (4 female and 6 male); 2.0 yr; ~6.5 kg entering the study
Calorimetry system characteristics	0.6×0.6×0.75 m; no extensive description of chamber enrichment
Adaptation details	introduced to the chamber at least 1 month before and for at least 3 consecutive days for acclimation
Measurement and treatments	food available at all times; 12 hr restriction and data extrapolated to 24 hr; monadic feeding but with baseline, food restriction and energy available periods
Energy expenditure	baseline = 1,258±33.7 kJ/d (~198 kJ/kg/d); weight loss = 1,050±65 kJ/d (~162 kJ/kg/d); weight regain = 1,047±23.3 kJ/d (~161 kJ/kg/d)
Outcomes	energy restriction results in a mass-adjusted decrease in EE in cats which is maintained after they have regained this weight
Wei *et al.* (2014)	
Population	9 adult male cats; 2.5-3.6 yr; ~5.2 kg entering the study
Calorimetry system characteristics	0.6×0.6×0.75 m
Adaptation details	introduced to the chamber at least 1 month before and for at least 3 consecutive days for acclimation
Measurement and treatments	monadic; food available at all times; 12 hr restriction and data extrapolated to 24 hr
Energy expenditure	pre-neutering 1,108±62 (~213 kJ/kg/d); post-neutering (day 3-4) 1,174±71 (~225 kJ/kg/d); post-neutering (day 9-10); 1,177±74 (~226 kJ/kg/d); RQ for all groups ~0.85
Outcomes	results suggest that post-neutering weight gain results from an increase in energy intake, not a decrease in EE
Wei *et al.* (2011)	
Population	20 cats, 5 male and 15 female; 4.3 yr; 6.0 kg
Calorimetry system characteristics	0.6×0.6×0.75 m
Adaptation details	introduced to the chamber at least 1 month before and for at least 3 consecutive days for acclimation

Study	Details[1,2]
Wei *et al.* (2011) Continued.	
Measurement and treatments	MP vs HP diet; food available at all times; 12 hr restriction and data extrapolated to 24 hr
Energy expenditure	Q = 1,120-1,155 kJ/d (~186-192 kJ/kg/d)
Outcomes	HP diet did not reduce food intake or promote weight loss in obese cats; cats fed HP diets had greater EE than cats fed MP and may help to promote weight loss when energy intake is reduced
Green *et al.* (2008)	
Population	5 males and 5 females; 3.2 yr; 4.52 kg
Calorimetry system characteristics	0.6×0.6×0.75 m
Adaptation details	introduced to the chamber at least 1 month before and for at least 3 consecutive days for acclimation
Measurement and treatments	MP and HP diets; food available at all times; 12 hr restriction and data extrapolated to 24 hr
Energy expenditure	910-1,090 kJ/d (~202-242 kJ/kg/d)
Outcomes	cats are able to modify protein oxidation when given protein above their requirement
Lester *et al.* (1999)	
Population	6 male neutered; 6 female spayed; 4 yr
Calorimetry system characteristics	50×50×60 cm, 125 l; air mixed with the help of a fan; temp maintained at 25 °C; daily calibration of analysers
Adaptation details	animals acclimated over one week for periods of 2, 4 and 6 hrs; animals assumed to be acclimated when they express normal behaviour
Measurement and treatments	HF and HC diets; 24 hr calorimetry; food available during calorimetry collections
Energy expenditure	39-60 kcal/kg/d (~160-250 kJ/kg/d)
Outcomes	24 hour EE was positively correlated with $BW^{0.66}$; data was similar when expressed as per kg lean body mass; increasing dietary fat increased fat oxidation, decreased carboxidation, but did not result in a change in protein oxidation
Gooding *et al.* (2014)	
Population	10 neutered or spayed males and females; 10 months; 3.6 kg
Calorimetry system characteristics	53.3×53.3×73 cm; outfitted with a shelf, feeder, water bowl, hammock, litter box, toy and free area with a fleece bed
Adaptation details	acclimation to chambers demonstrated in Gooding *et al.* (2012) and over a 12 week period; panel maintained with reoccurring exposures
Measurement and treatments	HF and HC diets; fasted and fed measurements
Energy expenditure	~80 kcal/$kg^{0.67}$/d (~146 kJ/kg/d)
Outcomes	no differences in EE between diets; changes observed in macronutrient oxidation and EE suggest that cats adapt whole bogy nutrient metabolism in response to diet, but may require longer than 14 d to adapt to the new macronutrient content

Study	Details[1,2]
Shoveller *et al.* (2014)	
Population	10 neutered males and 10 spayed females; 3 yr; 4.32 kg
Calorimetry system characteristics	53.3×53.3×73 cm; outfitted with a shelf, feeder, water bowl, hammock, litter box, toy and free area with a fleece bed
Adaptation details	acclimation to chambers demonstrated in Gooding *et al.* (2012) and over a 12 week period; panel maintained with reoccurring exposures
Measurement and treatments	diets with and without L-carnitine addition (100 mg/kg); fasted and fed measurements
Energy expenditure	lean cats – fasting Q = 50-55 kcal/kg/d (209-230 kJ/kg/d) –post-prandial Q = 45 kcal/d (186 kJ/kg/d) overweight cats – fasting Q = 43-50 kcal/kg/d (179-209 kJ/kg/d) – post-prandial Q = 34-38 kcal/kg/d (142-158 kJ/kg/d)
Outcomes	overweight, but not lean cats, have greater energy expenditure when fed diets with 100 mg/kg of supplemental L-carnitine
Berendt *et al.* (2014)	
Population	10 neutered males and 10 spayed females; 5 yr; 4.93 kg
Calorimetry system characteristics	53.3×53.3×73 cm; outfitted with a shelf, feeder, water bowl, hammock, litter box, toy and free area with a fleece bed
Adaptation details	acclimation to chambers demonstrated in Gooding *et al.* (2012) and over a 12 week period; panel maintained with reoccurring exposures
Measurement and treatments	3 commercial diets with varying carbohydrates ingredients ranging in glycaemic response
Energy expenditure	post-prandial 200 kcal/d; 41 kcal/kg/d (~172 kJ/kg/d)
Outcomes	Q was not different between diets, but RQ responded to the glycaemic index of the carbohydrates utilised
Center *et al.* (2012)	
Population	32 overweight cats; 3/7 yr
Calorimetry system characteristics	two cages with different dimensions (31.8×33×19.1 cm [~20 l] and 62.2×27.9×26.7 cm [~46 l]) were constructed to accommodate small to large-sized cats; each cage was configured to allow enough room for the occupant cat to sit or recline comfortably with their head oriented toward the air intake vents in the sliding door; dimly lit room with little to no traffic during collection periods
Adaptation details	2-3 week acclimation to the chamber; cats were observed for behaviour throughout collections
Measurement and treatments	weight loss with titrated levels of L-carnitine.
Energy expenditure	34-61 kcal/kg/d (~142-255 kJ/kg/d)
Outcomes	dietary L-carnitine supplementation (100 or 150 mg/kg) provides metabolic advantage in over-conditioned cats undergoing weight loss

[1] BCS = body condition score; BW = body weight; EE = energy expenditure; HP = high protein; HF = high fat; HC = high carbohydrate; MP = medium protein; RQ = respiratory quotient.
[2] When results were presented in kcal, mean values were multiplied by 4.184 to allow comparison across publications and presented in bold for differentiation.

Appendix 13B.

Estimates of energy expenditure (Q) (as measured by indirect calorimetry and influenced by diet composition), chamber design and acclimation to respiration chambers reported in the canine literature.

Study	Details[1,2,3]
Calorimetry chamber system	
Crist and Romsos (1987)	
Population	no. animals: not reported; source: laboratory; breed: Beagles; sex: F; BW: exp. 1 – Group 1 = 8.76±0.3 kg, Group 2 = 8.93±0.13 kg, Group 3 = 9.21± 0.01 kg; exp. 2 – BW not reported; age: 2-4 yr (exp. 1), 4-6 yr (exp. 2)
Calorimetry system characteristics	dimensions not reported, described as a stainless steel respirator chamber, with front and rear clear plastic doors, equivalent in size to their metabolism cage and located in the same room; CO_2 0.495±0.012%; flow rate 23.86±0.87 l/min; room temperature 22 °C
Adaptation details	authors report that dogs had been previously exposed to the chamber and calorimetry measurements were made the day prior to study
Measurement and calculations	measurements began at initiation of the morning meal for 24 h (an evening meal was given); Q = 3.866(O_2 l/d) + 1.2(CO_2 l/d) assume reported Q values represent 24 h mean, but not explained by the authors
Treatments	exp. 1: Group 1: dogs fed 2,926 KJ ME/d & maintained in a room at 22 °C for 8 wk; Group 2: dogs fed 627 KJ ME/d & maintained at 22 °C for 4 wk then 3 °C for the final 4 wk; Group 3: dogs fed 627 KJ ME/d & maintained at 3 °C for 4 wk then 22 °C for the final 4 wk exp. 2: Group 1: dogs were fasted for 6 days; Group 2: dogs were fed 2,508 KJ ME/d (maintenance) for 6 days; Group 3: dogs were fed 5,685 KJ ME/d (overfed) for 6 days
Outcomes	Q: exp. 1: Group 1 = 2,742 KJ/d *(539 KJ/(kg$^{0.75}$·d))*; Group 2 = 3,751 KJ/d *(711 KJ/(kg$^{0.75}$·d))*; Group 3 = 2,717 KJ/d *(500 KJ/(kg$^{0.75}$·d))* exp. 2: Group 1 = 2,173 KJ/d; Group 2 = 2,433 KJ/d; Group 3 = 2,692 KJ/d
Larsson *et al.* (2014)	
Population	no. animals: 11; source: client owned; breed: various; sex: 9 F & 2 M; BW: 15-41 kg; age: 4-11 yr
Calorimetry system characteristics	dimensions: 3.5 m^3; CO_2 0.3-0.4%; flow rate 42 l/min; room temperature 16.5 °C; humidity 60%
Adaptation details	not reported
Measurement and calculations	4-5.5 h ('depending on behaviour of dog'); Q calculated using the Brouwer equation assume measurements were taken in fasted state
Treatments	simultaneous measurement of ^{13}C-bicarbonate recovery; animals were grouped post-hoc based on their behaviour in the chamber; Group 1: resting, animals spent the majority of their time lying down; Group 2: limited description, 'more active than the rest group'
Outcomes	Q: Group 1 = 363 KJ/kg$^{0.75}$/d; Group 2 = 718 KJ/kg$^{0.75}$/d RQ: Group 1 = 0.77; Group 2 = 0.94

Study	Details[1,2,3]
McKnight *et al.* (2014a)	
Population	no. animals: 8; source: laboratory; breed: Beagles; sex: 7 F & 1 M; BW: 9.62±0.31 kg; age: 7.6±0.4 yr
Calorimetry system characteristics	dimensions: 76×53×61 cm; CO_2 0.4-0.8%; flow rate 9-18 l/min
Adaptation details	authors report that dogs were acclimatised over an 8 week period (minimum exposure of 1 h per week), to rest comfortably in the chamber with no excessive activity or movement and were discouraged from urinating or defecating in the chamber
Measurement and calculations	two fasting collections after which dog were fed a meal and measurements continued for 23 hrs; Q was calculated using the abbreviated Weir equation
Treatments	Group 1: low carbohydrate diet (32.2% protein, 27.9% fat, 30.6% carbohydrate) with a placebo supplement; Group 2: low carbohydrate diet; with a MH-containing supplement; Group 3: high carbohydrate diet (27.0% protein, 11.2% fat, 53.6% carbohydrate) with a placebo supplement; Group 4: high carbohydrate diet with a MH-containing supplement
Outcomes	Q (KJ/kg$^{0.75}$/d): Group 1 = 387.8 (fasted), 457.2 (fed 0-3.8 h), 443.3 (fed 5-9.6 h), 369.4 (fed 10.9-16.7 h) and 353.1 (fed 18-22.5 h); Group 2 = 360.8 (fasted); 464.7 (fed 0-3.8 h), 464.0 (fed 5-9.6 h), 369.3 (fed 10.9-16.7 h) and 339.1 (fed 18-22.5 h); Group 3 = 342.0 (fasted), 426.4 (fed 0-3.8 h), 446.7 (fed 5-9.6 h), 351.6 (fed 10.9-16.7 h) and 334.6 (fed 18-22.5 h); Group 4 = 366.5 (fasted), 437.5 (fed 0-3.8 h), 480.9 (fed 5-9.6 h), 366.1 (fed 10.9-16.7 h) and 335.9 (fed 18-22.5 h) RQ: Group 1 = 0.764 (fasted), 0.819 (fed 0-3.8 h), 0.834 (fed 5-9.6 h), 0.829 (fed 10.9-16.7 h) and 0.797 (fed 18-22.5 h); Group 2 = 0.759 (fasted), 0.824 (fed 0-3.8 h), 0.828 (fed 5-9.6 h), 0.825 (fed 10.9-16.7 h) and 0.793 (fed 18-22.5 h); Group 3 = 0.780 (fasted), 0.838 (fed 0-3.8 h), 0.914 (fed 5-9.6 h), 0.918 (fed 10.9-16.7 h) and 0.869 (fed 18-22.5 h); Group 4 = 0.803 (fasted), 0.851 (fed 0-3.8 h), 0.915 (fed 5-9.6 h), 0.920 (fed 10.9-16.7 h) and 0.880 (fed 18-22.5 h)
McKnight *et al.* (2014b)	
Population	No. animals: 6; source: laboratory; breed: Labrador Retrievers; sex: M; BW: 31.2 kg; age: 5.4 yr
Calorimetry system characteristics	dimensions: 76×53×61 cm; CO_2 0.4-0.8%; flow rate 22-26 l/min
Adaptation details	authors report that dogs were acclimatised over an 8 week period (minimum exposure of 1 h per week), to rest comfortably in the chamber with no excessive activity or movement and were discouraged from urinating or defecating; in the chamber
Measurement and calculations	two fasting collections after which dog were fed a meal and measurements continued for 12 h.; Q was calculated using the abbreviated Weir equation
Treatments	Group 1: control diet; Group 2: MH containing diet

Study	Details[1,2,3]
McKnight *et al.* (2014b) Continued	
Outcomes	Q (KJ/kg$^{0.75}$/d): Group 1 = 278.6 (fasted); Group 2 = 287.6 (fasted); Group 1 & 2: 398-481 (fed, range) RQ: Group 1 = 0.74 (fasted); Group 2 = 0.76 (fasted); Group 1 & 2: 0.72-0.90 (fed, range) as numerical values were not provided, rough estimates were made based on visual interpretation of graphical data
Pouteau *et al.* (2002)	
Population	no. animals: 6; source: laboratory; breed: Beagles; sex: 4 F & 2 M; BW: 18.3±1.3 kg; age: 5.2±0.3 yr
Calorimetry system characteristics	dimensions, CO_2 and room temperature not reported; flow rate 40 l/min
Adaptation details	not reported
Measurement and calculations	20 h with hourly calibration
Treatments	Group 1: 24 h food restriction (fasted); Group 2: consumption of a single meal (552 KJ/kg$^{0.75}$/d; 24.8% protein, 15.0% fat,; 6.3% ash, 6.4% total dietary fibre) (fed)
Outcomes	Q (KJ/kg$^{0.75}$/d): Group 1 (fasted) = 395±21 (0-4 h), 352±12 (4-8 h), 316±21 (8-12 h), 308±17 (12-16 h), 311±21 (16-20 h) and 338±14 (0-20 h); (Group 2 (fed) = 455±30 (0-4 h), 388±26 (4-8 h), 342±25 (8-12 h), 336±16 (12-16 h), 328±20 (16-20 h) and 379±25 (0-20 h) RQ: Group 1 (fasted) = 0.78±0.01 (0-4 h), 0.78±0.01 (4-8 h), 0.80±0.01 (8-12 h), 0.81±0.01 (12-16 h), 0.81±0.01 (16-20 h) and 0.79±0.01 (0-20 h); Group 2 (fed) = 0.89±0.01 (0-4 h), 0.93±0.01 (4-8 h), 0.95±0.01 (8-12 h), 0.92±0.02 (12-16 h), 0.87±0.02 (16-20 h) and 0.91±0.01 (0-20 h)
Richey *et al.* (2009)	
Population	no. animals: 12; Source: *assume laboratory*; breed: Mongrel dogs; sex: M; BW: 31.8±1.5 kg; age: 'adult'
Calorimetry system characteristics	dimensions not reported, described as a standard dog kennel adapted for calorimetry; CO_2, flow rate and room temperature not reported
Adaptation details	authors stated that dogs were acclimated over a 3 week period, but no description of acclimation was given
Measurement and calculations	4 h (06:30-10:30 h); animals were fasted 12 h prior to measurements and Q was calculated using the abbreviated Weir equation
Treatments	Group 1: oral placebo; Group 2: Rimonabant (1.25 mg/kg/d); measurements at baseline and 16 wk for the purpose of this review, only data from the control animals are presented
Outcomes	Q (KJ/d): week 0 = 6,836.6±266.7 *(511 KJ/(kg$^{0.75}$·d))*; week 16 = 6,818.2±585.7 *(510 KJ/(kg$^{0.75}$·d)*

Study	Details[1,2,3]
Yoo *et al.* (2006)	
Population	no. animals: 10; source: breeders/owners; breed: Labrador Retrievers; sex: M (intact); BW: 32.0±3.6 kg (LF), 31.6±3.6 kg (HF); age: 2.2±0.5 yr
Calorimetry system characteristics	dimensions: 1.3×0.8×1.0 m; CO_2 not reported; flow rate 25 l/min; room temperature 23 °C
Adaptation details	1 month acclimation to chambers which involved calorimetry recordings
Measurement and calculations	8 h (16:00-24:00 h); resting and physical activity monitored every 5 min; water provided *ad libitum*; animals were fasted 8 h prior to measurements and had free access to water; resting & physical activity were monitored every 5 min during measurements; Q was calculated using the abbreviated Weir equation
Treatments	Group 1: LF diet (21% protein, 5% fat, 55% NFE, 5% fibre, 4.6% ash, 12.9 kJ/g ME; 550.6±86.2 KJ/kg$^{0.75}$/d); Group 2: HF diet (25% protein, 18.4% fat, 38% NFE, 4.5% fibre, 6.3% ash, 15.8 kJ/g ME; 614.6±121.8 KJ/kg$^{0.75}$/d)
Outcomes	Q (KJ/d): Group 1 = 4,861±413 *(361 KJ/(kg$^{0.75}$·d))*; Group 2 = 4,940±361 *(370 KJ/(kg$^{0.75}$·d))*
Calorimetry face mask system used	
O'Toole *et al.* (2001)	
Population	no. animals: 7; source: staff owned; breed: various; sex: 4 F & 3 M (neutered); BW: 1.3-33.7 kg; age: 2-9 yr
Calorimetry system characteristics	flow rate 2-10 l/min; room temperature 22 °C
Adaptation details	10 min adaptation to wearing face mask; authors note that 'staff-owned dogs appeared relaxed for most of the readings'
Measurement and calculations	4 consecutive series of 16 min readings; animals were fasted prior to measurements; during measurements the authors state that no attempt was made to record the dogs activity (i.e. tail wagging, moving paws); Q was calculated using the abbreviated Weir equation
Treatments	measurements conducted on two separate days (day 1 and day 2) at the same time of day with the same research staff present
Outcomes	Q (KJ/d): Day 1 & 2 (range) 3,810-3,894 *(317-324 KJ/(kg$^{0.75}$·d))* as numerical values were not provided, rough estimates were made based on visual interpretation of graphical data

Study	Details[1,2,3]
O'Toole *et al.* (2004)	
Population	no. animals: 10; source: staff/client owned; breed: various; sex: 5 F & 5 M (neutered); BW: 21.1-33.5 kg; age: 2-9 yr
Calorimetry system characteristics	flow rate not reported
Adaptation details	10 min adaptation to wearing face
Measurement and calculations	15 consecutive 1 min readings on two separate days; Q was calculated using the abbreviated Weir equation
Treatments	control: healthy staff-owned dogs; medicine: client-owned dogs with a variety of medical illnesses; surgery: client-owned dogs recovering from a major surgical procedure; trauma: client-owned dogs in a post-trauma period for the purpose of this review, only data from the control animals is presented
Outcomes	Q (KJ/kg/d): 126-147 *(288-336 KJ/(kg$^{0.75}$·d))* as numerical values were not provided, rough estimates were made based on visual interpretation of graphical data
Mazzaferro *et al.* (2001)	
Population	no. animals: 27; source: client owned; breed: Beagles; sex: F (intact); BW not reported; age: 1 yr
Calorimetry system characteristics	room temperature 20-22 °C
Adaptation details	once weekly sessions for 8 consecutive weeks; for all evaluation periods each dog was allowed to acclimate to its environment, handling, and the collection mask for 5 to 15 min before sample collection
Measurement and calculations	minimal 10 min gas equilibration followed by 15 min measurements; animals were fasted overnight prior to measurements; Q calculated using the abbreviated Weir equation; authors note that all dogs were lying down (resting) during measurements
Treatments	control: healthy; Group 2: dogs with OSA before surgery; Group 3: dogs with OSA 24 h after surgery for the purpose of this review, only data from the control animals is presented
Outcomes	Q (KJ/kg$^{0.75}$/d): 320.7±34.8 RQ: 0.77±0.06

[1] Abbreviations used: F = female; M = male; BW = body weight; HF = high fat; LF = low fat; ME = metabolic energy; MH = mannoheptulose; OSA = osteosarcoma; RQ = respiratory quotient.

[2] Reported literature Q values were converted to units of KJ/(kg$^{0.75}$·d) (shown in italics) to enable comparisons among studies.

[3] Remarks in italics are by the author of this chapter.

Chapter 14: Computing energy expenditure from indirect calorimetry data: a calculation exercise

S.J.J. Alferink[1], M.J.W. Heetkamp[2] and W.J.J. Gerrits[1]*
[1]Animal Nutrition Group, Wageningen University, P.O. Box 338, 6700 AH Wageningen, the Netherlands; [2]Adaptation Physiology Group, Wageningen University, P.O. Box 338, 6700 AH Wageningen, the Netherlands; sven.alferink@wur.nl

Abstract

Energy expenditure (Q) can be accurately derived from the volume of O_2 consumed (VO_2), and the volume of CO_2 (VCO_2) and CH_4 (VCH_4) produced. When the measurements are performed using a respiration chamber, VO_2,VCO_2 and VCH_4 are calculated by the difference between the inflow (l/h) and outflow rates (l/h), plus the change in volume of gas in the chamber between successive measurements. There are many steps involved in the calculation of Q from raw data. These steps are rarely published in full detail, nor are they well documented for the training of students, researchers or technical staff. The objective of this chapter is to provide a complete calculation exercise for students at MSc level and researchers or technicians with little background in indirect calorimetry. Based on an example dataset and using a stepwise approach, the calculations used for calibrations, volumes of gas, Q, the respiratory quotient and activity related Q are explained and illustrated. Excel spreadsheet files are available online (http://www.wageningenacademic.com/doi/abs/10.3920/978-90-8686-809-4) to support this exercise.

Keywords: calibration, gas volume, physical activity, energy metabolism, gas exchange, respiration chamber

14.1 Introduction

In indirect calorimetry, energy expenditure (Q), or heat production, is derived from gaseous exchange measurements, using simple equations developed in the previous century, based on the general assumption that the heat produced in a chemical reaction is independent of the pathway between the initial and final states. Therefore, coefficients of these equations are based on the O_2 consumption (VO_2), CO_2 production (VCO_2), and the heat of combustion upon complete oxidation of substrates (see e.g. Brouwer, 1965, Gerrits *et al.*, 2015a). Consequently, using these equations, Q can be accurately derived from VO_2 and VCO_2, including some corrections for protein oxidation and methane production. When the measurements are performed using a respiration chamber, VO_2 and VCO_2 are calculated by the difference between the inflow (l/h) and outflow rates (l/h), plus the change in volume of gas in the chamber between successive measurements (l/h) (Van Milgen *et al.*, 1997; Verstegen *et al.*,

DOI 10.3920/978-90-8686-261-0_14, © Wageningen Academic Publishers 2015

1987). As the volume of gas is dependent on for example the ambient temperature, relative humidity and barometric pressure, all of these variables are either standardised, and/or continuously recorded (see Chapter 2 by Heetkamp *et al.*, 2015). Consequently, there are many steps involved in the calculation of Q from gaseous exchange data. These steps are rarely published in full detail, nor are they well documented for the training of students, researchers or technical staff.

In this chapter we detail the calculations we use at the calorimetry facilities of Wageningen University, the Netherlands. These facilities have been described in Chapter 2 by Heetkamp *et al.* (2015). The volume of a gas is dependent on its temperature, humidity and pressure. These parameters need to be measured and corrected for to arrive at standardised gas volumes. Calibration procedures are in place to correct for deviations in analysis equipment (e.g. gas volume meters) and drift of gas analysers. When using multiplexing systems, i.e. measuring gas concentrations in airflows originating from various respiratory chambers sequentially using the same analyser, time differences between the analysis of different air samples require interpolation. It should be noted that when the time between sequential measurements of gas concentrations is reduced to facilitate the monitoring of short term changes in Q, the precision of gas concentration measurement has to increase to maintain precision of the measurement of VCO_2 and VO_2. Alternatively, and specifically for high resolution data, modelling strategies have been developed to deal with the issues described above (Chapter 4 by Labussiere *et al.*, 2015; Van Milgen *et al.*, 1997).

The objective of this chapter is to provide a complete calculation exercise for students at MSc level and researchers or technicians with little background in indirect calorimetry. It is recommended that these exercises are accompanied by the reading of the principles of indirect calorimetry (Chapter 1 by Gerrits *et al.*, 2015a), and to be familiar with the general design of calorimetry facilities (e.g. Chapter 2 by Heetkamp *et al.*, 2015; Lighton, 2008). Using a stepwise approach, the calculations used for calibrations, volumes of gas, Q, the respiratory quotient (RQ) and activity related Q (AQ) are explained. The calculations can be performed in a pre-arranged MS Excel 2010 spreadsheet that is available online (http://www.wageningenacademic.com/doi/abs/10.3920/978-90-8686-809-4). The spreadsheet contains the formulas to perform the calculations on an example dataset. As a large part of the calculations is identical for O_2, CO_2 and CH_4, the individual calculation steps are only detailed for the calculation of VCO_2. A spreadsheet with answers is also available online to check whether the calculations that are performed are correct.

This dataset (from an experiment published by Bolhuis *et al.*, 2008) contains calorimetry measurements taken from a group of 12 pigs (in 2 pens) over a period of 24 h, using a multiplexing system with a 9 minute interval, a chamber volume of 80,000 l and a washout time τ of about 100 min (Figure 1). The pigs (25 kg BW) were housed in the chamber for 7 d, and fed at 06:30 h and 15:30 h. Lights were on between 06:00 h and 18:00 h. Physical activity was recorded using radar-Doppler devices. The time required for completion of the calculation exercises is estimated between 2 and 3h.

Figure 1. One pen of pigs, housed on a straw bedding in the respiration chamber (Bolhuis et al., *2008)*

14.2 General principles

For the calculation of Q (kJ) from VO_2 (l), VCO_2 (l), VCH_4 (l), and oxidation of amino acids, measured by urinary excretion of N, we use Brouwer's equation (Equation 1; Brouwer, 1965):

$$Q = 16.175\ VO_2 + 5.021\ VCO_2 - 5.99\ N - 2.167\ VCH_4 \tag{1}$$

As urinary nitrogen contributes very little to Q, and moreover, its variation is difficult to measure within the day, we neglect the urinary nitrogen contribution to Q in our calculations. VO_2, VCO_2 and VCH_4 are calculated as the sum of (1) the difference between the inflow and outflow rates, and (2) the gas volume change in the chamber between two successive measurements. The first part is referred to as the 'ventilation part', and the second part as the 'volume part'.

14.3 The calculations stepwise and in detail

14.3.1 Gas concentrations

The VO_2, VCO_2 and VCH_4 are calculated from gas concentrations measured in air that enters the chamber and in air that leaves the chamber. Continuously, sampled air is flowing through a series of gas analysers (para-magnetic for O_2 and non-dispersive infrared for CO_2 and CH_4). The gas concentrations are recorded every 9 min and the results are stored in a database. Once every 24 h a zero and a span calibration gas sample containing known concentrations of O_2, CO_2 and CH_4 is analysed. The data are stored and used afterwards to correct the measured gas concentrations. Potential changes in the measured calibration gas concentrations over the

24 h period are linearised, and an intercept and regression coefficient is calculated. These are used to calculate the calibrated concentrations of O_2, CO_2 and CH_4.

Calculation Step 1.

Open the MS Excel workbook 'Computing Energy Expenditure' and find 'Worksheet 1. Gas concentration correction'.

Calculate the regression coefficient a and intercept b for the CO_2 concentrations in columns N and O. Use the green coloured equations and variables in the header.

Calculate the corrected concentrations of CO_2 in ingoing and outgoing air in columns T and W. Use the blue coloured equations and variables in the header.

14.3.2 The ventilation part

The ventilation volumes of O_2, CO_2 and CH_4 are calculated by multiplying the calibrated gas concentrations of ingoing and outgoing sampled air by the volume of air flowing into and out of the chamber, respectively. When using a hypobaric system, the outgoing airflow is measured, in our case using a diaphragm type or rotor type gas volume meter. These gas meters are calibrated before and after every experiment using a gas meter calibrated to a high standard at the Dutch Metrology Institute (NMI Certin, Dordrecht, the Netherlands). The calibration equations are used to correct the measured air outflow. This correction is typically nonlinear, i.e. flow dependent and therefore a multipoint calibration is necessary. To correctly calibrate a diaphragm type gas volume meter a stepwise increasing flow followed by a stepwise decreasing flow is necessary. See the graph in 'Worksheet 8. Gas meter calibration curve' from which the polynomial equation is used in 'Worksheet 2. Ventilation and chamber volume' column H.

The outgoing airflow needs to be corrected for temperature, humidity and barometric pressure to obtain the standardized temperature and pressure dry gas airflow (standard temperature (273.15 K) and pressure (101.3 kPa/760 mm Hg) dry gas).

For these corrections the barometric pressure, air temperature and relative humidity inside the chamber, and the temperature inside the gas meter are measured every minute, averaged and stored for every 9 minute data point.

Calculation Step 2.

Use 'Worksheet 2. Ventilation & chamber volume' to calculate the pressure corrected ventilation in column K and the pressure + temperature corrected ventilation in column L. (Use the corrected ventilation rate (l/min) and the gas meter temperature).

When the VO_2 and VCO_2 are not equal (i.e. the RQ does not equal 1.00) and/or gases like CH_4 are emitted, the volumes of ingoing and outgoing air will differ. To correct for this, the volume of ingoing air is calculated as the outgoing STPD multiplied by a correction factor, commonly referred to as the 'nitrogen factor'. The nitrogen factor is the ratio between the concentration of nitrogen gas leaving and entering the chamber. As N_2 is assumed inert and not produced by subjects inside the chamber, it is assumed that the flow of STPD N_2 into the chamber equals the flow of STPD N_2 out of the chamber. The concentrations of N_2 are calculated as 100% minus the concentrations of O_2, CO_2 and CH_4.

Calculation Step 3.
Use 'Worksheet 3. Ventilation gas volume' to calculate the nitrogen factor in column K.

The outgoing STPD airflow is multiplied by the nitrogen factor to calculate the ingoing STPD airflow.

Calculation Step 4.
Use 'Worksheet '3. Ventilation gas volume' to calculate the ingoing STPD in column L.

The calibrated gas concentrations in inflowing STPD air are averaged over two subsequent 9 minute data points and multiplied with the STPD ingoing airflow for that 9 minute period. This gives the volumes of O_2, CO_2 and CH_4 entering the chamber in l/min.

Calculation Step 5.
Use 'Worksheet 3. Ventilation gas volume' to calculate the ingoing CO_2 in l/min in column M.

The calibrated gas in outflowing STPD air are handled in the same way. This gives the volumes of O_2, CO_2 and CH_4 leaving the chamber in l/min.

Calculation Step 6.
Use 'Worksheet 3. Ventilation gas volume' to calculate the outgoing CO_2 in l/min in column N.

The ingoing volumes of O_2, CO_2 and CH_4 are subtracted from the outgoing volumes and expressed in l/d.

Calculation Step 7.
Use 'Worksheet 3. Ventilation gas volume' to calculate CO_2 production in l/d in column O. (P.S. There are 1,440 minutes in a day).

14.3.3 The chamber volume part

Any change in the volume of O_2, CO_2 or CH_4 within the chamber between two 9 min measurements contributes to the consumption or production of O_2, CO_2 or CH_4. Such changes result from e.g. diurnal or postprandial changes in VO_2, VCO_2 and VCH_4, or are related to variation in chamber STPD volume related to variation in climate. To calculate this part of the gas exchange the known measured volume of the chamber (80,000 l) is expressed in litres of STPD. The STPD volume of the chamber needs to be calculated for every 9 minute datapoint because temperature, humidity and pressure change over time and will have an effect on the STPD volume of the chamber. In the setting described by Bolhuis *et al.* (2008), and used for this exercise, temperature and humidity were controlled within a rather narrow range.

Air pressure inside a chamber is directly depending on barometric pressure, automatically regulated in our example experiment between -50 and -100 Pa against barometric pressure. We assume this variable under pressure not to have a significant effect on our calculation. For example, a difference of 50 Pa has an effect of (100 – (101,300 – 50)/101,300 × 100 =) 0.05% reduction in air pressure.

Calculation Step 8.
Go back to 'Worksheet 2. Ventilation and chamber volume' to calculate the pressure corrected volume in column N and the pressure + temperature corrected volume in column O.

The calibrated gas concentrations of outflowing STPD are multiplied by the STPD chamber volume to get the STPD volume of CO_2, O_2 and CH_4 inside the chamber for each 9 minute data point.

Calculation Step 9.
Use 'Worksheet 4. Chamber gas volume' to calculate the CO_2 content in l in column K.

The STPD volume change over the 9 min time interval is then calculated for every 9 minute data point by subtracting the previous volume from the current volume.

Calculation Step 10.
Use 'Worksheet 4. Chamber gas volume' to calculate the CO_2 Volume change in column N.

An STPD volume change of the chamber between successive measurements, because of changes in temperature, humidity or pressure, will, due to a constant outflow, change the amount of air entering the chamber because pressure is maintained constant. In a hypobaric respiration chamber, where the gas flow meter measures the outgoing airflow, this change in airflow is not measured by the gas meter so a correction is necessary. To correct for this effect the average ingoing gas concentration of the 9 minute period is multiplied by the volume change of the respiration chamber. Note that because of the higher concentrations, the effect on O_2 is larger compared with CO_2 and CH_4.

Calculation Step 11.
Use 'Worksheet 4. Chamber gas volume' to calculate the ingoing CO_2 correction in column O.

This gas volume is then subtracted from the volume change of CO_2 inside the chamber to obtain the CO_2 production caused by volume changes in the chamber.

Calculation Step 12.
Use 'Worksheet 4. Chamber gas volume' to calculate the CO_2 production that results from volume changes in column P.

Finally the gas volume changes, calculated for each 9 min interval, are expressed in l of gas/d.

Calculation Step 13.

Use 'Worksheet 4. Chamber gas volume' to calculate the CO_2 production in l/d in column Q. (PS. Correction from per minute to per day can be done by dividing the outcome by 9 and multiplying by 1,440).

14.3.4 Calculating VO_2, VCO_2 and VCH_4

The total gas volumes for O_2, CO_2 and CH_4 are calculated by adding up the ventilation part and the volume part.

Calculation Step 14.

Use 'Worksheet 5. Energy expenditure' to calculate the total gas exchange for CO_2, O_2 and CH_4 in columns J, K and L.

14.3.5 Energy expenditure and respiratory quotient

Energy expenditure is calculated using Equation 1.

Calculation Step 15.

Use 'Worksheet 5. Energy expenditure' to calculate the Q in kJ/day in column N.

This Q is then corrected for the total metabolic weight of all animals in the respiration chamber to arrive at the energy expenditure in kilojoules per kilogram of metabolic weight per day (kJ/kg$^{0.75}$/day).

Calculation Step 16.

Use 'Worksheet 5. Energy expenditure' to calculate the Q in kJ/kg 0.75/day in column P.

RQ is calculated by dividing the VCO_2 by the VO_2. The RQ is an indicator of the type of substrate oxidized. In addition, VO_2, VCO_2 may be used for the estimation of the net rates of carbohydrate and fat oxidation. For details see Chapter5 by Van den Borne *et al.* (2015). The RQ for complete oxidation of carbohydrates is 1.00. For protein it is approximately 0.84 and for fat it is approximately 0.71. For details see Chapter 1 by Gerrits *et al.* (2015a).

Calculation Step 17.

Use 'Worksheet 5. Energy expenditure' to calculate the RQ in column Q.

14.3.6 Quantifying activity related energy expenditure

When timed recordings of physical activity are available, Q can be decomposed into energy expenditure on physical activity (AQ) and resting metabolism. Decomposition of Q is possible because the short term variation in Q is mainly due to changes in physical activity, which means that AQ can be estimated from the time correlation between the recorded intensity of physical activity and Q. For details, see Gerrits *et al.* (2015b). In the example dataset (see 'Worksheet 7. Activity'), physical activity of these group-housed pigs was recorded using radar-Doppler meters. These activity meters give a pulse signal to a counter device when a set threshold value is passed. The counts are summed per 9 min interval and stored with a time stamp in the computer database.

A linear regression of the activity counts versus the Q over the 24 h period is used to estimate the activity related AQ (Equation 2):

$$Q = A \times C + B \tag{2}$$

in which A = regression coefficient = energy costs per unit activity; C = activity counts (arbitrary units); B = constant (intercept) = resting energy expenditure.

Then, AQ can be calculated by multiplying the energy costs per unit activity (C) times the actual recorded number of counts in the corresponding 9 min interval. Please note that in this calculation it is assumed that the resting Q is constant. There are more advanced approaches to decompose Q into AQ, resting Q and the thermic effect of feeding. For further reading, the reader is referred to e.g. Chapter 8 by Gerrits *et al.* (2015b), Chapter 4 by Labussière *et al.* (2015) and Van Klinken *et al.* (2012).

> Calculation Step 18.
> Use the equation from the appropriate chart on 'Worksheet 6. Graph Q & activity' to calculate the AQ in column T of 'Worksheet 5. Energy expenditure'.

Results are shown in several graphs in 'Worksheet 6. Graph Q & activity'. Calculated nine minute and hourly data are presented for Q and AQ. Activity corrected Q, calculated as Q – AQ, is also presented. From this graph it is obvious that correction for AQ gives a clearer view on diurnal rhythm in Q related to the two meals and the subsequent thermic effect of feeding. The graph plotting 9-minute and hourly data of RQ against time shows that RQ on an hourly basis exceeds unity during substantial part of the day. For interpretation of these data, the reader is referred to Chapter 5 by Van den Borne *et al.* (2015), Chapter 4 by Labussière *et al.* (2015) and Bolhuis *et al.* (2008).

Supplementary material

The MS Excel worksheets used in this chapter can be found online at: http://www.wageningenacademic.com/doi/abs/10.3920/978-90-8686-809-4

References

Bolhuis, J.E., Van Den Brand, H., Staals, S.T.M., Zandstra, T., Alferink, S.J.J., Heetkamp, M.J.W. and Gerrits, W.J.J., 2008. Effects of fermentable starch and straw-enriched housing on energy partitioning of growing pigs. Animal 2: 1028-1036.

Brouwer, E., 1965. Report of the Sub-committee on Constants and Factors. In: Blaxter, K.L. (ed.) Proceedings of the 3rd Symposium on energy metabolism, Troon, Scotland. EAAP Publication no 11, Academic Press, London, UK, pp. 441-443.

Gerrits, W.J.J., Heetkamp, M.J.W., Labussière, E. and Van Klinken, J.B., 2015b. Quantifying physical activity heat in farm animals. In: Gerrits, W.J.J. and Labussière, E. (eds.) Indirect calorimetry. Techniques, computations and applications. Wageningen Academic Publishers, Wageningen, the Netherlands, pp. 155-170.

Gerrits, W.J.J., Van den Bornc, J.J.G.C. and Labussière, E., 2015a. Deriving heat production from gaseous exchange: validity of the approach. In: Gerrits, W.J.J. and Labussière, E. (eds.) Indirect calorimetry. Techniques, computations and applications. Wageningen Academic Publishers, Wageningen, the Netherlands, pp. 19-34.

Heetkamp, M.J.W., Alfcrink, S.J.J., Zandstra, T., Hendriks, P., Van den Brand, H. and Gerrits, W.J.J., 2015. Design of climate respiration chambers, adjustable to the metabolic mass of subjects. In: Gerrits, W.J.J. and Labussière, E. (eds.) Indirect calorimetry. Techniques, computations and applications. Wageningen Academic Publishers, Wageningen, the Netherlands, pp. 35-56.

Labussière, E., Dubois, S., Van Milgen, J. and Noblet, J., 2015. Modelling gas exchanges to partition heat production between fasting, thermic effect of feeding and physical activity in farm animals. In: Gerrits, W.J.J. and Labussière, E. (eds.) Indirect calorimetry. Techniques, computations and applications. Wageningen Academic Publishers, Wageningen, the Netherlands, pp. 73-86.

Lighton, J.R.B., 2008. Measuring metabolic rates, a manual for scientists. Oxford University press, Oxford, UK.

Van den Borne, J.J.G.C., Heetkamp, M.J.W., Alferink, S.J.J. and Gerrits, W.J.J., 2015. Moving from a complete energy balance towards substrate oxidation: use of stable isotopes. In: Gerrits, W.J.J. and Labussière, E. (eds.) Indirect calorimetry. Techniques, computations and applications. Wageningen Academic Publishers, Wageningen, the Netherlands, pp. 87-114.

Van Klinken, J.B., Van den Berg, S.A.A., Havekes, L.M. and Willems van Dijk, K., 2012. Estimation of activity related energy expenditure and resting metabolic rate in freely moving mice from indirect calorimetry data. PLoS One 7: e36162.

Van Milgen, J., Noblet, J., Dubois, S. and Bernier, J.F., 1997. Dynamic aspects of oxygen consumption and carbon dioxide production in swine. British Journal of Nutrition 78: 397-410.

Verstegen, M.W.A., Van Der Hel, W., Brandsma, H.A., Henken, A.M. and Bransen, A.M., 1987. The Wageningen respiration unit for animal production research: a description of the equipment and its possibilities. In: Verstegen, M.W.A. and Henken, A.M. (eds.) Energy metabolism in farm animals. Martinus Nijhoff Publishers, Dordrecht, the Netherlands, pp. 21-48.

Acknowledgements

The editors gratefully acknowledge the independent reviews that were performed on all book chapters conducted by:

Joost van den Borne	FrieslandCampina Innovation Centre, Wageningen, the Netherlands
Guido Bosch	Animal Nutrition Group, Wageningen University, Wageningen, the Netherlands
Anne Colin	INRA, UR83 Poultry Research, INRA, Nouzilly, France
Michael Derno	Leibniz-Institut für Nutztierbiologie, Dummerstorf, Germany
Patrick Even	CNRS, UMR INRA – AgroParisTech, Laboratory of Human Nutrition and Intestinal Physiology, AgroParisTech, Paris, France
Marcel Heetkamp	Adaptation Physiology Group, Wageningen University, Wageningen, the Netherlands
Kees de Lange	Department of Animal & Poultry Science, Guelph University, Canada
Jaap van Milgen	INRA, UMR INRA – Agrocampus Ouest, Physiology, Environment and Genetics for the Animal and Livestock Systems, Saint-Gilles, France
Cesar Pinares Patino	CSIRO Agriculture, Canberra, Australia
Chris Reynolds	Animal, Dairy and Food Chain Sciences, University of Reading, Reading, United Kingdom
Martin Verstegen	Animal Nutrition Group, Wageningen University, Wageningen, the Netherlands

Index

www.ingramcontent.com/pod-product-compliance
Lightning Source LLC
LaVergne TN
LVHW042256300926
849161LV00003B/618